Principles of Microprocessors

Principles of Microprocessors

IAN L. SAYERS
Motorola (UK) Ltd

ADRIAN P. ROBSON
Newcastle upon Tyne Polytechnic

ALAN E. ADAMS
University of Newcastle upon Tyne

E. GRAEME CHESTER
University of Newcastle upon Tyne

CRC Press
Boca Raton   Ann Arbor   Boston

Library of Congress Cataloging-in-Publication Data

Catalog record is available from the Library of Congress.

Direct all inquiries to CRC Press, Inc., 2000 Corporate Blvd., N.W., Boca Raton, Florida, 33431.

© 1991 by CRC Press, Inc.

International Standard Book Number 0-8493-8605-5

Printed in the United States

Contents

Preface, vii

1   Introduction to Microprocessors, 1

    1.1   Introduction, 1
    1.2   A brief computer genealogy, 2
    1.3   Basic computer concepts, 12
    1.4   Summary, 18
    1.5   Further reading, 19

2   Data Representation, 20

    2.1   Introduction, 20
    2.2   Number systems, 20
    2.3   Representing negative values, 32
    2.4   Overflow, 35
    2.5   Floating-point notation, 36
    2.6   Error detection and correction, 41
    2.7   Unweighted codes, 45
    2.8   Summary, 47
    2.9   Further reading, 48
    2.10 Problems, 48

3   Basic Logic Structures, 50

    3.1   Introduction, 50
    3.2   Boolean algebra, 50
    3.3   Gate symbology, 57
    3.4   Simple combinational design, 58
    3.5   Flip-flops, 65
    3.6   Design using medium-scale integration devices, 73
    3.7   Circuit technologies, 82
    3.8   Summary, 91
    3.9   Further reading, 91
    3.10 Problems, 91

4   Architecture of a Typical Microprocessor, 94

    4.1   Introduction, 94
    4.2   Machine instructions, 94
    4.3   Internal components, 102
    4.4   Instruction execution, 112
    4.5   Interrupts, 120
    4.6   Summary, 123
    4.7   Further reading, 123
    4.8   Problems, 123

5 Assembly-language Programming, 125

5.1 Introduction, 125
5.2 Programming model, 126
5.3 The NeMiSyS assembler language, 127
5.4 Programming the NeMiSyS microprocessor, 137
5.5 Summary, 153
5.6 Further reading, 154
5.7 Problems, 154

6 Memory Organization, 156

6.1 Introduction, 156
6.2 Semiconductor memory, 158
6.3 Semiconductor memory-system design, 177
6.4 Mass storage, 196
6.5 Summary, 213
6.6 Further reading, 214
6.7 Problems, 214

7 Input/Output Devices, 218

7.1 Connecting peripherals to the microprocessor, 218
7.2 Interface devices, 227
7.3 The user interface, 234
7.4 Data communication and storage, 244
7.5 Sensors and actuators, 246
7.6 Analog interfaces, 252
7.7 Summary, 263
7.8 Further Reading, 263
7.9 Problems, 264

8 High-level Languages, 266

8.1 Introduction, 266
8.2 The Pascal language, 270
8.3 The C language, 281
8.4 Interfacing high-level language to assembly code, 286
8.5 Summary, 290
8.6 Further reading, 290
8.7 Problems, 291

9 Practical Microprocessor Architectures, 292

9.1 Introduction, 292
9.2 8-bit processors, 295
9.3 The 16/32-bit processors, 311
9.4 RISC processors, 323
9.5 Single-chip computers, 329
9.6 Summary, 334
9.7 Further reading, 334
9.8 Problems, 334

Appendix 1: NeMiSyS Software, 338

Appendix 2: Character Sets, 340

Index, 344

Preface

The main aim of this book is to provide a single text that can be used for both undergraduate electronic engineering and computer-science/engineering courses which teach basic hardware and software design of microprocessor systems. The book can also be used as a supplementary text for teaching of inter- or cross-disciplinary courses where microprocessor techniques form only a part of the core curriculum. As the book contains a simulation of the processor described, it will be useful in remote-learning situations or for self study. The book is completely self contained, so students with no previous knowledge of digital hardware and/or software experience will be able to use the text with ease; for the more experienced students, selected chapters may be omitted without losing continuity.

The book covers the main characteristics of commonly available SSI and MSI chips and their use in implementing Boolean functions. This theme is continued in later chapters when the structure of LSI chips used in the design of complete microprocessor systems is presented. The techniques needed to implement correctly structured programs is also discussed. The use of high-level languages such as C and Pascal are presented in connection with their increasing use in programming microprocessor systems. An attempt has been made to achieve a balance between both hardware and software aspects of system design.

A unique feature of this text is that the description of the microprocessor is based on a software simulation provided with the book and designed to run on the most commonly available computer, the IBM PC and its derivatives. This "microprocessor" is used as a basis for describing the main design principles. Not only does the simulator provide a realistic model of a typical microprocessor but also the environment in which the students may find themselves when programming real systems. This allows the discussion to relate very closely to the microprocessor as the student will be working within very well defined constraints—a situation which is not possible with many hardware systems as peripheral hardware and assemblers may vary widely, even for machines with essentially the same internal processor. Using the simulator, it is also possible to demonstrate concepts, such as interrupts and low-level micro sequencing. These features of a microprocessor are not normally accessible in a hardware environment.

In order to support the student's activities in the real world, most chapters present manufacturers' data on actual chips, to illustrate the discussion or to demonstrate the trade-offs that are involved in any design. This theme is continued, in the final chapter, with the presentation of a series of overviews of real processor architectures in terms of the simulated processor. Using this technique a turgid and often lengthy discussion of a single processor is avoided.

Overview of the book

The book can be started from any chapter and followed through in sequence depending on the student's experience. Each chapter contains a problem section which allows the student to test his/her understanding of the ideas presented in that chapter. A solution book is available for supervisors.

Chapter 1 presents the genealogy of the modern microprocessor and thus the philosophy behind modern computer systems. An outline of the constituent parts of a computer system are presented as a lead-in to the remaining chapters.

Chapter 2 presents the fundamentals of number systems and binary arithmetic. This is used to demonstrate how binary arithmetic can perform all the operations normally associated with the decimal system. It also allows students to refresh ideas they may have been using already and perhaps to expand their knowledge.

Chapter 3 is to familiarize students from other than an electronics background with Boolean algebra and electronic digital systems in general. The concepts presented also illustrate how the basic building blocks can be produced for a digital computer. The chapter discusses the main logic families currently available and their use in modern processor design.

Chapters 4 and 5 describe the operations and programming of the NeMiSyS processor. These chapters cover all the basic aspects of microprocessor technology from the fetch/execute cycle to the structure of programs. Chapter 4 also presents the hierarchical view of a computer that allows orientation between the hardware and software aspects of a computer system.

Chapter 6 deals with the design of memory arrays for a microprocessor. The correct design of the memory array can make or break the overall design by improving or reducing the performance of the associated processor. This chapter begins by examining the low-level design of a memory chip and ends by discussing how memory can be used to increase the functionality of a microprocessor. Magnetic-media storage is also discussed as well as the emerging technology of optical media.

Chapter 7 looks at the external problems of interfacing a microprocessor to the real world. The control of external peripherals is probably one of the most useful features of a modern computer system. The chapter also looks at how people can interact with systems via VDUs and keyboards. The interaction of computers over large distances through networks is also discussed.

Chapter 8 presents high-level languages (HLL) in the context of microprocessor systems as an alternative to assembly-language programming. The use of C and Pascal in controlling external peripheral devices is discussed. As the power of a "simple" microprocessor approaches that of the traditional mainframe computer, HLLs are becoming the normal method of programming these machines, while assembly-language programming is becoming a more specialized environment for programming.

Chapter 9, as already mentioned, discusses real microprocessors currently available along with the different approaches taken by manufacturers to instruction sets and internal architectures.

Acknowledgements

The authors would like to express their thanks to all those who made this book possible—especially the many students who "class tested" much of the material. In particular the authors would like to acknowledge the resources provided by the Department of Electrical and Electronic Engineering at the University of Newcastle upon Tyne, UK, and the Department of Computer Science at Newcastle Polytechnic, UK. In addition the authors would like to thank Scott Midkiff of Virginia Polytechnic Institute, Virginia, USA, for his invaluable comments and extensive reviewing of both the draft manuscript and initial versions of the simulator software. Finally we are indebted to Navin Sullivan of CRC Press for his guidance and encouragement during the writing of this text.

August 1990

Ian L. Sayers
Adrian P. Robson
Alan E. Adams
E. Graeme Chester

1: Introduction to Microprocessors

1.1 Introduction

The microprocessor has revolutionized almost every aspect of our lives. The uses of the ubiquitous microprocessor are legion: almost every piece of modern electrical equipment now contains the device or has been designed with its assistance. The microprocessor has improved the functionality of existing systems and brought others into existence. The penetration of microprocessor-based hardware into the consumer market has largely been unnoticed by most people. How many end-users care whether their new purchase contains a microprocessor-based computer (*microcomputer*)? Usually their main concern is the cost and, ultimately, the reliability of the product. However, in the industrial and commercial sectors the microprocessor has had a profound effect on working practices, the robot and wordprocessor being, perhaps, the most notable manifestations of the new information-technology age.

The versatility of the microprocessor lends itself to a wide range of applications, old and new. Hence, the traditional layman's view of a "computer" as a massive data-processing engine is no longer valid. Undoubtedly, as the increasing density and complexity of integrated circuits makes the microprocessor even more powerful, the range of applications will increase markedly. Fortunately, for this text, the current range of computers, and microprocessors, all depend on certain fundamental principles which have remained constant since the earliest stored-program machines. However, previously held beliefs are being challenged as basic physical limits are reached in semiconductor technology. These limits make it increasingly difficult to extract more performance from a straightforward traditional microprocessor design. Several approaches have been taken to obtain the maximum performance from current semiconductor technologies.

The first approach involves an attack on the main underlying principles of processor design. Instead of continually increasing the complexity of the processor, thereby increasing the burden on the overworked silicon chip, the new approach simplifies the processor by reducing the complexity and simplifying certain aspects of the design. The reduction in complexity brings with it an increase in overall processor performance, at the cost of increasing the size of the actual stored program. The increase in program size would, normally, have been detrimental to the overall complexity of the computer, if it were not for the hand-in-hand development of semiconductor memory technologies. Nevertheless the new Reduced Instruction Set Computers (RISC), as the new processor types are called, are making massive inroads into the "traditional" processor market.

1

The second approach to computer design involves the technique of "divide and conquer," i.e. rather than use a single fast processor to execute the program, several processors are used to execute small sections of the program in parallel, increasing the overall execution speed at the expense of more hardware. This illustrates another feature of microprocessor design (as well as engineering in general), that hardware cost can usually be traded off for increases in speed. However, as hardware prices decrease, relatively, then a systematic approach to parallel computer architectures will become normal practice. Parallel computer architecture designs have been around for some time, mainly as research tools, but commercial processors are now available that are specifically designed to be connected together to allow parallel program execution. The next few years will see an increase in the use of parallel processors, with the probable introduction of personal computers/workstations based on this hardware, at which time software engineers will have to learn an almost complete set of new skills. As new developments in computer technology are announced almost daily, the future of computer engineering will be very interesting, just as is the history of computing.

1.2 A brief computer genealogy

The starting point for a discussion of the history of computers really depends on how you define the word *computer*. Today, the word computer refers to a machine (hardware), usually electronic, that is capable of executing a sequence of instructions from a stored program (software). This definition attempts to exclude the humble calculator: however the calculator is a simple extension of the human brain (which is run by "wet ware"!). A wider definition of a computer could be any machine that assists in data-processing operations, arithmetic or linguistic, by representing the data to be manipulated in some physical form. In giving the data a tangible form, processing of that data is greatly simplified.

The earliest, recorded, mechanical calculating aid is the abacus, invented in Babylonia in, about, 3000 BC. The first number systems were only used to record numbers and were not intended to be used for performing calculations. For example, Roman numerals are adequate for recording numerical data; however, it was almost impossible to perform calculations using them, e.g. MCMLXXXVIII multiplied by XIV. On the other hand, the abacus could be used to perform these calculations very easily. Although today the abacus is associated, almost exclusively, with the Orient, it was widely used in Europe by merchants and traders. These businessmen were generally uneducated; however, they were able to solve involved mathematical problems with just the simple abacus. Between the 13th and 17th centuries, in Europe, the old Roman system was slowly replaced by the Hindu–Arabic system of notation. In some countries the two systems coexisted for some time, with curious mixtures occasionally occurring. The Hindu–Arabic system used the nine digits familiar to the Europeans; it also introduced the concept of zero and the position dependence of the notation, which made "pen and paper" mathematics much easier. Eventually the abacus was abandoned, even though it could still cope adequately with the new number system. The introduction of positional notation into mathematics made the future development of computers a far easier task. However, the decimal system, which is easy for humans to understand, is not necessarily the best representation for the computer.

Snowballing European trade and commerce soon meant that mechanical means had to be found to take the drudgery out of the increasing amount of mathematical work. The first step towards a "mechanical" calculating machine was made by John Napier in the 17th century. Napier's discovery of logarithms was entirely original, based on his own endeavors. The principle of logarithms is quite straightforward, being based upon the fact that a number can be represented by a power to a particular base. Using logarithms multiplication and division could then be performed by simply adding or subtracting the powers, respectively. The answer could then be obtained by using antilogarithms to change the resulting power back into a number. The generation of the logarithm and antilogarithm tables sparked off a great deal of activity in the mathematical world. In the later years of Napier's life he produced a multiplication table laid out on a series of four-sided wooden rods. Using these rods, or Napier's bones, it was possible to perform quite complex multiplication and division operations. The most significant invention to arise from Napier's logarithms was the slide-rule, produced by William Ougthred in about 1622. The development of the slide-rule made multiplication and division a truly mechanical task, with a little helping hand. The slide-rule, in its various guises, became an indispensable tool to scientists and engineers alike.

The first truly mechanical calculator is often attributed to the French mathematician and philosopher Blaise Pascal, in about 1640. However, about twenty years before Pascal a German Professor, Wilhelm Schickard, designed and built the first working mechanical calculator. The calculating machine was based on Napier's logarithms and could therefore perform multiplication and division. A simple calculation could be performed using mental arithmetic; however, for larger calculations an adder/subtractor circuit was also provided, in the same machine, in order to produce the intermediate results from the main calculation. The adder/subtractor circuit used an ingenious arrangement of cogs to deal with carries. The machine was also believed to have contained a means of warning the user that an overflow or underflow had occurred in the calculation. Unfortunately this quite remarkable invention disappeared into oblivion during the Thirty Years' war, except for a few letters describing the machine. Consequently the enormous advances made by Schickard were lost and had to be rediscovered by Pascal 20 years later.

Pascal's calculating machine, called the Pascaline, is generally regarded as being the forefather of all other calculating machines, making a significant contribution to advancement in that field, unlike the Schickard calculator. The Pascaline was a simple adder, that could perform subtraction operations by using a method known as nine's complement. The Pascaline machine

3

introduced the concept of the carry mechanism to the world. Unfortunately the calculator had several drawbacks, the main one being that it was not reliable enough to gain general acceptance among its potential users. Another minor drawback of not being able to perform multiplication and division could be overcome by using Napier's logarithms or repeated additions/subtractions.

Gottfried Wilhelm von Leibnitz was the next major contributor to the mechanical-calculator field. Essentially the advance made by Leibnitz was the introduction of a special stepped wheel to allow the machine to perform multiplication operations. Again, due to the limitations of metal-working technology at that time, Leibnitz never managed to perfect the operation of his innovative machine. Leibnitz also made another significant contribution to computer design, being one of the first Western mathematicians to study the binary number system. This system is at the heart of all modern computers. Although Leibnitz's machine did not work, it spawned many imitations. The most successful of which was the Arithmometer, designed by the Frenchman Thomas de Colmar in about 1820. The Arithmometer was very reliable and easy to use. These features accounted for its immense popularity with over 1000 machines being sold in the thirty years that followed. The Arithmometer was perhaps the first mass-produced calculating machine.

The industrial revolution was just beginning when the Arithmometer was introduced. The rapid developments occurring at this time not only required mechanical calculators, but also accurate mathematical tables of all kinds. The traditional methods of calculating these tables by hand were inaccurate and prone to serious errors. In fact, errors in a set of navigation tables were blamed for several naval accidents. Furthermore some table publishers introduced deliberate mistakes to simplify the detection of copying by their competitors. The methods used to calculate the tables were well known and understood: they were ripe for mechanization, a fact not lost on the mathematician and inventor Charles Babbage. The Difference Engine, as Babbage called his machine, was designed to calculate the tables by means of constant differences. Not only would the machine generate the tables but it would also produce the metal plates for mass printing of the final results. Thus the engine would eliminate most, if not all, of the errors due to miscalculation and printing. Although Babbage was not the first person to conceive of this idea, he was the first to raise sufficient funds to build the machine. In this case most of the funding came from the British Government: they had realized the importance of good quality tables for navigation, especially now that the Empire was beginning to expand.

The main problem facing Babbage, at that time, was the shortage of high-quality metal-working machines to produce the precision gear mechanisms to complete the project. A great deal of project money and time was spent on designing new machine tools to build the required parts. This greatly enhanced the British tool market, but it delayed the complete project. Babbage was also a perfectionist, continually refining and redesigning parts of the Difference Engine. Changes in government, his main sponsors, and difficulties with Clements, his mechanical engineer, conspired against Babbage, ultimately leading to the demise of the Difference Engine. Only one small part of the calculator was built and worked perfectly. Although Babbage may not have completed his project, the publicity given to the Difference Engine spurred on the Swede, Pehr Georg Scheutz, to design his own Difference Engine, in about 1850. The Tabulating Machine, as Scheutz's machine was called, was the first demonstration of the power of mechanical calculators. Ironically, one machine was bought by the British Register General and heavily used for calculating insurance premiums.

During the construction of the Difference Engine, Babbage conceived an even more ambitious calculating engine. The Difference Engine was dedicated to one task and could not perform any other operations; therefore Babbage reasoned that a machine that could perform general calculations would be more useful and make the Difference Engine obsolete. The new machine, called the Analytical Engine, is regarded, by many, as the forerunner of the modern computer. The parallels between the Analytical Engine and a modern computer are quite striking. The design and refinement of the Analytical Engine stretched over several decades, from about 1836 until Babbage's death, in 1871. The engine was never built due to several reasons, not the least of which was the lack of funds; also the mechanical tolerances required for such a machine were far beyond those available from the tool makers of the time. Although the Analytical Engine is referred to as a single machine, Babbage in fact designed several Analytical Engines with increasing sophistication; this may have been another contributing factor to the eventual downfall of the machine.

The Analytical Engine was a testament to one man's insight into the problems of producing mechanical calculators, the Engines contained many features that would not be unfamiliar to a modern computer designer. The main computational section of the engine was the *mill*; today this would correspond to the arithmetic and logic unit (ALU) of a computer. There was also a storage section to contain the data to be manipulated. The internal operations of the complete unit were determined by a barrel arrangement containing metal slats. The slats engaged levers within the mill and store to determine the overall operation, e.g. division or multiplication, performed by the engine. Simply changing the barrel changed the operations that could be performed; therefore this section represents the control unit of the Analytical Engine. One further feature of the Analytical Engine was its ability to print out results.

The external program to be executed by the Analytical Engine was stored on punched cards, three different types of cards being used, each with independent readers. The idea of using cards to store the program came from Jacquard's automatic loom, which used punched cards to control the pattern to be woven. Although Babbage had sorted out most of the mechanical details of his design, the programming aspects remained less well defined. Only a few programs were written to exercise the "imaginary" machine. One further interesting feature of the engine was its ability to perform conditional branching to another part of the program, depending on the result of the current calculations; by today's standards the branching ability was very limited. However, the ability to redirect program flow by conditional tests on the results of calculations is now an accepted feature of any programming language.

Most of the information available about the programming of the Analytical Engine is entirely due to Lady Ada Lovelace, an amateur mathematician and daughter of Lord Byron. Ada became involved with Babbage after they met in 1833, making comprehensive explanatory notes on the principles of programming the engine. Many of her notes were published in widely circulated and popular journals and, consequently, Ada Lovelace is considered, by many, to be the first software engineer.

As precision metal working improved, more advanced mechanical calculators appeared on the market. The next step on the road towards the computer required a transition from mechanical to electrical means of calculation. The first step down this road is credited to an American, Herman Hollerith, who, in the 1880s, developed a machine that would ultimately be used in the preparation of the data from the 1890 US census. The Hollerith machine was based on punched cards, although originally Hollerith had used paper tape for the prototype design. The information from the census forms was initially transcribed onto the cards by punching holes in specific areas of the cards; the cards were then placed onto a special card reader. The reading device consisted of a press, with a series of pins which could stick through the holes in the cards to connect with an array of contacts underneath. The operator simply inserted the punched card into the press, those pins that connected with the underlying contact array then completed an electrical circuit which advanced an associated counter. Using this technique, even including the time to transcribe the data onto the punched cards, the whole census operation was greatly speeded up. With the great success of Hollerith's machine in the 1890 US census, many other countries ordered the equipment. Eventually, industry saw the advantages to be gained in using the machine. The company Hollerith formed to build the units, the Tabulating Machine Company, eventually became the International Business Machine Corporation, or more familiarly IBM. Hollerith's machine is most notable for its use of electricity and all the advantages that it endowed on the system.

The next significant phase in digital computer development did not occur until the mid 1930s, mainly due to the fact that a suitable technology was needed for any further advances to be made: that technology involved the use of electronics. The first "computer" was based on electromechanical relays and Boolean algebra—although the designer, Konrad Zuse, did not realize this fact at the time. Zuse wanted to improve his job, as an engineer, by taking the grind out of the mathematical calculations he had to perform. In designing his "calculator" Zuse made some very important decisions: the machine was to be capable of solving a range of calculations, it was to be based on binary, rather than decimal arithmetic, and it would use rules based on Boolean algebra for its operations. The first machine Zuse designed and

6

built, the Z1, was entirely mechanical and paid for out of his own funds. The next machine, the Z2, used electromechanical relays for the main calculating part of the machine, the memory remained mechanical: this curious mixture of technologies worked very well. The subsequent machine, the Z3 completed in December 1941, was based on relays and built to assist in aircraft design during World War II. The Z3 performed well and was, arguably, the first fully functional stored-program general-purpose calculating machine. As the war drew to a close Zuse designed the Z4, an even more powerful machine. However, this was not considered to be an important part of the German war effort and was, fortunately for the Allies, largely ignored. After the war the Z4 was used in Germany for many years, being the only machine of its type available that was not classified as top secret.

At this time several avenues of research were adding to the overall understanding of computing. In Britain, Alan Turing published his paper, "On Computable Numbers...", which considerably advanced the theory of computers. In America, Claude E. Shannon published his thesis on a systematic approach to switching circuit design, changing the design of relay logic circuits into a science instead of an art, as had previously been the case.

Totally independently of Zuse, American mathematicians were working along similar lines. The Havard Mark I, built by Howard Aiken between 1937 and 1943, was perhaps one of the most publicized machines, not least because the work was sponsored by IBM. The Mark I was such a long time in development that it was obsolete before it was finished. The final machine contained well over 0.75 million mechanical parts and weighed about 5 tons. As most people were unaware of the work by Zuse in Germany, the Mark I appeared to be very impressive.

Two major problems during the Second World War, more than anything else, spurred on the development of the electronic computer. In Britain the problem involved the requirement of Allied intelligence to decipher the coded German messages that were being transmitted with their Enigma machines. The information contained in the messages was vital to the Allied war effort. The American problem concerned the very important job of producing ballistic tables for the new armaments that were being designed. The usual way of producing these tables involved the use of an analog differential analyzer. This machine required several hours to set up before a calculation could begin. Consequently production of ballistic tables for new weapons began to seriously lag behind, delaying the introduction of new equipment.

The British Government attempted to solve the German code problem by bringing together A. Turing, I. J. Good and D. Michie to produce a machine which could automatically decipher the coded messages. The machine they designed and built, the Colossus, was operational by December 1943. The Colossus was built using about 2000 tubes (valves), a significant step in the history of the computer. The machine simply compared the message to be decoded with the known permutations that could be generated by the German Enigma machine, until a match was found. Once a match had been found the decoded message was printed out. The messages were all punched onto tape and read by sophisticated optical readers so that the tape did not wear out, since it might pass through the reader a number of times before a message was decoded. In all, ten Colossus-type machines were built and made an enormous contribution to the war effort. In order to keep their existence secret the information gained by their use had to be treated very carefully in case the Germans should realize that their secret messages were being decoded by the enemy. It is a tribute to the secrecy that surrounded this project that it was a long time after the end of the war before Colossus was acknowledged publicly. Although Colossus was not a true computer, it is significant because all the logic operations, etc., were carried out using electronic tubes.

In America the problem of producing ballistic tables was turned over to the Moore School of Engineering at the University of Pennsylvania. The team, led by J. W. Mauchly and J. P. Eckert, won this contract by demonstrating the potential speed that could be obtained from the calculator by using tubes in its design. Although the final speed obtained from the machine was less than that envisaged, it was still extremely fast. The project to build ENIAC (Electronic Numerical Integrator and Calculator) began in 1943 and was completed in 1946. ENIAC fell far short of being a fully fledged computer, but it was faster than any calculator then available. In the end the project was a complete success: it had shown the potential available to a future computer industry. The final machine was built from over 35,000 electronic components, 18,000 of which were tubes, consuming over 175 kW of power. To improve reliability, ENIAC was run continuously day and night; this reduced the problems of the tube heaters burning out or being stressed and then failing, due to high turn-on currents. Also the tubes were run at voltages well below their maximum value and air cooled by large fans. Setting the appropriate connections to program the ENIAC for a particular task was a tricky and painstaking job; however, the speed at which it could perform calculations far outweighed the non-"user friendly" mode of programming the machine. In fact ENIAC was completed too late to perform its assigned task; instead it was used in the cold war to "perfect" the H-bomb.

The program for ENIAC was stored externally, not internally as in a modern computer. Even as ENIAC was being built the idea of building a true stored-program computer was beginning to take shape in the minds of Mauchly and Eckert. At about this time, early 1944, John von Neumann joined the team, lending weight and influence to the project. Von Neumann helped Eckert and Mauchly refine their ideas about the new computer, EDVAC (Electronic Discrete Variable Computer). As is usual in a team effort, no one person can be credited with producing specific ideas; however, von Neumann undoubtedly influenced heavily the important design considerations. In 1945 von Neumann produced a report on the design of EDVAC entitled "Report on the EDVAC." This report contained a complete specification for the design of a general-purpose stored-program digital electronic computer. The concept of the central processing unit (CPU), random-access memory for program storage, and the use of binary arithmetic, were all embodied within this report. Although von Neumann left space to name his associates in the paper, this was never done and von Neumann was credited with conceiving the whole idea, much to the annoyance of his fellow workers. In 1952 the EDVAC computer was successfully completed, but by this time the honor of producing the worlds first stored-program computer had gone to Britain. So significant was the contribution made by von Neumann that all "normal" computer architectures are now generally referred to as "von Neumann" machines.

The report on the EDVAC and Turing's earlier paper "On Computable Numbers..." caused British scientists to consider building their own computer. They had already gained valuable experience with the highly secret Colossus machine. To build the new computer J. R. Womorsley, who had seen the ENIAC project and received a copy of the EDVAC report, began to assemble a development team at the National Physical Laboratory (NPL), England. The first member of this team was Alan Turing who, using his own work and the EDVAC report, formulated the design of the Automatic Computing Engine, ACE. Unfortunately due to bureaucracy only a very much reduced version of ACE was eventually completed in 1950. By this time a despondent Turing had left and joined Max Newman (who had began Colossus) at Manchester University, who was building the Manchester Mark I computer. The Mark I was the first fully operational electronic stored-program computer, executing its first program in June 1948. The Mark I was further developed by Ferranti Ltd., into the first commercial computer, the Ferranti Mark I. Another British computer was also under construction at this time at Cambridge University, called EDSAC (Electron Delay Storage Automatic Computer); it was also based on the EDVAC report. The rest, as they say, is now history.

The massive developments taking place in computer technology assured their continued existence. The main problems with these computers was their size, cost and, above all, reliability: the tube was causing serious problems and a successor had to be found to improve the situation. The replacement came in 1945 when J. Bardeen, W. H. Brattain and W. Schockley, of Bell Laboratories, invented the transistor. The could be used as a transistor could be used as a logic switching element in the design of a new generation of computers, replacing the tube. The transistor allowed a massive improvement in reliability and speed, while reducing the overall power consumption of the computer.

Even as the transistor was revolutionizing the computer industry, moves were occurring in England that would profoundly affect the way computers were built. An English engineer, W. A. Dummer, realizing that as the size of electrical equipment was reduced then the reliability increased, proposed the design of the first integrated circuit (IC) to increase reliability even further. Again the ability and technology to undertake this work was not yet available and the idea was dropped in England. However American engineers took up the challenge and in 1958 Jack Kilby, working for Texas Instruments, produced the first IC. Other engineers working on the

9

development of integrated circuits slowly refined the different fabrication methods. This resulted in a considerable increase in the complexity of the available ICs. The computer could now be built from even more reliable and smaller components.

The next step in this saga involved the lateral thinking of one man, Ted Hoff, an engineer at Intel, who conceived the first CPU on a single chip, the 4004, in the early 1970s. The 4004 is regarded as the first microprocessor (CPU) and was the forerunner of today's modern microprocessors. The advent of the microprocessor brought with it the promise that, in the near future, an individual would be able to own a powerful computer, rather than a large company-run computer system.

The pattern for future computers had been set by the Ferranti Mark I and ENIAC machines. They were large and required the resources of a major company to buy and operate. Indeed, these large *mainframe* computers needed their own minor workforce to keep them operational: the "inevitable incidentals" that come with owning such a machine. As the engineers and scientists began to use computers in their day-to-day work, the demand for such machines also began to grow. However, remember only large companies could afford such machines. Developments in time-sharing computer systems allowed the computer owners to sell time on their machines to other users. Thus many bureau computing companies sprang up to service the demands of this burgeoning market, fuelling even more demand. This view of mainframe computers remained dominant for many decades, the blinkered view of mainframe manufacturers preventing many from seeing the next development in computer hardware.

The next step in computer technology came from the need for smaller and more accessible machines. The increasing availability of integrated circuits meant that the physical bulk of a computer could be reduced markedly, a fact that was recognized by a, then, small computer company, Digital Equipment Corporation (DEC), in the early 1960s. Kenneth Olsen of DEC had realized that most computer users did not require the power available from a mainframe to perform their routine tasks. The first successful machine (although not the first machine) to be produced by DEC was the PDP-8 (Programmable Data Processor Model 8), which was about the size of a filing cabinet and so the *minicomputer* was born. The PDP-8 was not as powerful as current 8-bit microprocessors, even though it used 12-bit arithmetic. However, it was cheap, in comparison to mainframe computers. The PDP-8 found many uses with scientists and engineers, who could now transfer their computations from remote computers to local machines that did not require large numbers of people to support its operations. The new minicomputers found many applications in real-time control, from oil refineries to chemical plants. Many other computer manufacturers emerged with minicomputer products, for example Data General Corporation and their Nova. It was probably the PDP-8 that did more to change attitudes towards computers than any other machine. In 1977 DEC continued their innovative success by introducing one of the first *superminis* in the form of the VAX 11/780. This machine is still used as a yardstick by many other manufacturers when comparing the performance of their machines with competitors.

As has already been discussed, the first microprocessor appeared around 1971; however, this processor was not powerful enough to support any real computation. It was not until the 8008 was introduced that developments began to take place. The minicomputer had already shown that the demand for low-cost computing was healthy and increasing. The microprocessor should have given immediate impetus to the established computer manufacturers to continue the trend they had already begun. However, this was not the case. Most manufacturers could not see why an individual would want to own a *personal computer*. Even DEC hesitated when faced with the possibility of introducing a cut-down version of their successful PDP-8 (the PDP-8 was not based on a microprocessor) as a contender in the personal market.

The reluctance of computer manufacturers to enter the personal-computer market was more than matched by the enthusiasm of individuals. Undoubtedly, many people were considering building a personal computer at the same time, but the first description of such a machine appeared in the July 1974 issue of Radio Electronics and was due to an electronics graduate, J. A. Titus, from Virginia Polytechnic Institute, Blacksburg. His machine was based on the 8008 and could contain up to 16 Kbytes of memory, when expanded. The Mark-8, the name given to the machine, was a tremendous success, with many readers purchasing the design manual and the circuit board on which the computer could be built. The interest in personal computers generated by the Mark-8 was channelled into the formation of user groups, which would be instrumental in further developments of these machines. The next development came hard on the heels of the Mark-8. The Altair 8800, designed by E. Roberts of Micro Instrumentation and Telemetry Systems Inc. (MITS), was a more substantial machine based on the 8080 and capable of supporting 64 Kbytes of memory. Again design information about the Altair was published in a magazine, Popular Electronics. The orders for the Altair computer flooded into MITS, overwhelming their ability to supply the machine quickly. The Altair was also significant in that it was one of the first machines, of its type, to use a BASIC interpreter. The interpreter was written by William Gates and Paul Allen who later went onto found the Microsoft Corporation.

The Altair computer spurred many other, would-be, designers to consider building their own computer. This was further encouraged by the formation of many user groups across America. One of these groups, called the Homebrew Computer Club in Silicon Valley, began as a very small group of less than 20 people, but the membership rapidly grew to well over 400. Stephen Wozniak, a founder member of this group, had a profound effect on the personal-computer scene, by designing one of the first commercially available ready-built machines. In conjunction with colleagues he designed and built a complete computer based on the 6502 processor; the new machine even incorporated a BASIC interpreter. One of his friends, Steven Jobs, then persuaded Wozniak to form a partnership to sell the new machine and the Apple

Computer Company was formed. The first machine lacked the sophistication of a keyboard, visual display unit (VDU) and external storage media, but it was extremely popular at such a low price of only a few hundred dollars. The second machine designed by Apple, the Apple II, included most of these missing features, or the ability to include them. The Apple II was an immense success, making its founders millionaires after only a few years.

In the early days of personal computers the main problem was the lack of standardization between different manufacturers. This included the method of storing the data on floppy disks, graphics capability, expansion capabilities and even the dialect of used to program the computer (BASIC had been settled upon as easy to use and implement in this new type of machine). Another area that required some standardization was the operating system that allowed the user/programmer to interact with the computer: this is still true today. The scene altered considerably when IBM entered the personal-computer market with the PC. The new computer became an immense success with huge sales worldwide. The success of the PC attracted many imitators who produced compatible machines, usually at lower cost. This further enhanced the size of the market, virtually sweeping aside any competitors that were producing non-PC-compatible machines. With one type of computer, although from different manufacturers, software authors now concentrated on producing high-quality programs that could be used on a single, widespread, computer architecture. As microprocessor technology improved, IBM upgraded the original PC to give it more features; however, compatibility was maintained with the older machines, thereby continuing the dynasty. This approach also meant that the MSDOS operating system, produced by Microsoft for the PC, became a standard. If it were not for this consistent approach to hardware and software then this text would not have been possible. The floppy disk provided with this book is designed to run on the PC, purely because of its widespread use.

1.3 Basic computer concepts

The purpose of a computer is to execute a series of stored instructions, the program, to perform functions on a set of data. This ad-hoc definition of a computer highlights the various components that now constitute a modern computer system. As shown by Figure 1.1 there are five main hardware components, the memory, the input device, the output device, the microprocessor and a master clock to synchronize all the computer's actions. The main component of the computer shown in Figure 1.1 is the microprocessor (the topic of this book). The microprocessor is the part of the computer that performs all the data manipulation and controls the rest of the system. Within a normal processor there are usually three subsections.

1 Register file: this holds the internal data that the processor is currently using while executing the program. This information may include the state of the processor and a program counter to allow the processor to keep track of its current position in the program.

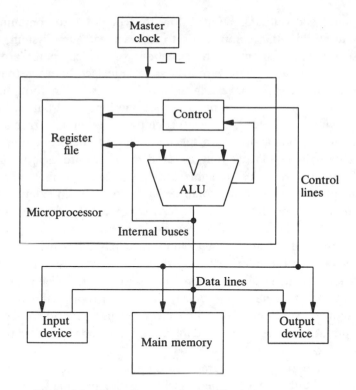

Figure 1.1. Block diagram of a typical microcomputer.

2 Control section: this section of the processor determines which functions are to be performed on the data. This information is obtained by decoding the incoming program instructions. The controller is also responsible for redirecting program flow, depending on the changing state of the processor. The orchestration of the actions performed by hardware external to the processor is also undertaken by the controller.

3 Arithmetic and logic unit (ALU): this section of the processor is responsible for performing the actual data-manipulation operations, such as addition, subtraction and, in some cases, multiplication.

The processor deals with data as a collection of bits. The number of bits operated on by most processors is fixed. However, microprocessors capable of dealing with 1 to 32 bits are available, the usual bit sizes being 1, 4, 8, 16 and 32 (at the moment). The master clock is responsible for synchronizing all the events that are taking place within the computer. The frequency of this clock determines the fundamental speed at which the microprocessor can execute an instruction.

The main memory is used to store the program to be executed and the data, in some cases, that the program will require, e.g. variables, etc. The design of the main memory can have a dramatic effect on the performance of the complete computer system. Originally memory, suitable for interfacing to a microprocessor, was not very large; today memory containing over four million locations is available.

The input and output sections allow the computer to communicate with the outside world. If a computer is to interact with any system, human or otherwise, then the communication sections of the computer are vital. In general the input/output hardware acts as an interface between the processor and the outside world, converting signals into a suitable form for both to work properly. In a microcomputer system the input and output hardware is usually provided by special, dedicated, chips, however in larger systems the input/output may be handled by a special dedicated microcomputer system.

The above discussion has only covered the hardware aspects of the computer. The software is less tangible but is just as important. The software covers the programming of the computer and, to a certain extent, the method used to represent the data. In a modern computer the data are represented by binary values. The binary values can represent any information that is being processed at the time, for example if a payroll is being dealt with by the computer then the internal binary data will be representing numeric data and, possibly, character information that is to be printed on the pay documents. On the other hand, the binary data could represent a pictorial scene if the computer were performing image-processing operations. Consequently the interpretation placed on the data is entirely dependent on the program being executed.

When computers first appeared the only way they could be programmed was to enter the 0s and 1s (called machine code) that made up the program and data directly into the memory, usually by hand. Obviously this state of affairs could not continue with the increasing size of the computer. The next step involved the use of mnemonics that represented, in a readable form, the instructions that could be executed by the computer. The programmer could then simply write the program using the mnemonics. Another piece of software, called an assembler, would then read this information and automatically produce the binary patterns representing the instructions. This method was more reliable, as human operators are considerably better at remembering textual information (the mnemonics) than numeric information (the machine code). The mnemonic program is, normally, called the source file and the machine code is called the object code. The object code can be loaded into the memory of the computer for execution; the source file cannot be executed and is merely there to assist the programmer in performing his/her tasks more efficiently.

Development of more efficient programming languages involved the design of still higher levels of specifying the actions of a computer. One of the first computer languages to be developed, in the late 1950s, was FORTRAN; the name was an acronym for FORmula TRANslation. The FORTRAN language used concepts that were easy to understand and were quite divorced from the actual computer that was running the program. The assembly-language programs were specific to the machine they were running on; if the machine changed then the entire program would have to be rewritten using the new machine's mnemonics. The FORTRAN program could not be executed directly on the computer; it first had to be converted into machine code.

However, the conversion from the FORTRAN source program could be performed automatically by another piece of software called a compiler. The operations performed by a compiler are complex and can vary between different implementations. The main tasks of the compiler are to check that the program is syntactically and semantically correct and to then generate the machine code for the particular computer. Some compilers generate assembly-language programs which can be converted into machine code by an assembler. Unfortunately the compiler cannot determine whether or not the program is logically correct. Many compiled languages have followed in the footsteps of FORTRAN, these include Pascal (after Blaise Pascal), ADA (after Ada Lovelace), C and Algol (ALGOrithmic Language). The main drawback with using compiled languages is that they tend to produce inefficient machine code, in terms of both size and speed of execution. However, in most applications these drawbacks are unimportant; the ease with which the program can be written and maintained, once it has been perfected, tend to outweigh any disadvantages. Also the program is not dependent on the computer being used. Hence, it should be possible to transfer the program to any other computer with a compiler for that language, making the programs portable. In practice this can be a difficult task, even though most languages have an agreed international standard defining their syntax. The problems arise due to the fact that many computer manufacturers tend to produce extensions to the standard language to exploit features on their machine. Therefore, unless a programmer scrupulously avoids these extensions, the program may be impossible to transfer to any other machine without major modifications, at which point it is usually better to start again.

Another form of computer language is represented by BASIC (Beginners' All-purpose Symbolic Instruction Code). In this case the language is interpreted rather than compiled, that is, instead of producing machine code via a compiler, there is a program on the computer which reads the BASIC program and performs the indicated operations directly, i.e. the software interprets the BASIC program for the computer. The use of an interpreter cuts out the compilation phase of running the program. However, since the program being executed is not in the "native tongue" (machine code) of the computer it can take a long time to execute. Professional programmers tend to prefer compiled languages. All of the above languages are termed high-level languages (HLL).

The development of HLLs has certainly made programming computers a far easier task. Unfortunately the languages are a long way from being ideal. It would be much better if the computer could be programmed in something resembling natural language. The problem is that natural language is very complex and can be difficult for even humans to decipher at times. However, research is currently underway to produce programming languages that are closer towards this goal. The research is also aimed at producing machines which use artificial intelligence to solve problems. The current computer hardware lacks the power to deal with such a complex task, but hardware research is also improving this aspect of the problem.

In the early days of microcomputers the only programming method available was machine code; on machines with a small memory this was fine. As microprocessors grew in power and could handle larger memories, the techniques of the mini and mainframe computer worlds were slowly transferred to the microcomputer scene. Hence, it is now possible to program microprocessors using most of the HLLs. This is not totally unexpected as the power of today's microprocessor is usually equal to or greater than that available from "yesterday's" mainframe.

1.3.1 An example of a microprocessor computer system

The last section discussed the structure of a computer in general terms. The block diagram in Figure 1.2 gives a more concrete example of a computer in the form of a personal computer. The components illustrated are typical and may not be found in every personal computer.

The main component of the example machine is the microprocessor: any one of the currently available types could be used for this purpose. However, today the computer would most likely be based on the 16/32-bit processors, such as the MC680X0 family (MC68000, MC68020, MC68030 and MC68040) or 80X86 family (80186, 80286, 80386 and 80486) (the processors chosen for many IBM PC compatibles). The 16/32-bit processors have the ability to support co-processors that can alleviate the main processor of the need to perform the more computationally intensive tasks, such as floating-point arithmetic operations (8086 + 8087 and MC68020 + MC68881) and memory-management chores (MC68020 + MC68451). The co-processors are optional and not essential for the successful operation of the computer. Currently the trend is to integrate more of the "co-processor" functions onto the actual microprocessor, negating the need for external co-processing elements.

The main memory of the computer is made up from random-access memory (RAM) and read-only memory (ROM). The ROM is used to store the program that the microprocessor will execute when powered up or after a reset: usually this is referred to as the boot ROM. The boot program must initialize any input/output devices and load in the operating system, if it is stored on an external medium, e.g. a floppy disk. The program may also perform some self-test operations to detect any faults in the hardware. The RAM is used to store the operating system, when it is loaded, and any programs that are executed plus their associated variables. In some computer systems the integrity of the data stored in the RAM may be protected by the use of parity codes. Using such codes it is possible to detect when a particular bit of the RAM is faulty: this fault information can then be relayed to the user and repair action initiated.

The computer user can interact with the machine via the keyboard and visual display unit (VDU). The keyboard allows the input of commands to the operating system and the writing of source files for new programs. The VDU is used to display the textual information and return results from the executing programs. Some computers may use the VDU to represent

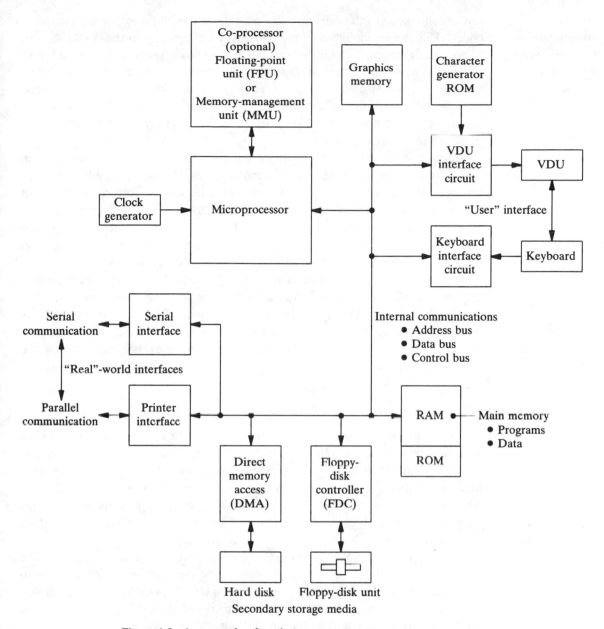

Figure 1.2. An example of a microcomputer system.

pictorial information in the form of graphs or drawings. Associated with the VDU-interface hardware there is usually a section of RAM. This RAM may be part of the general-purpose main memory or specialized video RAM (VRAM), which are designed to optimize the performance of the graphics display. In both cases the RAM is used to store the information that is being displayed on the VDU. The VDU is like a television screen and needs to be refreshed from the RAM to retain the image. Another piece of hardware linked with the VDU circuits is the character-generator ROM. The ROM is used to store the graphic representation of the character set. The characters

17

are usually stored internally in the computer as an 8-bit sequence character code. This is not enough to directly draw the character on the VDU screen, hence, the character sequence decodes a particular section of the ROM to obtain the necessary information to produce the image on the screen. Alternatively, the RAM contents may directly represent the picture on the screen with 0s and 1s turning off and on each element of the image. This is becoming a more common method of producing the image on the screen as the power of the machines increases.

The computer usually stores important data, such as the operating system and programs, on external magnetic media. The two most popular forms of magnetic media are the floppy disk and the hard disk. Both these pieces of hardware require an interface to the microprocessor, represented, in the figure, by the floppy-disk controller (FDC) and direct memory access chip (DMA) for the hard disk. Other forms of storage such as tape or optical disks are slowly making inroads into the personal-computer market.

On some personal computers provision is made for directly interfacing a printer to the system. The printer allows programs to be listed or, if a word processor is being used then the documents generated can be printed. A serial interface may also be provided on the computer to allow it to be connected to other computers: information can then be exchanged between the two computers.

1.4 Summary

The intention of this introductory chapter has been to present the historical developments that culminated in the design of the modern computer. In particular the ideas and the people behind the most important concepts have been highlighted. The names of the most conspicuous engineers/scientists have become synonomous with particular aspects of a computer design. However, as with all technological advances there is usually an army of other engineers/scientists whose contribution is just as important.

The main components of a modern computer have been identified in the later sections of this chapter. In the rest of the book chapters are devoted to each of the main components, giving complete explanations of all design aspects. The example computer system discussed in Section 1.3.1 serves to illustrate how a real computer could be built.

Unlike most texts on microprocessor principles no single commercial processor has been chosen as the discussion model. Instead a general-purpose microprocessor architecture has been developed that avoids most of the pitfalls associated with choosing one specific commercial processor. The processor architecture to be discussed is supplied as a software simulator on the floppy disk provided; therefore the computer is now acting as the teacher. Programs can be run on this simulated processor in the same way as a normal processor. Simulating the processor allows more information to be obtained than from a black-box approach; it also means that this microprocessor is definitely accessible. The knowledge gained from this simulated processor is then used as the model for explaining the operation of several commercially available processors, covering the MC6809, Z80, MC68000, ARM RISC processor and the MCS-48 single-chip microcontroller. Consequently it should be much easier to understand the architecture of any new processor that may be encountered, a much better situation than understanding a single commercial processor in complete depth, especially when learning about microprocessors.

1.5 Further reading

1 Augarten, S. (1984). *Bit by Bit—An Illustrated History of Computers*, Ticknor and Fields.
2 Forester, T. (1980). *The Microelectronics Revolution*, Blackwell Scientific Publications.
3 Hodges, A. (1983). *Alan Turing—The Engima of Intelligence*, Counterpoint.
4 Hyman, A. (1982). *Charles Babbage: Pioneer of the Computer*, Princeton University Press.
5 Metropolis, N., Howlett, J. and Rota, G.-C. (1980). *A History of Computing in the 20th Century*, Academic Press.
6 Randell, B. (ed) (1937). *The origins of digital computers*, Springer-Verlag. (Contains many of the original papers.)
7 Roberts, H. E. and Yates, W. (1975). Altair 8800: The most powerful minicomputer project ever presented—can be built for under $400, *Popular Electronics*, Jan–Feb, 33–58.
8 Shannon, C. E. (1938). Symbolic analysis of relay and switching circuits, *Trans. Amer IEE* **57**.
9 Titus, J. A. (1974). Build the Mark-8: your personal computer, *Radio Electronics*, July, 29.
10 Turing, A. (1937). On computable numbers with an application to the Entscheidungs problem, *Proc. London Math. Soc.* **42**, 230–265.

2: Data Representation

2.1 Introduction

The topic of this chapter is data representation. We shall deal with two forms of data, numerical data and non-numerical data. At first glance these two types of data appear to be represented, in the computer, in very similar fashions. However, the similarity is only superficial and it is most important to distinguish between numerical and non-numerical data types. In general a code representing a numerical value will be a weighted code. This means that the various components of the code have values, or weights, associated with their position within the code. Non-numerical values are represented by unweighted codes: in an unweighted code each individual component of the code has no more significance than any other part; it is simply the overall pattern of the code that is important.

2.2 Number systems

The vast majority of humans use the same number system to represent magnitudes. The number system so employed is, of course, the decimal system. The reasons why humans use the decimal system are fairly obvious and reside at the ends of your arms. The ability to count up to ten with considerable facility has been codified and refined over centuries and has led to the development of a set of symbols to represent relative magnitudes between zero and nine. We are of course referring to the symbols 0, 1, 2, 3, 4, 5, 6, 7, 8, 9. If, when using the decimal system, we wish to represent values greater than nine then we have to use combinations of these symbols. For example, there is no special symbol in the decimal system for the value fifteen so we represent this value as 15, the 1's position in front of the 5 showing that it has more significance, that it is of greater value, than the next symbol to the right. The 1 thus has a greater weight (in this case ten times the weight) associated with it than the 5.

The above may seem very pedantic to some readers, but one of the problems when dealing with number representation is that we are so used to dealing only in the decimal system that we come to believe that there is something special about using ten symbols, that this is the way the universe is built, that this is the "natural" way of doing things. This is, of course, not the case. There is nothing special about the decimal system other than it fits in with our own physique; the physique of the digital computer is somewhat different from our own and thus requires a different number system to represent its values.

For technical reasons, associated with the electronic circuitry used in their implementation, digital computers effectively have only two fingers, rather than ten, on which they can count. The two states which exist within the computer may be referred to as "true" and "false," or "high" and "low," or 1 and 0. This last form, using ones and zeros, is often the most convenient and this is the representation we shall be using most. Since there are only two symbols, not ten, we obviously cannot employ the decimal system within the computer but must use a number system that has only two symbols. Such a system is known as the binary system and is said to have a radix of two. The radix of a system refers to the number of symbols used (the number of different values that can be represented before it is necessary to use multiple symbols). Thus the radix of the decimal system is ten.

2.2.1 The binary system

As we have only two symbols, combinations of these symbols must be used for the representation of both numerical and non-numerical data. Initially we will concentrate on the representation of numerical values within the computer. In the decimal system the significance of the symbols increases by a power of ten as we move to the left. For example:

$$279_{10} = 2 \times 10^2 + 7 \times 10^1 + 9 \times 10^0.$$

Notice that now we have written this number in full, the position of the digits is no longer significant and the order of writing the right-hand side of the above equation could be altered. Note also the subscript 10 to the figure 279; this is to show that this figure represents a value using the decimal system. As we are about to use several different number systems, which have symbols in common, it is important to know just which system is being employed; the placement of a subscript is the conventional way of defining the radix of the system being used.

In the binary system the meaning of position within a number follows the same pattern as for the decimal system. However, as we now have only two symbols we must move to the next column when these two symbols have been exhausted. This means that the significance (or weight) of each column will increase as a power of two as we move leftwards across the symbols (in any number system the significance of each symbol will change by a multiple of the radix of the system as we move across the number). Suppose then that we have a 4-bit (bit is a corruption of binary digit, i.e. one binary symbol) binary value, 1101, then the magnitude that this value represents could be more formally expressed as:

$$1101_2 = 1 \times 2^3 + 1 \times 2^2 + 0 \times 2^1 + 1 \times 2^0$$

$$= 13_{10}.$$

Note the use of the subscript 2 to show that we are using the binary system; the symbol B could also be used here.

Using 4 bits we can represent values in the range zero to fifteen, as shown in Table 2.1.

Table 2.1.

Decimal	Binary	Decimal	Binary
0	0000	8	1000
1	0001	9	1001
2	0010	10	1010
3	0011	11	1011
4	0100	12	1100
5	0101	13	1101
6	0110	14	1110
7	0111	15	1111

Similar principles to those described above are also applied to the representation of fractional values, e.g.

$$0.315_{10} = 0\times10^0 + 3\times10^{-1} + 1\times10^{-2} + 5\times10^{-3}$$

$$0.1101_2 = 0\times2^0 + 1\times2^{-1} + 1\times2^{-2} + 0\times2^{-3} + 1\times2^{-4}$$

$$= 0 \quad + \tfrac{1}{2} \quad + \tfrac{1}{4} \quad + 0 \quad + \tfrac{1}{16}$$

$$= 0.8135_{10}.$$

If we use 4 bits as fractional values then we can represent values in the range zero to fifteen-sixteenths as shown in Table 2.2.

Table 2.2.

Decimal	Binary	Decimal	Binary
0.0	0.0000	0.5	0.1000
0.0625	0.0001	0.5625	0.1001
0.125	0.0010	0.625	0.1010
0.1875	0.0011	0.6875	0.1011
0.25	0.0100	0.75	0.1100
0.3125	0.0101	0.8125	0.1101
0.375	0.0110	0.875	0.1110
0.4375	0.0111	0.9375	0.1111

The number of different values that may be represented by a number will depend upon the radix of the system and the number of digits employed. In general, using n digits in a number system of radix R, we can represent R^n different values. If we restrict ourselves to integer values then the largest number that could be expressed in this format will be $R^n - 1$, since one of the R^n values will be zero. For example, if we use three decimal digits we can represent one thousand integer values, the largest of which will be $10^3 - 1$ (999_{10}). If we use eight binary digits then we can express 2^8 (256_{10}) combinations of digits, the largest integer value being $2^8 - 1$ ($1111\,1111_2 = 255_{10}$).

2.2.2 Conversion between binary and decimal formats

It is usual for most computing systems to allow data entry in decimal format, the initial conversion to binary taking place automatically at the keyboard, keypad or selector switch. Similarly numerical data is most often output by the computer in decimal format. However, it may be necessary, particularly with small dedicated microprocessor-based systems, for the user to work directly in binary. The conversion process is quite straight forward and, although it may consume time, it is not difficult to implement.

The general technique that is used to convert integer values from a system using one radix to a system employing a different radix, is to continually divide the original value by the radix of the new system. The equivalent value in the new radix system is formed by the remainders developed during the division operations. As an example consider converting the value 268_{10} into binary.

	Remainder
268/2 = 134	0
134/2 = 67	0
67/2 = 33	1
33/2 = 16	1
16/2 = 8	0
8/2 = 4	0
4/2 = 2	0
2/2 = 1	0
1/2 = 0	1

Thus $268_{10} = 100001100_2$. Note that the last remainder forms the most significant bit (msb—the leftmost bit) of the answer.

A similar method is used to convert fractional values to a system employing a different radix. In this case we multiply by the radix of the new system. The required value is formed from the integer part of the result of the multiplication. Further parts of the fraction may be formed by continuing to multiply the fractional result of previous operations.

Suppose we wish to produce the binary equivalent of 0.719_{10}.

	Integer part
0.719×2 = 1.438	1
0.438×2 = 0.876	0
0.876×2 = 1.752	1
0.752×2 = 1.504	1
0.504×2 = 1.008	1

Thus $0.719_{10} = 0.10111_2$, or nearly so. Notice that the fractional part of the last multiplication is not zero and thus the equality quoted is not exact; indeed it is often the case that fractional values in one number system do not have an exact equivalent in another system. It is usual in these cases to agree an acceptable number of significant figures that are to be used. Notice also that it is the first integer value produced that forms the most significant bit of the result.

A further point that is worth mentioning at this stage is that within the digital computer there is no formal way of representing the binary point ".". Remember that we only have two symbols within the computer and these are being used to represent 0 and 1. The position of the binary point is assumed to reside at a predetermined position within the stored word; just where depends upon the design of the particular machine being used and upon the type of algorithm being implemented at the time.

In the previous examples we have been converting from decimal to binary. The same techniques could be used to convert to decimal from binary, although we would now have to work in binary arithmetic. However, it is probably more efficient simply to consider the weights of the ones within the binary number, in terms of their decimal equivalent, and to form the new decimal value from the summation of these values.

For example, let us determine the decimal equivalent of 101101.011_2. Taking the integer part first, we have one values with weights:

$$2^0 = 1, \qquad 2^2 = 4, \qquad 2^3 = 8, \qquad 2^5 = 32,$$

a total of 45. For the fractional part we have values with weights:

$$2^{-2} = 0.25, \qquad 2^{-3} = 0.125,$$

a total of 0.325. Thus $101101.011_2 = 45.325_{10}$.

2.2.3 Fundamental arithmetic operations in the binary system

ADDITION

As there are only two symbols used in the binary system there can be only four possible combinations of these symbols in any addition operation. Thus all possible single-bit addition operations can be summarized in four lines, namely:

	Sum bit	Carry bit	
$0+0 =$	0	0	
$0+1 =$	1	0	
$1+0 =$	1	0	
$1+1 =$	0	1	The result here is two, but to represent the value two we have a zero result in the sum column and carry one to the next most significant column.

All addition operations in binary will be based upon combinations of the four options shown above. Let us consider the addition of two 4-bit words:

```
    1 0 1 0
  + 1 0 1 1
```

	1	$0 + 1 = 1$
	<u>1</u> 0	$1 + 1 = 0$ with carry 1 to the next column
	1	$0 + 0 + 1 = 1$
<u>1 0</u>		$1 + 1 = 0$ with carry 1 to next column.

```
  1 0 1 0 1
```

Note that the addition of the two 4-bit numbers has yielded a 5-bit result. In general the addition of two n-bit numbers will produce a result that is $n+1$ bits long.

SUBTRACTION

As with addition, the subtraction of one binary digit from another can occur in only four possible combinations:

	Difference bit	Borrow bit	
$0 - 0 =$	0	0	
$0 - 1 =$	1	1	Here we borrow from the most significant column. This has a value of two compared to the difference column, giving a difference signal of one.
$1 - 0 =$	1	0	
$1 - 1 =$	0	0	

We can apply these principles to determine the result of subtractions between multi-bit words.

```
    1   0   1   0
  - 0   1   0   1
```

	<u>−1</u> 1	Here we borrow two from the next column, leaving a difference of one. Note the −1 to compensate in the next column.
	0	
<u>−1</u> 1		As above we must borrow from the next column.
0		

```
    0   1   0   1
```

It should be noted that subtraction, as a distinct operation, rarely takes place in digital computers but is replaced by the addition operation—this is discussed further in a later section which deals with the representation of negative values.

MULTIPLICATION

The multiplication of binary numbers is carried out in a very similar fashion to that of the multiplication of decimal values and is, in fact, rather simpler since the multiplying values are either 1 or 0.

```
        1 0 1 1
      × 1 1 0 1
```

1 0 1 1	First partial product formed by multiplying multiplicand by least significant bit of the multiplier.
0 0 0 0	Second partial product formed by multiplying the multiplicand by the second bit of the multiplier. Note shift to the left because of the increased significance of the multiplying bit.
1 0 1 1	Third and fourth partial products formed in a similar manner.
1 0 1 1	
1 0 0 0 1 1 1 1	Final result formed by summing partial products.

Note that an 8-bit result is produced by the multiplication process. In general the multiplication of two n-bit numbers will produce a result that is $2n$ bits long. Similarly if the multiplier is n bits long and the multiplicand is m bits long then the result will be $n+m$ bits in length.

Until relatively recently many computer systems did not employ special hardware circuitry for multiplications and the process was implemented in software. One of the algorithms employed for this process, the shift-and-add technique, used a sequence of operations very similar to the hand multiplication technique illustrated above. It will be readily appreciated that the software method of multiplication was a long process (especially if the word-lengths were long). With the falling cost of hardware it is now common to find specialist multiplication circuitry included in even the smallest of computers, allowing the operation to be carried out in just a few clock cycles.

DIVISION

The division of binary numbers is carried out in a similar manner to the division of decimal numbers, the divisor being divided into the dividend to form a quotient and remainder. Consider $1101 \div 0011$.

```
              0 1 0 0
      0 0 1 1 )1 1 0 1
              1 1
              0 0 0
              0 0 0 1   remainder 1
```

As with multiplication, division used to be carried out as a software operation, often involving repeated subtraction (modified addition) of divisor from dividend, and although many different algorithms have been developed in order to speed up the process, division using software is still an excessively time-consuming business. With the advent of lower hardware costs and particularly cheap memory structures, it is now relatively easy to produce fast division circuits; often these are based around the computer's specialist multi-plier circuits with a look-up table to form the reciprocal of the divisor.

2.2.4 Other number systems

OCTAL

Writing down long strings of binary digits is tedious and prone to error. What is required is some form of shorthand notation. We have seen that, although it is not difficult to convert between binary and decimal it can be a time-consuming process. The main reason why the binary system and the decimal system do not fit together in a simple fashion is because there is not a very direct relationship between the radices of the two systems. Suppose we consider another system whose radix is an integer power of two: suppose we chose a radix of eight (2^3). Such a system is known as the octal system and will use eight symbols, $0, \ldots, 7$. Larger values are, of course, represented by combinations of these symbols, the significance of each digit increasing by a power of eight as we move to the left.

$$672_8 = 6 \times 8^2 + 7 \times 8^1 + 2 \times 8^0$$
$$= 110\,111\,010_2$$
$$= 442_{10}.$$

To convert between binary and octal we could adopt the same techniques that we used when converting between binary and decimal. However, because of the relation between the radices of the binary and octal systems, a simpler and faster method exists which makes the octal system a very convenient system for the recording of binary patterns. If we consider the range of values that can be represented by a single octal digit, i.e. eight, we can see that this exactly matches the range of values that can be represented by three binary digits. Thus the translation between binary and octal may be achieved simply by splitting the binary word into groups of three bits and then replacing each group by a single octal digit.

$$101\,100\,011_2 = 101 \quad 100 \quad 011$$
$$= 5 \quad 4 \quad 3$$
$$= 543_8.$$

Similarly, to convert from octal to binary we need only to replace each octal digit by a group of three binary digits.

$$107_8 = 1 \quad 0 \quad 7$$
$$= 001 \quad 000 \quad 111$$
$$= 1000\,111_2.$$

Fractional values may be dealt with in a similar manner.

$$436.25_8 = 4 \quad 3 \quad 6 \quad . \quad 2 \quad 5$$

$$= 100 \ 011 \ 110 \ . \ 010 \ 101$$

$$= 100\,011\,110.010\,101_2.$$

HEXADECIMAL

Earlier in this section we chose a radix of eight for the number system we wished to use because eight is an integer power of two. Of course there are other integer powers of two that could be chosen. If we select a number system which has a radix of sixteen (2^4) then we are in fact selecting the hexadecimal number system. This system employs sixteen different symbols and so can represent values ranging from zero to fifteen before it is necessary to use multiple symbols and hence include positional information. The symbols used are the conventional decimal symbols $0, \ldots, 9$ to represent values up to nine; for values between ten and fifteen then the first six letters of the alphabet are employed. Thus the full sixteen symbols of the hexadecimal system are 0, 1, 2, 3, 4, 5, 6, 7, 8, 9, A, B, C, D, E, F.

$$2AF_H = 2 \times 16^2 + A(\text{ten}) \times 16^1 + F(\text{fifteen}) \times 16^0$$

$$= 687_{10}.$$

As we have already said, one hexadecimal digit can represent any value between zero and fifteen, the same range of values that may be represented by four binary digits. This allows a similar ease of conversion between binary and hexadecimal as between binary and octal. However each hexadecimal digit is now equivalent to four binary digits. The conversion between binary and hexadecimal thus becomes simply a matter of relating single hexadecimal digits to groups of four binary digits.

$$B75_H = B \quad 7 \quad 5$$

$$= 1011 \ 0111 \ 0101$$

$$= 1011\,0111\,0101_2;$$

$$1100\,0010\,1101.1010\,01_2 = 1100 \ 0010 \ 1101 \ . \ 1010 \ 0100$$

$$= C \quad 2 \quad D \quad . \quad A \quad 4$$

$$= C2D.A4_H.$$

Generally speaking, hexadecimal is much more commonly used for the recording of binary patterns than octal. The reason for this is that most digital computers use wordlengths (the number of bits that are usually treated together, as they often represent one item of data) of 8 bits, or multiples of 8 bits. If we use octal to record our 8-bit patterns then we must employ three octal digits, one of which will only be representing two bits and thus only have a range of zero to three. If we use hexadecimal however we need employ only two digits, each of which will represent a group of 4 bits.

In the octal and hexadecimal numbering systems we have convenient ways of representing the binary patterns that exist within the digital computer in a convenient form. However we, as humans, are, generally speaking, no more familiar with these two number systems than we are with the binary system itself. An alternative number system takes a different approach to the problem of interfacing between our, usually decimal, environment and the binary universe of the computer. This system, known as binary-coded decimal or BCD, effectively forces the computer system to adapt to a decimal world rather than requiring us to change our way of working to match that of the computer.

In the BCD system there is a straight correlation between each decimal digit and a group of binary digits just as there was for the octal and hexadecimal systems. As there are ten decimal symbols we require at least four binary digits to allow sufficient bit combinations for each digit, and BCD does indeed represent each decimal digit as a group of 4 bits, as shown in Table 2.3.

Table 2.3.

Decimal	BCD	Decimal	BCD
0	0000	5	0101
1	0001	6	0110
2	0010	7	0111
3	0011	8	1000
4	0100	9	1001

Conversion between decimal and BCD is thus as simple a process as converting between hexadecimal and straight binary, each decimal digit being replaced by a group of 4 bits, or vice versa.

$$153_{10} = \quad 1 \quad\quad 5 \quad\quad 3$$
$$= 0001 \quad 0101 \quad 0011$$
$$= 0001\,0101\,0011_{BCD};$$
$$1001\,0010\,0100_{BCD} = 1001 \quad 0010 \quad 0100$$
$$= \quad 9 \quad\quad 2 \quad\quad 4$$
$$= 924_{10}.$$

It appears, at first sight, that with the BCD system we have solved all our problems in relation to matching our decimal environment with the binary universe of the computer. However, this is not the case and, except in circumstances requiring constant updating of data via an operator (for example on some types of numerically controlled machinery), BCD is little used. There are two major problems with BCD. Firstly, the code is redundant in that we are using 4 bits to represent only ten values, whereas in straight binary 4 bits could be used to represent sixteen values. Secondly, and as a direct consequence of the unused six combinations, any arithmetic operations using BCD are much more involved, and thus slower, than equivalent operations using straight binary. Consider the addition $762_{10} + 849_{10} = 1611_{10}$ using BCD.

```
        0111    0110    0010
  +     1000    0100    1001
        ────    ────    ────
        1111    1010    1011
                       +0110
                ─────  ─────
                  +1    0001
               ─────
                1011
               +0110
        ─────   ─────
          +1     0001
        10000
       +0110
       ─────
      1  0110
  0001  0110    0001    0001
  ────  ────    ────    ────
   1     6       1       1
```

Notice that the sum of the least significant grouping gives a code (1011) which does not exist in BCD, it is one of the missing 6 states, to correct for this we must add an extra 6 to the result. This gives a carry one to the next group. This in turn must have a +6 correction factor applied, and so on until the addition is complete. If the 4-bit group sum is within the range 0 to 9 however no correction need take place.

Notice that correction factors for one group of 4 bits must be added before the addition for the next 4 bits can be completed. Thus there is a cascade of delays across the word and the addition takes a long time to complete.

EXCESS-THREE CODE (XS3)

Excess-three code, usually written as XS3, is a code that was developed to try and retain the advantages of BCD while minimizing the resulting arithmetical problems. XS3 is very similar to BCD in that a group of 4 bits is used to represent each decimal digit. However, compared to the BCD representation each value has an extra (an excess) of three added to it. Thus the XS3 codes for the ten decimal symbols are as shown in Table 2.4.

Table 2.4.

Decimal	XS3	Decimal	XS3
0	0011	5	1000
1	0100	6	1001
2	0101	7	1010
3	0110	8	1011
4	0111	9	1100

Again we are using 4 bits to represent only ten different combinations. However, since each code has an excess of three, when we add two XS3 numbers together we will already have the extra +6 correction factor that we had to employ with BCD addition. The need for further correction is not eliminated but the overall addition is speeded up since with XS3 this can be left to the last stage and completed simultaneously for all 4-bit groups.

Consider the XS3 addition of $148_{10} + 527_{10} = 675_{10}$.

```
          0100    0111    1011 XS3
    +1    1000    0101    1010 XS3
    ──────────────────────────────
                   +1     0101
          +0      1101
          1100
        −0011   −0011   +0011        Correction factors all being
    ──────────────────────────────   applied at the same time.
          1001    1010    1000 XS3
    ──────────────────────────────
           6       7       5
```

Notice that in the addition of the rightmost group of four the excess of six has effectively been lost by the carry of one to the next most significant group. Thus the correction factor for this least significant group of four must be plus three—to get back to XS3 format. However, for the other two groups of four, the carry out is zero and thus they retain an excess of six. In these cases the correction factor must be minus three to return to XS3 format. The judgment as to whether to add or subtract three makes XS3 arithmetic complicated, but the fact that all the corrections can be carried out at the same time and that earlier operations are not waiting for the completion of corrections of lesser significance does mean a considerable increase in speed over the equivalent operation in BCD, a speed advantage that increases as the wordlength grows.

Note that XS3 is fundamentally an unweighted code, since we cannot associate a particular arithmetic value with each digit position. However, we are still able to carry out arithmetic operations with XS3 because of its very close relationship to a weighted code (BCD). In general, unweighted codes are unsuitable for arithmetic operations.

OTHER WEIGHTED CODES

The most efficient coding, in the sense of being able to represent the maximum range of values using the fewest number of bits, will always be achieved by using straight binary. However, there may be occasions when, because of particular system considerations it may be advantageous to define column weights in other than a 1:2:4:8:16, etc., sequence. The use of other weightings is quite acceptable so long as consistency is maintained within the operation of the machine. An example of a weighted code, other than straight binary, is the 2:4:2:1 code. You will see that two columns in this code are given the same weight. The full code is listed in Table 2.5.

This code has the advantage that the value nine is represented by the code 1111 and thus there will be a natural carry out of one and a return to 0000 on a count of ten. This means that counters designed to count in 2:4:2:1 code will naturally divide the results of their counting into groups of four bits, each bit representing units, tens, hundreds, etc., in decimal.

Table 2.5.

Decimal	2:4:2:1 Code	Decimal	2:4:2:1 Code
0	0000	5	0101
1	0001	6	0110
2	0010	7	0111
3	0011	8	1110
4	0100	9	1111

2.3 Representing negative values

When we wish to distinguish between positive and negative values we usually precede the numbers we are interested in by "+" and "−", although the positive sign is often omitted in common usage. However, no such symbols exist within digital computers so we must find some other way to define the sign of the values we are using. Two general techniques are employed for this purpose, although one is much more widely used than the other.

2.3.1 Signed binary

The first technique, known as signed binary or sign-magnitude, appears, at first sight, to be the most obvious method but is, in fact, the least used of the two techniques. In the signed binary system the first (leftmost) bit of each word is used not for magnitude information but to represent the sign of the value (whose magnitude is defined by the other bits in the word). The symbols 0 and 1, in this position, are used as direct replacements for + and −. Zero is almost universally used for the + sign; however, a system which uses 0 as the negative sign is perfectly feasible so long as consistency is maintained. An example of signed binary, for an 8-bit machine might be the following.

$$
\begin{array}{ccc}
 & \text{Sign bit} & \text{Magnitude} \\
+7_{10} = & 0 & 000\,0111_2 \\
-7_{10} = & 1 & 000\,0111_2
\end{array}
$$

This is a very simple system to understand but there are certain penalties concerned with its operation. Firstly we are reducing the range of numbers that we are able to represent. If we use all 8 bits for value representation then we can represent values in the range 0 to $2^8 - 1 = 255$. However, by using one of the digits as a sign symbol the magnitude range is reduced to a limit of $2^7 - 1 = 127$, although now both positive and negative representations of this value are possible. As it happens, this is a disadvantage that is shared by the alternative technique for representing negative values and is one which cannot be avoided. Using signed binary also has consequences for the hardware design of the computer in that, if signed binary is employed, it is necessary to include circuitry that is capable of carrying out the process of subtraction. We shall see that this is not necessary if an alternative method of representation is employed.

2.3.2 Two's complement representation

Consider an 8-bit machine where the digit pattern 00101101 is added to 11010011:

$$
\begin{array}{r}
0010\,1101 \\
+\,1101\,0011 \\
\hline
0000\,0000
\end{array}
$$

1

↑
This digit lost

If we allow ourselves only an 8-bit answer then the final carry from the addition is lost since there is nowhere left to record its value. Thus as far as the computer is concerned the result of adding 00101101 to 11010011 is 00000000. Since thc only number that may be added to a value to form the result of zero is the negative of the original value then it appears, in computer terms, that 11010011 is the negative of of 00101101. In fact 11010011 is called the two's complement of 00101101. Two's complement notation, as it is called, is by far the most common way of representing negative values within modern digital computers.

The two's complement of a number may be formed in the following manner.

1 All positive numbers should start with at least one leading zero (thus the maximum positive number in an 8-bit machine is 01111111, as for signed binary).

2 Invert all bits, i.e. $1 \rightarrow 0$ and $0 \rightarrow 1$.

3 Add 1.

The final number formed is the two's complement of the original value, i.e. its negative as far as the computer is concerned. If we consider our original example, starting with the positive value 00101101, we invert all the digits to form the pattern 11010010, adding 1 wc gct 11010011 which we have already seen represents the negative of our original value in an 8-bit machine. Note that since all positive numbers must start with a zero all their two's complements will begin with a leading 1. The leading digit thus gives an immediate indication of the sign of the value. This is similar to the situation with signed binary. However, the two systems must not be confused since, in signed binary, sign and magnitude information are distinct, whereas with two's complement notation the two types of information are closely integrated. Note also that the two's complement representation of the same value will change if we change to a different wordlength. If, for example, our original value of 00101101 were to be placed in a 16-bit machine there would be more leading zeros, thus 0000000000101101. When we come to form the two's complement of this value all these leading zeros will become ones and so the negative of this number, in a 16-bit system, will be 1111111111010011. Note that two's complement notation forms a proper weighted code with the weight of the leftmost bit being -2^{n-1} for an n-bit word.

The two's complement of a value is easily formed within a machine, the complements of digits are readily available (the circuitry used to hold values within the central processing unit presents complements of bits automatically as a natural consequence of its structure) and adding in the extra one presents no difficulty and usually incurs no time overhead. As well as ease of formation, two's complement notation has a further significant advantage in that it does not require subtraction circuitry to be provided within the computer, since the process of subtraction is carried out by the addition of the two's complement. To illustrate this, consider the following examples (based upon a 4-bit machine for convenience). Suppose, using two's complement notation, we wish to carry out the simple subtraction, seven minus three. The two numbers represented as positive values will be:

$$7_{10} = 0111_2 \quad \text{and} \quad 3_{10} = 0011_2.$$

To represent minus three we form the two's complement of the pattern for plus three. Thus, inverting and adding one:

$$-3_{10} = 1101_2.$$

Thus $7-3$ is equivalent to:

$$
\begin{array}{r}
0111 \\
+ \ \ 1101 \\
\hline
1 \ \ 0100
\end{array}
$$

The final carry of one is lost so that the answer formed is 0100 (four), as we would expect. The system is consistent as we can see if we subtract seven from three. First, forming the two's complement of seven we get 1001, and so $3-7$ is equivalent to:

$$
\begin{array}{r}
0011 \\
+ \ \ 1001 \\
\hline
1100
\end{array}
$$

As the result starts with a one, we know that it represents a negative value. To determine its magnitude we must reverse the process by which we form the two's complement of a number, i.e. subtract one and then invert all the bits. This gives us a result of 0100. So our final answer above was minus four. This last operation of converting back from the two's complement of a value does not, of course, normally take place within the machine, since this is the way the machine represents negative numbers, but is just done for clarity here.

As well as two's complement notation it is also possible to use one's complement notation to represent negative values. This is very similar to two's complement notation. The one's complement of a binary number is formed simply by inverting all the digits. The one that is added to form the two's complement is omitted; however, when carrying out arithmetic using one's complement notation an extra one must be introduced during the arithmetic operation.

2.4 Overflow

We have already noted that the number of bits used to represent values within a particular computer will be fixed, indeed the use of two's complement notation depends upon there being such a fixed wordlength. However, a consequence of a fixed wordlength is that it restricts the range of number representation. Some arithmetic operations may produce results which exceed this limitation and it is important to recognize when these events occur. Consider the two examples of simple summations, using two's complement notation, shown below.

(a)   00110
  + <u>01000</u>   Carry into msb of 0.
       01110   Carry out from msb of 0.

(b)   01100
  + <u>01011</u>   Carry into msb of 1.
     <u>10111</u>   Carry out from msb of 0.

All the operands in each of the two examples are positive since they all having leading zeros. In example (a) the result is also positive: it too starts with a zero and, of course, we would expect a positive result if we add two positive values together. However, if we examine the result formed in example (b) we see that it begins with a one, that is, as far as the machine is concerned it is a negative value. It appears, therefore, that the addition of two positive numbers has yielded a negative result. This is clearly an error situation and is known as overflow. We now consider two further examples using negative operands.

(c)   11111
  + <u>11100</u>   Carry into msb of 1.
       11011   Carry out from msb of 1.

(d)   10000
  + <u>11110</u>   Carry into msb of 0.
     <u>01110</u>   Carry out from msb of 1.

Example (c) gives the correct result, the addition of two negative values producing a result that is also negative. Example (d), however, shows the addition of two negative numbers giving a positive result. Again this is an error situation owing to overflow.

Overflow and underflow occur because the magnitude of the results of arithmetic operations have become too large for the fixed wordlength of the computer to represent them properly: the results have gone out of range. It is very important to recognize when these conditions occur, as any subsequent manipulation of the data will be based upon incorrect values. Most computing machines set an error flag when overflow happens. The condition for setting the flag is based upon a comparison of the sign bits of the operand and the sign bit of the result. In Boolean terms (see Chapter 3) this can be written as

$$F = \bar{A}.\bar{B}.S + A.B.\bar{S},$$

where A and B represent the sign bits of the operands and S represents the sign bit of the result. In essence what the above equation is saying is that if the signs of both operands are the same as each other but different from the sign of the result then an error has occurred (overflow can never occur when the operand signs are different). It may be that in some arithmetic units, where one of the operands is lost as part of the addition process, it is not possible to implement the above checking structure; however, an alternative method of generating an error flag is available. If we examine the carries generated in examples (a) through (d) we can see that in cases where no overflow occurs the carry into the msb (C_m) is the same as the carry out of the msb position (C_{m+1}), whereas when an out-of-range condition does occur the value of the carry into the msb position is different from that of the carry out of the msb. Thus when these two carries are different we can set the error flag. This condition can be expressed succinctly as:

$$F = \overline{C_{m+1}}.C_m + C_{m+1}.\overline{C_m}.$$

2.5 Floating-point notation

As has already been stated, within any real computer there is a limited word length. This, in turn, imposes a limitation on the range of values that can be represented. So far we have always used what is known as fixed-point notation. This implies a fixed position for the binary point so that each bit position always has the same absolute weighting. Although, using fixed-point notation, the position of the binary point will not change, the choice of its position is somewhat arbitrary but its placement will determine the range of values that can be expressed. Consider a 16-bit word in a machine that employs two's complement format for the representation of negative values. One possible position for the binary point is immediately to the right of the least significant bit (lsb—the rightmost bit). This means, of course, that there is no fractional part to our number, i.e. we will only be able to represent integer values. In this case the range of values, other than zero, that can be represented are as follows.

Largest positive value:

$$0111\,1111\,1111\,1111. = 2^{15} - 1 = 32\,767.$$

Smallest positive value:

$$0000\,000\,000\,000\,001. = 1.$$

Largest negative value:

$$1000\,0000\,0000\,0000. = -2^{15} = -32\,768.$$

Smallest negative value:

$$1111\ 1111\ 1111\ 1111. = -1.$$

Thus the range is $+32\,767$ to $+1$, 0, -1 to $-32\,768$.

This is the greatest range of values that can be represented by 16 bits using fixed-point notation. If we devote some of the word to the representation of fractional values then the overall range will be reduced. As an example, suppose we place the binary point to the left of the fourth lsb (note that no binary point exists as a distinct symbol within the machine, it is only an assumed position that is being used). The range of values with this structure will be the following.

Largest positive value:

$$0111\ 1111\ 1111.1111 = (2^{11}-1).(1-2^{-4}) = 2047.9375.$$

Smallest positive value:

$$0000\ 0000\ 0000.0001 = 2^{-4} = 0.0625.$$

Largest negative value:

$$1000\ 0000\ 0000.0000 = -2048.$$

Smallest negative value:

$$1111\ 1111\ 1111.1111 = -0.0625.$$

Thus the range is $+2047.9375$ to $+0.0625$, 0, -0.0625 to -2048.

For many scientific and engineering calculations the range of values achievable using fixed-point notation is simply not adequate. To support the demands to be able to represent much larger, and smaller, values it is common for computers to offer another form of number representation known as floating-point representation. Floating-point format splits the binary word up into two main sections: a mantissa (or argument or fraction or significand) M, and an exponent (or characteristic) E: together these two sections represent the value V, where

$$V = M \times 2^E.$$

A binary-point position is assumed for the mantissa. However, the absolute weightings associated with the bits making up M will vary as the value of E changes—hence the term floating point.

The use of floating point greatly extends the range of number representation. If we take our 16-bit word and devote 10 bits to the mantissa, with a binary point placed to the right of the leftmost bit, and the remaining 6 bits to the exponent (both using two's complement notation for negative-number representation) then the range of values will be as follows.

Largest positive value:

$$0.1111\ 11111 \times 2^{01\ 1111} = 0.998 \times 2^{31} = 2.143 \times 10^9.$$

Smallest positive value:

$$0.0000\,00001 \times 2^{10\,0000} = 0.00193 \times 2^{-32} = 4.494 \times 10^{-13}.$$

Largest negative value:

$$1.0000\,00000 \times 2^{01\,1111} = -2.147 \times 10^9.$$

Smallest negative value:

$$1.1111\,11111 \times 2^{10\,0000} = -4.494 \times 10^{-13}.$$

Thus the range is $+2.143 \times 10^9$ to $+4.494 \times 10^{-13}$, 0, -4.494×10^{-13} to -2.147×10^9.

The penalty paid for this increase in range is some loss of precision, since some of the bits are now being used for the exponent, and an increase in the complexity of arithmetic operations. Figure 2.1 shows a typical structure for a floating-point word.

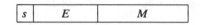

s represents the leftmost bit of the mantissa.

Figure 2.1.

This structure may, at first sight, appear rather strange, particularly with the mantissa being split into two parts; however, there are good reasons behind the use of such a format. The positioning of the sign bit of the mantissa (which is the sign of the overall floating-point word) in the leftmost position means that it occupies the same position as the sign bit of a fixed-point word, thus enabling the same circuitry and algorithms to be used for sign manipulation and checking in both formats. Putting the rest of the mantissa into the rightmost section of the word means that the lsb of the mantissa occupies the rightmost position, the same position as the lsb of fixed-point words. Again, this means that the same bit position in the word may be used to accept carry-in signals in both formats. Also putting the exponent to the left of the mantissa means that we have a continuing increase in significance as we move leftwards across the word. The same situation exists for fixed-point words and thus the same techniques for such operations as magnitude comparison can be used in both formats.

If we wish to be able to represent both very large and very small values then we must allow for both positive and negative exponent values. One way of doing this is to use two's complement notation for the representation of negative values, just as we did in our earlier example. However, a somewhat better alternative is to use what is known as a biassed exponent. With a biassed exponent a constant is added to each exponent such that, internally, all exponents appear positive (i.e. two's complement notation is not used). This would mean that with our 6-bit exponent, for internal representation, we would add 2^5 to the true value. The smallest exponent would thus be 00 0000 and the largest 11 1111; 10 0000 would represent a true exponent of zero. Using a biassed exponent has two advantages:

1 the absence of negative exponents provides some simplification during arithmetic operations;

2 a better representation for the value zero.

It might appear that the value of the exponent when the mantissa is zero is quite immaterial, since 0 times anything is zero. However, matters are not quite as simple as this. If we leave the exponent of a word at some arbitrary value when the mantissa becomes zero we have what is known as a dirty zero. As some arithmetic operations require us to shift mantissa values until exponents match, dirty zeros may result in an unnecessary loss of significance. This problem is overcome if we ensure that all zeros are clean zeros, that is, that the exponent part of the word is set to the smallest exponent value. Using a biassed exponent, this smallest value is itself zero. It is very easy to implement a sequence such that when the mantissa becomes zero the whole word is cleared, automatically setting the biassed exponent to its smallest value and producing a zero. A biassed exponent also means that the clean zero representation of a clean zero in floating point has exactly the same form as a fixed-point zero and this has obvious advantages for system design.

There are several different ways in which a floating-point number may be represented in decimal, for example 0.438×10^5 is the same as 0.00438×10^7. However, within our computer we only have a limited number of bits to use for our mantissa and to waste them on unnecessary leading zeros means that we are reducing the precision of our value representation. Thus it is standard practice to adjust the exponent value to ensure that the leading bit of the mantissa (other than the sign bit) is always 1: this process is called normalization and it is usual to conclude each part of an arithmetic operation with this activity (then known as post-normalization) to avoid a progressive loss of precision as the calculation continues.

2.5.1 Floating-point arithmetic

Floating-point arithmetic is more complicated and thus slower than equivalent fixed-point operations, the reason being that we must now deal with the exponent as well as with the mantissa.

FLOATING-POINT ADDITION AND SUBTRACTION

Before the addition or subtraction of floating-point numbers can begin it is first necessary to make the exponents of the operands agree. This is achieved by shifting the mantissas and making compensating changes to the exponents. It is usual to change the smallest exponent until it matches the largest exponent.

Consider the addition of $0.1101\,1000 \times 2^{0101}$ and $0.1000\,1000 \times 2^{0111}$.

Mantissa	Exponent
0.1101 1000	0101
0.1000 1000	0111

Shifting the first mantissa two places right to match exponents gives the following.

$$0.0011\,0110 \qquad 0111$$
$$0.1000\,1000 \qquad 0111$$

Adding mantissas, we now have

$$0.1011\,1110 \qquad 0111.$$

Thus the result is $0.1011\,1110 \times 2^{0111}$.

The shifting operation will lead to the loss of some precision if ones are lost from the smallest operand as part of the shifting process; indeed, if the difference in the original exponent values is greater that the number of bits in the mantissa, then all the bits of the smaller number will be lost and the result of the addition will be equal to the larger of the two operands.

Consider the following example:

$$0.1101\,0101 \times 2^{1100} + 0.1010\,1010 \times 2^{0011}.$$

To match the exponent of the second operand (0011) to the larger exponent of the first operand (1100) we must shift the mantissa of the second operand by $(1100 - 0011)$ nine places. As there are only 8 bits (apart from the sign bit) in the mantissa the resulting addition becomes:

$$0.1101\,0101 \times 2^{1100} + 0.0000\,0000 \times 2^{1100} = 0.1101\,0101 \times 2^{1100}.$$

The importance of not retaining dirty zeros becomes apparent if we consider the addition of a dirty zero to another value with a smaller exponent:

$$0.0000\,0000 \times 2^{1110} + 0.1111\,1111 \times 2^{0010}.$$

After shifting to match exponents this will become:

$$0.0000\,0000 \times 2^{1110} + 0.0000\,0000 \times 2^{1100} = 0.0000\,0000 \times 2^{1100}.$$

The presence of the dirty zero means that precision has been lost unnecessarily. If a clean zero had been used as one of the original operands then it (by definition) would have had the smallest exponent and the shifting operation would have resulted only in the loss of zeros.

It is quite common for floating-point arithmetic units to include special circuitry to determine, early on in the operation, if one of the mantissas is zero, in which case the result can be set to the value of the other operand immediately, and also if the difference in the exponent values is greater than the number of bits in the mantissa, in which case the result can be set to the value of the largest operand immediately. In other cases the exponent-matching operation is followed by a conventional addition of the mantissa parts. If this results in an overflow then the result mantissa is shifted one place to the right and the result exponent increased by one.

Floating-point subtraction may employ two's complement notation or signed binary. The only significant differences between subtraction and addition are that, if two's complement notation is used, when shifting a negative mantissa to the right in order to match exponents, ones will be introduced into the left-hand side of the word in order to preserve its sign. After the subtraction it may be necessary to shift the result mantissa to the left (modifying the exponent as a consequence) in order to get rid of unwanted leading bits (post normalization).

FLOATING POINT MULTIPLICATION AND DIVISION

In some ways floating-point multiplication and division are easier than floating-point addition and subtraction, since with multiplication and division there is no need to match exponents prior to the operation. However, exponent manipulation is still required. In the multiplication of two floating-point values the exponents of the operands will be summed to form the exponent of the result, the product of the mantissas being formed as for fixed-point working. For floating-point division the exponent of the divisor is subtracted from the exponent of the dividend before the mantissas are operated upon. In both cases a normalization process should be implemented after the main operation. The detection of an all-zero operand is particularly important for multiplication and division algorithms as the detection of a zero in either of the operands in a multiplication means that time can be saved by setting the result to zero at once, the detection of a dividend of zero in a division operation can be used to the same end whereas the detection of a divisor of zero is usually used to set an error condition.

FLOATING-POINT FORMAT

There are no absolute reasons why any particular number of bits should be devoted to the exponent (with the remainder left to the mantissa); the greater the number of bits given to the exponent the greater the range of values that may be represented by a defined wordlength, but also the lesser the precision of that representation. Until relatively recently it was usual to find different computer manufacturers providing different formats for the representation of floating-point numbers within their machines and this could lead to complications when tasks were moved from one machine to another. However, in the early 1980s the IEEE Computer Society produced a standard, number 754, for floating-point arithmetic. This standard defines word structures for 32-bit and 64-bit wordlengths used for floating-point numbers. The 32-bit wordlength, for example, uses bit 0 (the leftmost bit) for the sign bit, bits 1 to 8 for the exponent and bit 9 to 31 for the mantissa. The introduction of this standard has done much to ease the communication problems between machines when using floating-point representation.

2.6 Error detection and correction

Although digital systems are usually very reliable, errors can occur because of faulty components, poor connections or the effect of electrical noise. An error in a digital system will result in a 1 being replaced by a 0, or a 0 being replaced by a 1. It is essential that any such errors are detected so that action may be taken to rectify them. Several different systems have been developed to assist in the detection and even the self-correction of errors. However, all these different systems have one thing in common and that is that they are based on bit redundancy. In essence this means that some bits in the word are used for checking purposes rather than for data. As the sophistication of the error-detection/correction system grows, a greater proportion of the word is devoted to check bits rather than to data bits and thus the efficiency of data representation diminishes. The choice of the correct level of coding is thus an important factor at the design stage: too many check bits and we will be wasting system capacity that could be used for holding data bits; too few check bits and we may not be able to recognize that incorrect data is being produced.

2.6.1 Special codes

Special codes are those which have particular characteristics that enable errors to be readily detected as deviations from those characteristics. An example of such a code is BCD which uses only 10 of a possible 16 combinations; should one of the six unused combinations occur then an error must have appeared. However, BCD is not a very good code for error-detection purposes since many single errors may occur which will simply cause a change to another of the ten used combinations, thus leaving the error undetected. An example of another, special code that is a little more suitable for error detection is the 2-out-of-5 code. This is listed in Table 2.6.

Table 2.6.

Decimal	2-out-of-5	Decimal	2-out-of-5
0	00011	5	01100
1	00101	6	10001
2	00110	7	10010
3	01001	8	10100
4	01010	9	11000

The origin of the code's name is obvious from Table 2.6: each of the five bit combinations has two ones; any code with other than two ones will be seen as an error. Note, however, the increase in the inefficiency of our data representation, as we are now using five bits, which are capable of representing 32 different values to represent only ten different values. The 2-out-of-5 code suffers from a further disadvantage in that it is an unweighted code and thus must be converted into another code prior to any arithmetic operations.

2.6.2 Parity checking

By far the most common form of error detection used in digital systems is the single-bit parity check. This form of checking involves the addition of one more bit alongside the data bits. The data bits themselves are unaltered and thus if they were weighted it is still possible to involve them in arithmetic operations.

There are two forms of single-bit parity check: odd parity and even parity, the names referring to the number of ones in the overall word (including the parity bit) when it is error free. As an example, suppose we have a data word 01100. In an odd parity system this would become 011001, the parity bit making the overall number of ones in the word odd, whereas if we use even parity the code becomes 011000, with an even number of ones in the overall word. Errors are detected when we find a word with an even number of ones in an odd parity system, or a word with an odd number of ones in even parity system. Notice that if a double error occurs then the errors will go undetected because we will detect the correct number of ones in the word. Single-bit parity codes are distance-two codes: this means that changing two bits of a correct code will move us to another apparently correct code and the errors will go undetected.

There is little to choose between odd and even parity. However, odd parity tends to be more widely used because every word will contain at least one 1 and, since most machines use an even number of bits in their words, odd parity will show up errors that cause all the bits in a word to be set to all ones or all zeros. Parity-bit generation and checking are very easily implemented within the system by the use of exclusive-OR gates (see Chapter 3 for further details on these gates).

2.6.3 Error-correcting codes

HAMMING CODES

We mentioned above that a single parity-check digit produced a distance-two code which was capable of detecting one error but which would miss two errors. Also, with a single parity check, once the error is detected, the only remedial action that can be taken is to ask for a repeat of the message or to report the error to the operator. If, however, we add more parity bits to the word we can increase the distance of the code, allowing us to detect more errors, at the cost, of course, of increased redundancy in terms of data representation. If we add sufficient extra check bits then we can produce codes which can not only detect errors but also correct them. One class of error-detecting codes are known as Hamming codes after Richard Hamming, who did much of the early work on their development (the distance of a code is sometimes known as the Hamming distance for the same reason).

Consider a 7-bit word arranged as shown below.

Column number	1	2	3	4	5	6	7
Weight	P_1	P_2	8	P_3	4	2	1
Code example	1	0	1	1	0	1	0

This particular code is based on even parity, columns 3, 5, 6 and 7 representing the data bits with standard binary weightings. We can see that the value ten is being represented here. Columns 1, 2 and 4 are three parity-check bits. The use of three parity bits produces a distance-three code which can be used to detect up to two errors or to reliably correct a single error (if we use the code for error correction and two errors occur we will "correct" to the wrong value).

To see how the error correction works, suppose that the above code is received but that column 5 has been corrupted and changed from a 0 to a 1. Thus the codes we receive is

Column number	1	2	3	4	5	6	7
Code	1	0	1	1	1	1	0

On receipt of the code we carry out three parity-check operations, in this case looking for even parity. The first check is made on columns 4, 5, 6 and 7. As there are an odd number of ones in these columns, this shows an error. We record this error as a 1. The second check is made on columns 2, 3, 6 and 7. This shows even parity in these columns and thus that no error is present. This is recorded as a 0. The third check is made on columns 1, 2, 5 and 7. This check again shows an error as there are an odd number of ones. The error is recorded as a 1.

We can see that our checks reveal that the error is present in columns 4, 5, 6 and 7 and in 1, 2, 5 and 7 but not in 2, 3, 6 and 7. The only column that fits with this arrangement is column 5; so this is the column we must correct to produce the true data. However, it is not necessary to consider which column fits in some sets and not in others. If we consider the error/no error codes we recorded during our three checks, we can form a 3-bit word. This 3-bit word forms the column address of the bit that is in error, 101 in this case. If there are no errors the checks will produce an all-zero address (000); if there are two errors a non-zero address will be formed, but it will be the wrong address; more than two errors could produce an all-zero address and thus go undetected. Notice that this code is using 7 bits which are capable of representing 128 different patterns but is using only sixteen of them, as only four of the bits are data bits.

The number of parity bits, P, required for single-error correction may be determined from the following formula:

$$2^P \geqslant P + n + 1,$$

where n is the number of data bits in the word, e.g. for $n = 4$ as in the previous example, P must be at least three. The parity-check bits are then placed in columns whose addresses are integer powers of two, that is, columns 1, 2 and 4 in the previous example. Each parity-bit value will be determined by a parity check on all those columns whose column address, in binary, contains a one in a position corresponding to the (single) one in the address of the check digit itself; e.g. in the previous example check digit P_2, placed in column 2 (010) will be formed by a parity check on all columns whose address has a one in the middle bit position, that is, columns 2 (010), 3 (011), 6 (110) and 7 (111).

An alternative mechanism for error correction is based upon chain codes. Chain codes are easily formed by circuits known as shift registers (see Chapter 3 for more details on these circuits). Exclusive-OR gating is applied to these shift registers which converts them into a special form called linear feedback shift registers (LFSR). The data is fed into the shift registers and then a clock pulse is applied which shifts the data along and also generates new digits; these new digits are themselves shifted, producing further new digits. All the digits produced are related to each other and to the original data, the relationship being known from the design of the LFSR, and this relationship can be used to recover data lost through errors.

The number of check digits is large compared with the number of data bits, for example, a 4-bit data word will produce a total wordlength of 15 bits (11 check bits) and a 7-bit data length will generate a 127-bit word (120 check bits). Thus these codes are highly redundant, but they do have large Hamming distances. The 15-bit code has a Hamming distance of 8, for example, allowing for the detection of 7 errors, or the correction of 3 errors. Cross-correlation techniques are used at the receiver in order to determine the correct code. Because of their high redundancy these codes are not commonly used within digital computers but are used in transmission systems where very high reliability is required and where the transmission channel is noisy and the received signal weak. Such conditions occur when deep-space probes communicate with their ground stations.

2.7 Unweighted codes

So far in this chapter we have been much concerned with representing arithmetic values and carrying out mathematical operations and, with the exception of XS3 which is a rather special case, we have used weighted codes. However, it is often the case that we wish to represent other than arithmetic values or that we require a particular arrangement of bit patterns in some circumstances; we are likely then to resort to the use of unweighted codes, codes in which we cannot associate a particular value with a particular bit position but where the overall pattern of the bits is the important factor.

2.7.1 Cyclic codes

A particularly important group of unweighted codes are the cyclic codes. These codes are so called because of their wrap-around nature, where the bit pattern representing the largest value may be seen to be only one change away from the bit pattern representing the smallest value. These codes have particular application in some types of instrumentation devices.

The codes we have encountered so far have all involved multiple bit changes as we moved from one value to the next. For instance, consider the change from seven to eight in BCD: the code change will be from 0111 to

1000, a change which involves all four digits. Such multiple bit changes can lead to problems in some applications where the transitions may not take place simultaneously and thus several false codes sensed during the transition period. In order to overcome this problem we can use a cyclic code which also has the property of being a unit-distance code. The term unit-distance means that each code differs from its neighbors by only one change of digit and thus that false codes, resulting from multiple bit changes during transitions, will not occur. The most common application of these codes, usually referred to as Gray codes, is in shaft converters. A shaft converter is a transducer which translates the angular position of a shaft into an electrical signal, in our case a pattern of binary digits. They are often used in numerically controlled machines and robots. Figure 2.2(a) shows the structure of a 4-bit shaft encoder based on a straight binary weighting, whereas Figure 2.2(b) shows a similar shaft encoder using a Gray code. Each track on the disc will have a separate sensor (usually a light-emitting diode/phototransistor pair).

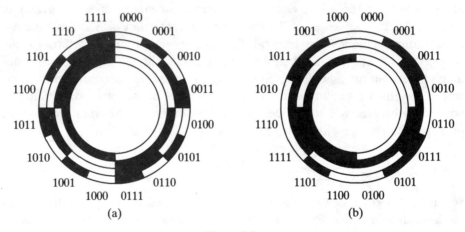

Figure 2.2.

It can be seen that an angular change on the binary-coded disc may result in several track changes; on the Gray-coded disc only one track changes state at any one time, thus eliminating the possibility of false readings. Practical shaft encoders will have many more tracks than the four shown in the Figure 2.2. This will allow a much greater angular precision; for example, a ten-track shaft encoder will contain 2^{10} different codes, an angular precision of a little over one-third of a degree. A full listing of the 4-bit Gray code is given in Table 2.7.

Table 2.7 shows a further property of these codes in that all but the most significant digit of the code is reflected about the center of tabulation; indeed, these codes are sometimes referred to as reflected codes. Notice also that cyclic codes of less than 2^n states may be derived from a full 2^n-state code by selecting the required number of states, centered on the center of tabulation. For example, in Table 2.7, by taking five states from each side of the center of tabulation, we produce a ten-state code that is itself of unit distance and cyclical in nature.

Table 2.7.

Decimal	Gray code		
0	0000		
1	0001		
2	0011		
3	0010		
4	0110		
5	0111		
6	0101		
7	0100	Center of tabulation	Ten-state code
8	1100		
9	1101		
10	1111		
11	1110		
12	1010		
13	1011		
14	1001		
15	1000		

2.7.2 Character codes

Studies have shown that computers spend most of their time in moving and storing data and that relatively little time is spent in arithmetic and logical processing. Most of the data presented to computers is in the form of records, usually entered as alphanumeric characters. Because of this the representation of alphanumeric characters within computers is well formalized with well-defined patterns of ones and zeros being used to represent letters, numerals, punctuation characters and control codes. It is obvious that if computers are to be able to communicate reliably with each other then the same bit patterns must be used to represent particular characters in each machine. The most commonly used convention for the representation of characters is known as ASCII (American Standard Code for Information Interchange). This code uses 7 bits to represent each character—though it is often transmitted as an 8-bit code, the extra bit being a parity-check bit. As an example "A" in ASCII is coded as 100 0001 and lower case "a" is 110 0001. A full listing of the ASCII code is given Appendix 2.

The second common code for character representation is EBCDIC (Extended Binary-Coded Decimal Interchange Code). This is an 8-bit code, often used in IBM machines. In EBCDIC the pattern 1111 1001 represents the "9" character, and 0101 1011 represents the "$" sign. A full listing of the EBCDIC code is given in Appendix 2.

2.8 Summary

Chapter 2 has considered the way in which magnitudes and symbols may be represented within the binary universe of the digital computer. The relationship between binary and other codes was discussed, together with the way in which arithmetic is carried out in the binary system. The importance of error

47

checking and correction to the reliable operation of digital systems has been outlined. It has been the intention of this chapter to give the reader an understanding of two-state systems; an appreciation of the limitations of such systems is a necessary precursor to many of the topics developed in later chapters.

2.9 Further reading

1 Bannister, B. R. and Whitehead, D. G. (1983). *Fundamentals of Modern Digital Systems*, MacMillan.
2 Cooke, D. J. and Bez, H. E. (1984). *Computer Mathematics*, Cambridge University Press.
3 Downton, A. C. (1984). *Computers and Microprocessors*, Van Nostrand Reinhold.
4 Floyd, T. L. (1982). *Digital Fundamentals*, Merrill.
5 Lewin, D. (1985). *Design of Logic Systems*, Van Nostrand Reinhold.
6 Peterson, F. J. and Hill, G. R. (1987). *Digital Systems*, Wiley.

2.10 Problems

2.1 Translate the following decimal numbers into binary:

(a) 14_{10}; (b) 175_{10}; (c) 2563_{10}.

2.2 Translate the following binary numbers into decimal (assuming that all numbers are positive):

(a) 10110011_2; (b) 10101010_2; (c) 01111110_2.

2.3 Represent the following values in both binary and decimal:

(a) 73_8; (b) 145_8; (c) $3F_H$; (d) $A4C_H$.

2.4 Carry out the following additions in binary:

(a) $82_{10}+74_8$; (b) 16_8+F7_H.

2.5 Carry out the following subtractions in binary:

(a) 75_8-14_{10}; (b) AA_H-25_{10}.

2.6 Carry out the following binary multiplications:

(a) $1010_2\times1001_2$; (b) $1110_2\times11.01_2$.

2.7 Carry out the following binary divisions:

(a) $11011_2\div11_2$; (b) $100001_2\div1000_2$.

2.8 Carry out the following in the home-base notation:

(a) 273_8+141_8; (b) $F03_H+1BA_H$.

2.9 Repeat the operations of question 2.8 by translating the values into binary and then working in the binary system. Verify your answers by converting them back to the number system used in question 2.8.

2.10 Represent the following values in binary-coded decimal (BCD):

(a) 349_{10}; (b) 174_8; (c) 10011001_2.

2.11 Represent the following values in excess-three code (XS3):

(a) 596_{10}; (b) $D3_H$; (c) 10011001_{BCD}.

2.12 Carry out the following additions in BCD:

(a) $141_{10}+591_{10}$; (b) 141_8+591_H.

2.13 Repeat question 2.12 using XS3 instead of BCD.

2.14 Assuming a machine with an 8-bit wordlength, represent the following values using two's complement notation.

(a) 27_{10};   (b) -19_{10};   (c) -63_8;   (d) $-0F_H$.

2.15 Repeat question 2.14 using signed-binary notation.

2.16 Carry out the following arithmetic operations by translating the values into binary and using two's complement notation. Assume a machine with an 8-bit wordlength.

(a) $82_{10} - 17_{10}$;   (b) $126_8 - 64_8$;   (c) $3B_H - 4A_H$.

2.17 Explain what is meant by overflow and how the condition might be recognized in a digital machine.

2.18 A computer uses a 32-bit wordlength. If only integer values can be represented, determine the range of values that can be represented using fixed-point notation. Assume the use of two's complement notation for negative numbers.

2.19 The 32-bit machine mentioned in question 2.18 can also use floating-point format. 22 bits are used for the mantissa (all of it fractional apart from the msb) and the remaining 10 bits are used for the exponent. The exponent is always an integer. Both exponent and mantissa use two's complements notation. Determine the range of values that can be represented by this structure.

2.20 Determine the bit pattern that would represent the following values in the floating-point system described in question 2.19:

(a) 0.015625_{10};   (b) $-7.392812_{10} \times 10^7$.

2.21 A system using the form of Hamming code described in section 2.6.3 receives the following bit pattern:

$$0100011.$$

Determine if the pattern is in error and, if so, what the most likely correct code is.

2.22 Form a 12-state cyclic Gray code.

3: Basic Logic Structures

3.1 Introduction

In Chapter 2 we discussed data representation and arithmetic as they relate to the two-state environment of the digital computer. We will now go on to consider how general conditions in a two-state system can be described, that is, we will discuss the algebra of such a system. This chapter will then go on to discuss the gating structures that can be used to fulfill digital design requirements and finally to describe some of the more common circuit technologies that are used to implement these gates.

3.2 Boolean algebra

The type of algebra used to describe two-state systems is known as Boolean algebra after George Boole, an English mathematician who, in the middle of the 19th century, did much to develop the axioms of the algebra. However it was not until the 1940s that any general application was found for his work.

The facility to manipulate and minimize Boolean expressions was once an essential skill for the designer of digital systems. Minimization of expressions usually meant that a particular design could be implemented with the least amount of hardware, and when hardware costs were the dominant factor in the determination of a product's price, the saving of just a few logic gates could be important. With improved technology, however, the cost of hardware is no longer so significant and design time may be more a critical factor in cost terms than producing a minimal gate structure. Nevertheless, a knowledge of Boolean algebra and the ability to interpret Boolean expressions is still important to anyone involved with digital systems since Boolean statements are still widely used on data sheets to describe device functions and to communicate ideas between designers and users.

To those being exposed to Boolean algebra for the first time it can appear to be very involved. However, the reader will quickly see that this algebra is much less complicated than "normal" linear algebra. As with the binary number system the important thing to keep in mind is that all defined variables in Boolean algebra may take only one of two values: true or false, high or low, one or zero.

3.2.1 Fundamental operations

There are only three basic operators used in Boolean algebra.

1 Complementing, sometimes called inversion, negation or the NOT operator. This operation is applied to a single entity and simply inverts the original value of that entity. The complement of an entity is shown by the placing of a bar over the symbol representing the entity. For instance, suppose we have a variable A, then the complement of this variable is shown as $\bar{A}$: $\bar{A}$ will take the opposite value to that of A, i.e. when A is true $\bar{A}$ is false.

This relationship can also be described by a truth table. A truth table lists all possible variable combinations and the result of any relevant functions formed by the combination of those variables. It is possible to produce this listing for Boolean variables since each variable can have only two values and hence the number of variable combinations is limited to 2^n, where n is the number of variables. In the case of the complement function there is only one variable (A) and so the truth table will have only two rows:

A	$\bar{A}$
0	1
1	0

2 The second fundamental Boolean operation combines two variables and is known as the OR operator (sometimes called the inclusive-OR operator to distinguish it from a similar function we will come across later). The result of the OR operator will be true if either, or both, of the two input variables are true. The symbol for the OR operator is a + sign. Do not confuse this with conventional arithmetic addition which employs the same symbol. In Boolean algebra,

$$f = A + B$$

is read as f equals A or B. The truth table for this function will be as follows.

A	B	f
0	0	0
0	1	1
1	0	1
1	1	1

3 The final fundamental Boolean operator is the AND function. The result of an AND function is true only if all the input variables to the function are true; otherwise the result of the function is false. The AND operator is formally shown by a **.** between the variables, but often the **.** is omitted and the variables are simply written next to each other. Thus

$$f = A.B = AB.$$

This statement would be read as f equals the AND of A and B or simply f equals A and B. The truth table for this function will be as follows.

A	B	f
0	0	0
0	1	0
1	0	0
1	1	1

The OR and the AND functions are not limited to pairs of variables but may be expanded. Thus

$$f = A + B + C$$

and

$$f = A.B.C.D = ABCD$$

are perfectly valid statements. In the first case f will be true when A or B or C, or any two, or all three of A, B and C are true. In the second case f will only be true when A and B and C and D are all true at the same time.

All more complex Boolean statements are built up by combining these three fundamental operations.

A particular operation is so common that it has its own symbol, $\oplus$, and is known as the exclusive-OR function, or Ex-OR. The result of this function is true if either, but not both, of the input variables is true, i.e.

$$f = A \oplus B$$

would be read as f is the exclusive OR of A with B. The truth table for the function is as follows.

A	B	f
0	0	0
0	1	1
1	0	1
1	1	0

However, the Ex-OR is not a fundamental Boolean operation since it can be described in terms of more simple operations, i.e.

$$f = A \oplus B = \bar{A}B + A\bar{B}.$$

3.2.2 Axioms and laws

The basic axioms of Boolean algebra are as follows.

$A.A = A$ The function is true when A and A are true and is thus only dependent upon the state of A.

$A.\bar{A} = 0$ A and $\bar{A}$ can never be true at the same time and so the result of the function is always false.

$A.0 = 0$ ANDing anything with something that is always false will yield a result that is always false.

$A.1 = A$ The 1 part of the function is always true and so the result depends only upon A.

$A + A = A$ The two parts of the OR function are true and false at the same time; thus the result only depends upon whether A is true or false.

$A + \bar{A} = 1$ Whenever A is false $\bar{A}$ is true and vice versa. Thus one part of the function is always true so the result is always true.

$A+0 = A$ Whether the result is true or false depends solely upon A.

$A+1 = 1$ Since the 1 part of the function is always true the result of the function is always true.

We now list the basic laws of Boolean algebra.

(a)
$$A+B = B+A,$$
(the commutative laws)

(b)
$$A.B = B.A;$$

(c)
$$A+(B+C) = (A+B)+C,$$
(the associative laws)

(d)
$$A.(B.C) = (A.B).C.$$

These two sets of laws state that the order of variables and the order of operation do not matter in Boolean algebra (they do matter in vector algebra, for example).

(e)
$$A.(B+C) = A.B+A.C. \qquad \text{(distributive laws)}$$

$$(A+B).(A+C) = A.A+A.B+A.C+B.C$$

$$= A+A.B+A.C+B.C$$

$$= A.(1+B+C)+B.C.$$

Thus

(f)
$$(A+B).(A+C) = A+B.C.$$

These two laws show how the AND and OR operators may be manipulated.

$$A+\bar{A}.B = A.(B+\bar{B})+\bar{A}.B$$

$$= A.B+A.\bar{B}+\bar{A}.B$$

$$= A.(B+\bar{B})+B.(A+\bar{A})$$

$$= A.1+B.1.$$

Thus

(g)
$$A+\bar{A}B = A+B. \qquad \text{(absorption law)}$$

This law shows that when a variable is ORed with the AND of its complement and another variable the complement may be eliminated.

(h)
$$\overline{A+B} = \bar{A}.\bar{B},$$
(De Morgan's laws)

(i)
$$\overline{A.B} = \bar{A}+\bar{B}.$$

(Note the bar over the whole term means that the final result of the operation is inverted rather than individual variables being inverted prior to the operation taking place.)

These last two laws can be very useful in the manipulation of Boolean expressions. The veracity of the statements may be shown using a truth table. (This is sometimes known as proof by perfect induction.)

A B	$\bar{A}$	$\bar{B}$	$A+B$	$\overline{A+B}$ (1)	$\bar{A}.\bar{B}$ (2)	$A.B$	$\overline{A.B}$ (3)	$\bar{A}+\bar{B}$ (4)
0 0	1	1	0	1	1	0	1	1
0 1	1	0	1	0	0	0	1	1
1 0	0	1	1	0	0	0	1	1
1 1	0	0	1	0	0	1	0	0

Since the results in column 1 match those in column 2,

$$\overline{A+B} = \bar{A}.\bar{B}.$$

Similarly since the results in column 3 match those in column 4,

$$\overline{A.B} = \bar{A}+\bar{B}.$$

These laws and axioms can be used to modify and manipulate Boolean expressions. There may be two main reasons for wanting to do this: firstly to check if two statements are equivalent and secondly to simplify an expression to its minimal form and hence reduce the amount of hardware required for its implementation. For example, we may be required to minimize the expression:

$$f = ABCD + AB\bar{C} + ABC\bar{D} + A\bar{B}\bar{C} + \bar{A}BCD$$

$$= (A+\bar{A})BCD + A\bar{C}(B+\bar{B}) + AB(C\bar{D}+CD+\bar{C})$$

$$= 1.BCD + A\bar{C}.1 + AB(C+\bar{C})$$

$$= BCD + A\bar{C} + AB.1$$

$$= BCD + A\bar{C} + AB,$$

i.e. by using manipulative techniques an expression originally involving five AND functions, three of which would require four-input gates, and a five-input OR function, has been reduced to an equivalent form which only needs three AND functions and a three-input OR gate.

Useful though the above techniques can be, it is easy to make mistakes. If the number of variables is not large then a graphical method may be used to simplify functions. This makes use of the Karnaugh map.

3.2.3 The Karnaugh map

The easiest way to appreciate the structure of the Karnaugh map is to see how it is built up from very simple beginnings. The Karnaugh map represents the whole of the Boolean universe under consideration as a rectangle. In the one-variable case the rectangle is split into two halves. One half represents the the condition of the variable being true, the other half represents the condition of the variable being false, as in Figure 3.1 (a). To represent a function on the map a 1 is placed in the region where the function is true, whereas a 0 is placed in the region where the function is false. Thus, if $f = A$ then the map representing f would be as in Figure 3.1 (b), and if $f = \bar{A}$ then f would be shown as in Figure 3.1 (c).

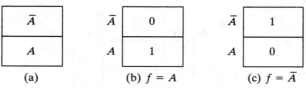

Figure 3.1. Single-variable maps.

If two variables are involved then the rectangle is split along its other axis to form four areas, as in Figure 3.2 (a). The areas on the map represent every possible combination of the two variables, e.g. the function $f = A \oplus B$ would be represented as in Figure 3.2 (b).

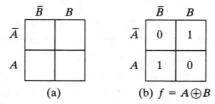

Figure 3.2. Two-variable maps.

If three variables are to be included then the third variable is introduced by further dividing the map as in Figure 3.3.

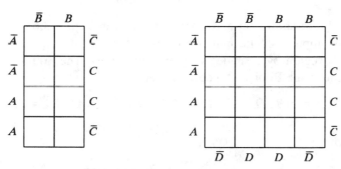

Figure 3.3. Three-variable maps   Figure 3.4. Four-variable maps

The two middle rows of the map represent regions where the third variable is truc. The upper and lower rows represent regions where the third variable is false. A fourth variable is introduced by dividing the map in a similar way as was done for the third variable but with the split being made along the other axis, as in Figure 3.4. The order in which the split is made is immaterial. It is also more usual to declare variables for each axis at the top left hand corner of the map and to show them as being true or false for each row and column by a 1,0 code. Thus the most usual representation for a four-variable map would be as in Figure 3.5.

The way in which the map has been split allows one region for each possible combination of variables, thus there are sixteen regions on a four-variable map. More than this, however, the arrangement of the split means that each

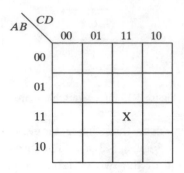

Figure 3.5. Region $ABCD$.

region differs from its immediate neighbors by only one variable. Consider the variable combination $ABCD$; this occupies the region of the map marked by an X in Figure 3.5. The regions adjacent to this area are $ABC\bar{D}$, $AB\bar{C}D$, $A\bar{B}CD$ and $\bar{A}BCD$, i.e. they all differ from $ABCD$ by the status of only one variable. This means that adjacent regions can be combined and the variable by which they differ eliminated from the description of the combined region, hence yielding a simpler expression.

Consider the function

$$f = ABCD + ABC\bar{D}.$$

This would be shown on a Karnaugh map as in Figure 3.6. Since two adjacent ones appear on the map we can combine them and describe the larger (outlined) region by a single, simpler expression. The graphical technique is equivalent to the Boolean-algebra sequence:

$$f = ABCD + ABC\bar{D} = ABC(D + \bar{D}) = ABC.1 = ABC.$$

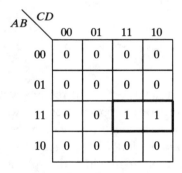

Figure 3.6. $f = ABCD + ABC\bar{D} = ABC$.

Adjacent ones may be grouped in twos, fours, eights and sixteens (which would cover the whole of a four-variable map). The larger the grouping the simpler the expression required to describe the group. In general, it is better to include a region in several different groups if it will produce larger groupings and hence simpler expressions. Consider the expression we manipulated

56

earlier using Boolean algebra:

$$f = ABCD + AB\bar{C} + ABC\bar{D} + A\bar{B}\bar{C} + \bar{A}BCD.$$

This would be drawn on a map as is shown in Figure 3.7.

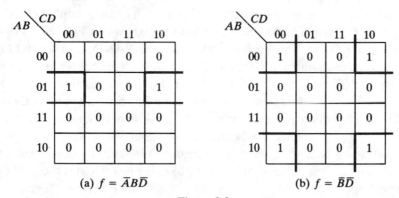

Figure 3.7. $f = AB + A\bar{C} + BCD$.

From the map we can see that there are two groups of four, one in the third row (AB) and one in the bottom left-hand corner ($A\bar{C}$). There is also one group of two (BCD). Thus from the map:

$$f = AB + A\bar{C} + BCD.$$

With practice deriving expressions from the map can be a very rapid process.

It should be noted that there are external adjacencics on the map, that is, in terms of making groupings, the left-hand edge of the map must be considered to be contiguous with the right-hand edge and the top edge of the map to be contiguous with the bottom edge. Thus the simplest representation for the function shown in Figure 3.8(a) is $\bar{A}B\bar{D}$ and that for the expression shown in Figure 3.8(b) is $\bar{B}\bar{D}$.

(a) $f = \bar{A}B\bar{D}$ (b) $f = \bar{B}\bar{D}$

Figure 3.8.

3.3 Gate symbology

One of the main reasons for using Boolean algebra is that it allows an abstract representation to be made of the operations that take place within

57

digital systems. The eventual systems themselves will be constructed from electronic circuitry which will carry out the defined Boolean operations, the voltage levels in the system representing the state of the variables in the Boolean description. A real system is likely to consist of hundreds, if not thousands, of similar circuits, each circuit carrying out a particular Boolean function. It is very unusual to describe the structure of such systems using normal circuit symbols for resistors, capacitors and transistors; rather the system is described at the higher level of logical function. Several different logic-symbol sets are in use today but perhaps the most widespread of these are those defined in IEEE Std. 9/ANSI Y32.14.

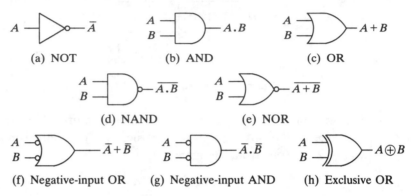

Figure 3.9. Gate symbols.

The symbols for the complement (NOT), AND and OR operations are shown in Figures 3.9(a), (b) and (c). For practical, electronic reasons, AND and OR gates are not the simplest reliable gates that can be produced but these functions followed by a complement operation. The AND-NOT combination is known as a NAND function ($\overline{AB}$) and hence the electronic circuits used to implement this function are known as NAND gates. Similarly the OR-NOT combination is known as the NOR function ($\overline{A+B}$) and is implemented by a NOR gate. The symbols for these functions are shown in Figures 3.9(d) and (e). Figure 3.9(f) shows a different symbol for the NAND function and Figure 3.9(g) a different symbol for the NOR function. The equivalence between 3.9(d) and 3.9(f), and 3.9(e) and 3.9(g) is, in fact, a graphical statement of De Morgan's laws. While two symbols for the same function are not strictly necessary, they can be very useful in communicating a designer's ideas to others. The use of symbol 3.9(d) would imply that the designer was looking for signals A and B both to be high simultaneously in order to initiate the next event. Symbol 3.9(f) implies that either A or B (or both) going low initiates the next event. The logical operation is the same but the emphasis is different. Figure 3.9(h) shows the symbol most commonly used for an Ex-OR gate.

3.4 Simple combinational design

The design of simple combinational circuits (the term combinational is used to distinguish these circuits from sequential circuits described later in this chapter) can follow a fairly standard pattern, utilizing some of the techniques described earlier. Let us consider the design sequence that would produce what is a very common structure, that of the half adder. A half adder will accept two binary signals and produce the arithmetic sum of those signals. Two outputs are required, these are known as the "sum" and the "carry", the carry signal having twice the significance of the sum signal. The half adder may thus be represented by the symbol, labelled HA, shown in Figure 3.10 (a). The half adder may form the least significant section of a larger, parallel adder; the more significant sections will need to be full adders (FA). Full adders accept three input signals and produce their arithmetic sum. They are thus able to accept as an input the carry signal from a stage of lesser significance, as illustrated in Figure 3.10 (b).

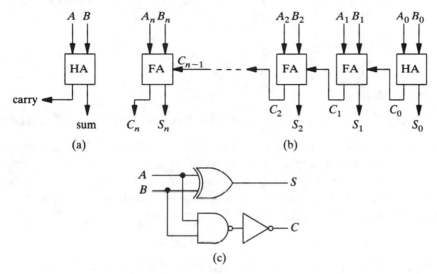

Figure 3.10. Adder structures.

For the simple structure of the half adder it is very easy to list the required output states for each input combination.

Inputs		Outputs	
A	B	Carry	Sum
0	0	0	0
0	1	0	1
1	0	0	1
1	1	1	0

We can now describe the outputs as Boolean functions of the inputs by ORing those rows of the table for which the output is 1. Each of the outputs is considered separately.

$$S = \bar{A}B + A\bar{B} = A \oplus B; \qquad C = AB.$$

Normally the next operation would be to simplify these Boolean functions, perhaps using a Karnaugh map. However, in this case, the functions cannot be simplified and the remaining operation of forming the function in terms of logic gates may be implemented immediately, see Figure 3.10 (c).

The above technique may be employed whenever it is possible to closely describe the outputs in terms of the inputs and hence form a truth table. Consider a further example, that of generating a parity signal for a 4-bit data word so that the 5-bit word formed has odd parity.

Inputs				Output	Inputs				Output
A	B	C	D	P	A	B	C	D	P
0	0	0	0	1	1	0	0	0	0
0	0	0	1	0	1	0	0	1	1
0	0	1	0	0	1	0	1	0	1
0	0	1	1	1	1	0	1	1	0
0	1	0	0	0	1	1	0	0	1
0	1	0	1	1	1	1	0	1	0
0	1	1	0	1	1	1	1	0	0
0	1	1	1	0	1	1	1	1	1

From the above,

$$P = \bar{A}\bar{B}\bar{C}\bar{D} + \bar{A}\bar{B}CD + \bar{A}B\bar{C}\bar{D} + \bar{A}BC\bar{D} + A\bar{B}\bar{C}D + A\bar{B}C\bar{D} + AB\bar{C}\bar{D} + ABCD.$$

Examining the Karnaugh map of this function, Figure 3.11 (a), appears to show that no group may be formed and that the function is in its simplest form. This is true in a strictly algebraic sense; however, the checkerboard pattern exhibited on the map is characteristic of a set of Ex-OR functions and these can be used to implement the function. Thus,

$$P = \overline{A \oplus B \oplus C \oplus D}$$

and the gating structure required to generate P will be that shown in Figure 3.11 (b).

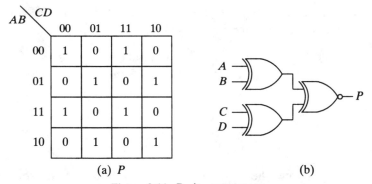

(a) P (b)

Figure 3.11. Parity generator.

3.4.1 NAND and NOR implementation

Because the simplest practical form of gates are NAND and NOR gates, rather than AND and OR gates, it is important to be able to implement functions using NAND or NOR structures. One approach to this form of implementation is to modify the Boolean expression describing the function so that it is in the form of a set of NAND or NOR operators.

Consider the function:

$$f = AB + CD.$$

This could be implemented using two AND gates and an OR gate, as shown in Figure 3.12 (a). However, by manipulating the expression using De Morgan's laws, thus:

$$f = \overline{\overline{AB + CD}} = \overline{\overline{AB}.\overline{CD}},$$

we can express f as a collection of three NAND terms and hence implement f with three NAND gates, as in Figure 3.12 (b).

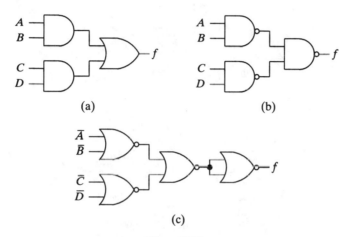

(a) (b)

(c)

Figure 3.12.

Similarly, by changing the expression in a different way, thus:

$$f = \overline{\overline{\overline{AB + CD}}} = \overline{\overline{\overline{A} + \overline{B}} + \overline{\overline{C} + \overline{D}}}$$

we can implement the function as an inverter (itself formed from a NOR gate) and three NOR gates, as in Figure 3.12 (c).

In practice it is not necessary to manipulate the Boolean expression into a final NAND or NOR form since the mechanism of translating between AND/OR and NAND or NOR can be formalized as follows:
1 set out the function as alternating layers of AND and OR gates: for conversion to NAND the final gate should be an OR gate, for conversion to NOR the final gate should be an AND gate; the final gate is level one, the preceding level of gates forms level two and so on;
2 translate all AND and OR gates to NAND or to NOR gates;
3 invert all variables which are connected to the circuit at odd levels.

As an example consider the function

$$f = A(C + D) + B(C + D) + E.$$

Figure 3.13 (a) shows this function as successive layers of AND and OR gates. Applying the rules above to form a NAND implementation, we form the circuit shown in Figure 3.13 (b). If we wish to create the equivalent NOR circuit we first need to modify the original AND/OR arrangement by adding an extra (dummy) AND gate to make the output-level gate an AND gate, see Figure 3.13 (c). This then translates into the NOR circuit shown in Figure 3.13 (d).

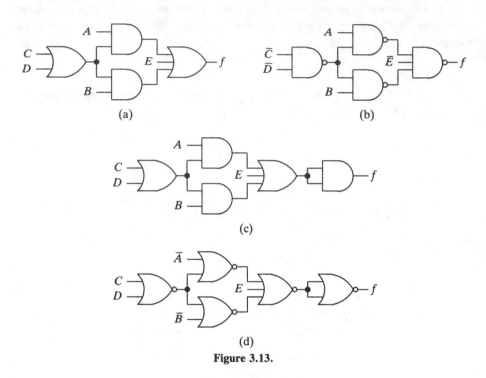

Figure 3.13.

3.4.2 "Don't care" states, decoders and code converters

So far, when we have been using the Karnaugh map, each area of the map has been declared as being either true or false. However, in practice, some input combinations never occur and thus there is no need to define an output condition. It is usual then to enter an x in the relevant area of the map, the x is said to declare a "don't care" state. This phrasing is somewhat unfortunate since it is often the case, not that we are uncaring of the output state for that input combination but simply that it will never appear. Properly exploited, the presence of "don't care" states can simplify a design and hence lead to lower implementation costs.

In order to illustrate this, let us consider the design of a decoder. A decoder is a circuit that will accept several input signals and will activate one of a number of output lines when a particular pattern appears at the input. The output line activated thus signals the presence of a given input code. Examples of the need for such structures appear in many instrumentation systems, for example, when it is necessary to recognize alarm conditions from the outputs of several sensors. In order to illustrate the technique we will consider decoding four signal lines which carry a BCD code. There will thus be ten distinct signal patterns and ten output lines, 0 to 9; see Figure 3.14 (a).

One way of decoding the output is always to sense all four input signals with a single gate, i.e. output line 9 would be derived from a four-input AND structure, the input signals being $W\overline{X}\overline{Y}Z$. However, if we note that in BCD only ten of the possible sixteen codes appear, we can develop a simpler decoder. First we must plot the code on a Karnaugh map. On the map we will declare the active output line for each input combination, entering x for those six combinations which do not appear in BCD; see Figure 3.14 (b). Since these combinations will never occur we can treat the value of the x areas as 1 or 0, whichever is more convenient at the time. This can lead to larger groupings and thus to simpler expressions for the description of active output conditions.

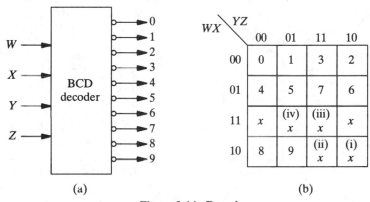

(a)   (b)

Figure 3.14. Decoder

The output signals for the decimal values 0 and 1 cannot be grouped with any of the "don't care" states and so we are still required to use all four input signals to decode each of these states, i.e. the code to activate the output line 0 will be $\overline{W}\overline{X}\overline{Y}\overline{Z}$, that for output line 1 will be $\overline{W}\overline{X}\overline{Y}Z$. However, for output line 2 we can group the 2 area with area (i) and describe the combined areas with only a three-variable function, $\overline{X}Y\overline{Z}$. Similarly, the active state for output line 3 can be described as $\overline{X}YZ$ by including area (ii). For an active output on line 9 we need only detect the input combination WZ because we can group area 9 with areas (ii), (iii) and (iv). Adopting this approach yields the following input combinations for each output line.

Output line	Active for input code	Output line	Active for input code
0	$\overline{W}\overline{X}\overline{Y}\overline{Z}$	5	$X\overline{Y}Z$
1	$\overline{W}\overline{X}\overline{Y}Z$	6	$XY\overline{Z}$
2	$\overline{X}Y\overline{Z}$	7	XYZ
3	$\overline{X}YZ$	8	$W\overline{Z}$
4	$X\overline{Y}\overline{Z}$	9	WZ

Thus by exploiting the "don't care" states our decoder will consist of 2 four-input gates, 6 three-input gates and 2 two-input gates, rather than the 10 four-input gates that would otherwise have been required.

The technique used in the design of decoders may be extended to the design of code converters. These are circuits that will accept input codes in one format and translate them to present equivalent values, but in a different format, as the output code. An example might be the conversion of BCD to XS3; see Figure 3.15(a).

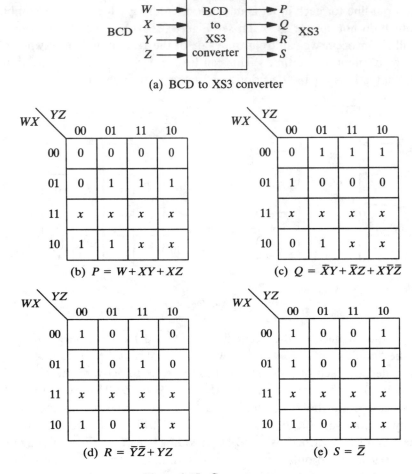

(a) BCD to XS3 converter

(b) $P = W + XY + XZ$

(c) $Q = \bar{X}Y + \bar{X}Z + X\bar{Y}\bar{Z}$

(d) $R = \bar{Y}\bar{Z} + YZ$

(e) $S = \bar{Z}$

Figure 3.15. Converter maps.

The sequence for the design of a code converter begins by listing the output codes required for each input-code condition.

BCD (input)				XS3 (output)				BCD (input)				XS3 (output)			
W	X	Y	Z	P	Q	R	S	W	X	Y	Z	P	Q	R	S
0	0	0	0	0	0	1	1	0	1	0	1	1	0	0	0
0	0	0	1	0	1	0	0	0	1	1	0	1	0	0	1
0	0	1	0	0	1	0	1	0	1	1	1	1	0	1	0
0	0	1	1	0	1	1	0	1	0	0	0	1	0	1	1
0	1	0	0	0	1	1	1	1	0	0	1	1	1	0	0

Each output line is treated separately and a Karnaugh map drawn to describe the required output condition for each input combination, "don't care" states being entered for each unused input combination; see Figures 3.15(b), (c), (d) and (e). Using the "don't care" states to best advantage, the outputs expressed as the simplest possible combination of input variables are:

$$P = W + X(Y+Z), \qquad Q = \bar{X}(Y+Z) + X\bar{Y}\bar{Z}, \qquad R = \bar{Y}Z + YZ, \qquad S = \bar{Z}.$$

By implementing these expressions with logic gates we will create an efficient BCD to XS3 code converter.

3.5 Flip-flops

So far in this chapter we have been concerned only with combinational circuits. Combinational circuits are those circuits whose outputs at any time depend only upon the circuit inputs at that time, that is, knowing the structure of a combinational circuit we can predict with complete accuracy, the output status for any defined input condition. However, a whole family of circuits exist for which the above rule does not hold true. For these circuits the output status depends not only upon the present inputs but also upon the history of those inputs. Such circuits are known as sequential circuits. An essential characteristic of all sequential circuits is that they contain feedback. It is highly unlikely that any digital system of any complexity will not contain sequential elements.

The simplest sequential element is also one of the most commonly used: this is the flip-flop. The simplest form of flip-flop is the "set-reset", or SR, flip-flop. The logic symbol for this device is shown in Figure 3.16(a). It may be constructed from cross-coupled NOR gates, as shown in Figure 3.16(b). Figure 3.16(c) is simply a restructuring of the previous diagram in order to show more clearly the feedback path inherent in the flip-flop's structure. The SR flip-flop may also be constructed from cross-coupled NAND gates with only minor changes in operation.

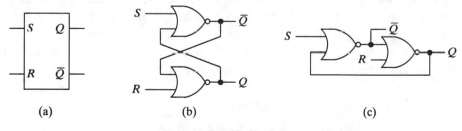

(a)　　　　　(b)　　　　　(c)

Figure 3.16. SR flip-flop.

Note that, because of the structure of the circuit, both true (Q) and complement ($\bar{Q}$) forms of the output are available simultaneously.

Because of its sequential nature the description of the flip-flop's operation is not as straightforward as that for a combinational circuit, as we must now take into account what has happened to the device prior to the presentation of the current inputs. Perhaps the best way to convey the response of the flip-flop is in the form of a truth table as shown below. Columns S and R show the status of the signals presently applied as inputs. Column Q shows the status of the output prior to the application of the present inputs, while Q^* shows the resulting value of Q after the presentation of the input signals. Note that, when Q is high ($\overline{Q}$ low) the flip-flop is said to be set, whereas when Q is low the flip-flop is said to be reset.

S	R	Q	Q^*	
0	0	0	0	No-change
0	0	1	1	
0	1	0	0	Resetting inputs
0	1	1	0	
1	0	0	1	Setting inputs
1	0	1	1	
1	1	0	-	Not-allowed inputs
1	1	1	-	

With this type of flip-flop $S = R = 0$ causes no change in the output condition, i.e. if the flip-flop was reset it remains reset and if it was set it remains set. $S = 0$, $R = 1$ causes the output to become, or remain, reset, whereas $S = 1$, $R = 0$ causes the output to become, or remain, set. $S = R = 1$ is known as a not-allowed condition; if we apply these inputs to the cross-coupled NOR structure then both the Q and the $\overline{Q}$ outputs will go low simultaneously, i.e., the $\overline{Q}$ line is no longer the complement of the Q line. Problems arise if the inputs then change to $S = R = 0$, since the output will be unpredictable as it will depend upon the relative switching speeds of the transistors making up individual circuits.

An important use of the SR flip-flop is illustrated by Figure 3.17. Note that at time (a) and at time (b) the input conditions are identical but the output states are different. By examining the inputs alone it is not possible to determine which of S or R last went high. However, by examining the output line we can tell that prior to time (a) S was high after R, whereas prior to time (b) R was high after S. What this means is that the flip-flop is capable of remembering signal conditions after those conditions have disappeared. In fact, a single flip-flop is capable of storing one bit of data.

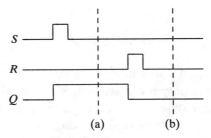

Figure 3.17. Flip-flop timing.

66

A supplementary, but very widespread, use of the SR flip-flop is to debounce mechanical switches. Mechanical switches do not present nice, clean signals as they are turned on and off; rather, as the pole of the switch leaves a contact sparking usually occurs (even at low voltages) and, as the pole closes with a new contact, it is likely to bounce and result in multiple on-off-on transitions. The spurious signals produced are immaterial if the the purpose of the switch action is just to illuminate or extinguish an indicator lamp, say. However, if the signal from the switch is being sent to a counter then false values will be recorded. By following the switch with an SR flip-flop, as in Figure 3.18, this problem can be eliminated, the switch is then said to be debounced.

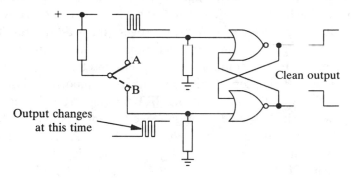

Figure 3.18. Switch debounce.

The operation of the debounce circuit is as follows: as the pole leaves the top contact, A, multiple transitions will be seen at the input to the upper NOR gate. However, the output from this gate will not change owing to these transitions since it is held low by the high signal on its lower input. Eventually the pole will leave contact A permanently and move across to hit contact B. On the first contact with B the output from the lower NOR gate will go low, sending a low signal to the upper NOR gate. The upper NOR gate now has two low inputs and so its output will go high. This, in turn, will present a high signal to the lower NOR gate. Further transitions owing to switch bounce on contact B will not be seen at the output since the output signal is kept low by the high signal from the upper gate. Note that for reliable operation of the debounce circuit the switch should be of the break-before-make type, i.e. the wiper must leave its previous contact before hitting the next contact.

3.5.1 Triggered flip-flops

THE SRT FLIP-FLOP

The simple SR flip-flop may be modified to form a more sophisticated device, the SRT flip-flop. The T stands for triggered and the symbol for this device is shown in Figure 3.19. The connection shown between the S and R inputs is the trigger input. The arrow head symbol shows that this input is sensitive

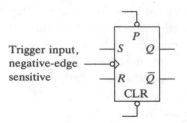

Figure 3.19. SRT flip-flop.

to changes (edges) in the connected signal and not to the steady-state condition of that signal (high or low). The small symbol against the arrow head shows that the device is sensitive to negative edges (high-to-low transitions) rather than positive edges (low-to-high transitions). The output from the SRT flip-flop responds to the status of the set and reset inputs in just the same way as did the SR flip-flop, except for one very important difference: the conditions on the S and R inputs of the SRT flip-flop are only registered by the flip-flop when (in this case) a negative edge appears on the trigger input. At all other times the status of S and R are ignored. If it is required to change the output independent of the signal on the trigger line then the preset (P) and clear (CLR) inputs may be used. Normally these lines will be held high but a low signal on P will force Q high immediately, whereas a low signal on CLR will force Q low immediately. P and CLR should not be taken low at the same time. P and CLR inputs are not always provided for SRT flip-flops as they consume pin locations on the integrated circuit.

The presence of an edge-sensitive input allows the development of the circuit shown in Figure 3.20: this is known as a toggle circuit.

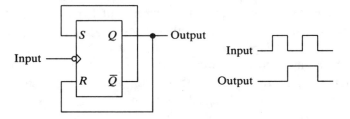

Figure 3.20. Toggle circuit.

Alternate transitions on the trigger line cause a change in the output level. This means that the output frequency is half that of the input signal, i.e. the circuit divides by two. By connecting several such circuits in cascade we can form a binary counter. Such a counter will have 2^n states, where n is the number of toggle circuits, or stages, in the counter.

Figure 3.21 (a) shows a three-stage counter and Figure 3.21 (b) shows the output waveforms generated as the input pulses appear. This type of counter is known as an asynchronous, or ripple-through counter. This is because the effect of input changes ripple through earlier stages and then go on to affect

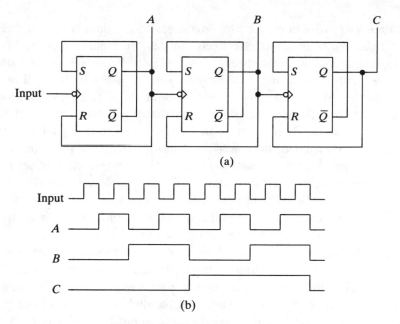

(a)

(b)

Figure 3.21. Three-stage counter

later stages, that is, the stages do not change together (synchronously) as a result of an input pulse but change one after the other. This can lead to decoding problems as transitory false states appear between valid states and also means that the counter is slow to respond, particularly if there are many stages. The maximum propagation delay between cause and effect for an n-stage asynchronous counter will be nt, where t is the propagation delay of a single flip-flop. Thus the delay increases as the size of the counter increases. We will later examine another type of counter that does not exhibit this property.

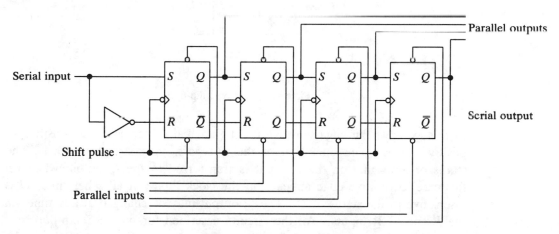

Figure 3.22. Shift register

A further very useful type of circuit that can be formed with the SRT flip-flop is the shift register. A four-stage shift register is shown in Figure 3.22.

69

Each time a negative edge appears on the shift line, data is moved one stage to the right. Note that this effectively divides the binary value by two. By placing conditional gating between the stages it is possible to develop the design of Figure 3.22 so that data may be shifted to the left as well as to the right. Shifting to the left effectively multiplies the data by two. An important application of the shift register is to translate between serial and parallel modes of operation. It is common to use both modes in digital systems. In parallel mode each bit of data in a word has its own data path; this leads to high-speed operation but is relatively expensive to implement. In serial mode data bits queue to use a single data path; this is slower but cheaper than parallel operation. Serial data transmission is often used over long distances, where the cost saving can be significant, or in applications where speed of data transfer is not important, e.g. interfacing to a keyboard.

The shift register can be used to translate between parallel and serial modes by preloading the register with parallel data via the *P* and CLR inputs. Pulsing the shift line will then cause the data to appear in serial form from the last stage of the register. Serial to parallel translation is achieved by accepting data one bit at a time on the serial input line, with one shift pulse occurring for each data bit. When the register is full data may be read out in parallel by sensing the outputs from each stage simultaneously.

THE *D*-TYPE FLIP-FLOP

There are several forms of triggered flip-flop. One of the most popular is the *D*-type. The symbol for this device is shown in Figure 3.23.

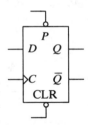

Figure 3.23. *D*-type flip-flop.

In this example the edge-sensitive input (labelled *C*, for clock) is sensitive to positive edges. The operation of the *D*-type is very straightforward. The status of the signal on the *D* input is transferred to the *Q* output when (in this case) a positive edge appears on the clock input; at all other times *Q* is insensitive to the status of *D*. The *Q* output may be changed at any time via the *P* and CLR inputs. All the circuits described for the SRT flip-flop may be implemented using *D*-type flip-flops. For example, Figure 3.24 shows the toggle circuit for the *D*-type. By connecting several such stages in cascade, counters may be produced.

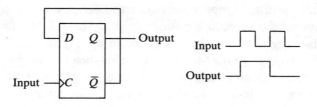

Figure 3.24. *D*-type toggle circuit.

THE *J-K* FLIP-FLOP

The third very common flip-flop type is the *J-K* flip-flop. This is very similar to the SRT flip-flop with the *J* input being equivalent to the *S* input and the *K* input being equivalent to the *R* input. The major difference, compared to the SRT flip-flop, is that $J = K = 1$ is an allowed condition. Under these circumstances the output of the flip-flop toggles for each active transition on the trigger input. This means that an asynchronous counter may be constructed as in Figure 3.25. Notice that the *J* and *K* inputs of each stage are tied permanently high so that each flip-flop changes state when a negative edge appears on its trigger line.

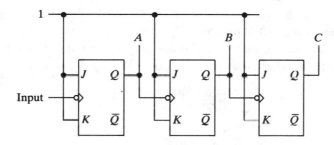

Figure 3.25. *J-K* asynchronous counter.

Another type of counter may be formed, however, which does not suffer from the ripple-through delays of the asynchronous structure. This is the synchronous counter. A 4-bit synchronous counter is shown in Figure 3.26.

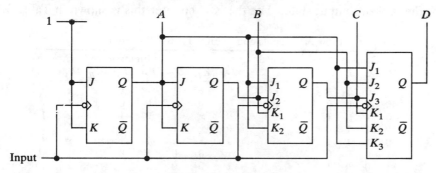

Figure 3.26. *J-K* synchronous counter.

71

Notice that all the trigger inputs are connected directly to the primary input signal and not to the preceding counter stage. Thus all the flip-flop transitions take place together and there is no build up of delay as changes ripple through earlier stages to affect later ones. This means that the overall delay in a synchronous counter does not increase as the size of the counter increases. There is a limit to the size of synchronous counters, however, and this is set by the loading effects on the clock line; intermediary buffers will be required for very long counters. Another point to note from Figure 3.26 is that some flip-flops are shown with multiple J and K inputs. This is quite common: effectively the multiple inputs are ANDed together to form the final J and K signals which determine the flip-flop's switching.

FORMAL DESIGN OF SYNCHRONOUS COUNTERS

A well-established technique exists for the design of synchronous counters. It is particularly useful when it is required to produce a non-binary count. The technique lists all required output states and alongside them the required input conditions to produce the next output state when the next active edge appears on the trigger line. Using a Karnaugh map to simplify functions, these inputs to the flip-flops can be derived from the present flip-flop outputs.

To illustrate this technique let us consider the design of a synchronous XS3 counter, using J-K flip-flops. The first things to consider are the transitions demanded of an individual flip-flop and the inputs required to produce those transitions. There are only four possible output changes.

Output transition required $Q \rightarrow Q^*$	Prior input state JK
$0 \rightarrow 0$	$0x$ (01 or 00)
$0 \rightarrow 1$	$1x$ (10 or 11)
$1 \rightarrow 1$	$x0$ (10 or 00)
$1 \rightarrow 0$	$x1$ (01 or 11)

Using these facts we can construct the table listing inputs against outputs. Remember the inputs are those required to produce the next output state, not the present output state. For the XS3 counter this is shown in Table 3.1.

Table 3.1. Inputs and outputs for the XS3 counter.

D	C	B	A	Decimal	J_D	K_D	J_C	K_C	J_B	K_B	J_A	K_A
0	0	1	1	0	0	x	1	x	x	1	x	1
0	1	0	0	1	0	x	x	0	0	x	1	x
0	1	0	1	2	0	x	x	0	1	x	x	1
0	1	1	0	3	0	x	x	0	x	0	1	x
0	1	1	1	4	1	x	x	1	x	1	x	1
1	0	0	0	5	x	0	0	x	0	x	1	x
1	0	0	1	6	x	0	0	x	1	x	x	1
1	0	1	0	7	x	0	0	x	x	0	1	x
1	0	1	1	8	x	0	1	x	x	1	x	1
1	1	0	0	9	x	1	x	1	1	x	1	x
[0	0	1	1]									

Using the data in the table, the J and K inputs can be expressed as functions of the present output signals A, B, C and D. Using Karnaugh maps at this stage leads to simplified expressions for the input functions: the Karnaugh maps for the XS3 counter are shown in Figures 3.27(a) to (f). In this case there is no need to construct maps for J_A and K_A as we can see from the listing that these lines are always 1 or x, and therefore they can be tied permanently high. Each J and K input having been expressed as a function of the current output values, it is now possible to generate the required counter structure. This involves the introduction of the necessary gating (as defined by the functions for the J and K inputs) to produce the required input signals for each stage of the counter. The input signals to this gating are derived from the current output signals of the flip-flops, as shown in Figure 3.27(g).

3.6 Design using medium-scale integration devices

The simple gates and flip-flops that have been considered so far in this chapter would be classed as small-scale integration (SSI) devices. It would be very unusual for a modern-day design to be implemented entirely from SSI devices: rather the designer would employ SSI devices to link together larger logic elements of medium-scale integration (MSI) devices. We will now examine four such more highly integrated structures, an arithmetic and logic unit, a multiplexor/demultiplexor (these two devices are really borderline SSI/MSI devices), a read-only memory and a programmable logic array.

3.6.1 Arithmetic and logic unit

Every computer and hence every microcomputer contains an arithmetic and logic unit (ALU) of some form. In many ways the ALU, together with the control unit, can be considered to be the heart of the processing system. The ALU accepts data from the rest of the system, performs basic arithmetic and logical operations upon the data and then returns the result back to the other units of the computer. The number of possible operations will vary with the sophistication of the system. Addition, subtraction, ANDing, ORing and inversion will be found even in the simplest structure; more complex units will provide further functions such as multiplication.

Figure 3.28(a) shows a basic ALU structure. Two input registers (basically sets of flip-flops) are used for the temporary storage of input operands; a further register is shown which will hold the result of the ALU operation (though this register may not be strictly necessary in all situations). These registers take data from, and send data to, the bi-directional databus. Access to and from the bus is determined by timing signals from the control unit. The output register of the ALU will not be the only source of data to the data bus. It is thus important that this register only sends its signals to the bus when the bus is not being used by other units. In order to achieve this, the output register (and all other data sources to the bus) must use tri-state drives. These can be switched to a non-active third state known as a

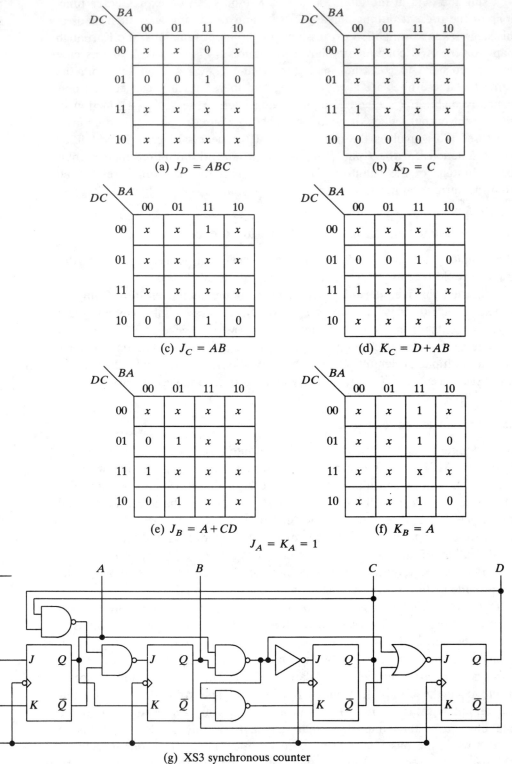

(a) $J_D = ABC$

(b) $K_D = C$

(c) $J_C = AB$

(d) $K_C = D+AB$

(e) $J_B = A+CD$

(f) $K_B = A$

$J_A = K_A = 1$

(g) XS3 synchronous counter

Figure 3.27. XS3 synchronous counter.

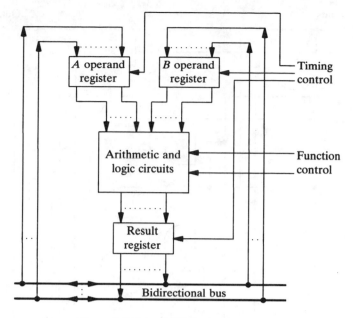

Figure 3.28 (a). ALU data paths.

high-impedance state when other devices are driving the bus (more details of tri-state drives are given in Section 3.7.1).

The core of the ALU will be based around a parallel adder structure similar to that of Figure 3.10 (b). However, control of the inputs signals together with further function-control signals from the control unit mean that much more than basic addition can take place using this circuit.

Consider Figure 3.28 (b), which shows a more detailed view of the central part of a 4-bit ALU. This circuit is shown with three function-selection inputs S_0, S_1 and S_2. With all three of these function-control inputs low, the true values of the input operands, A_0 to A_3 and B_0 to B_3, arc fed to the full adders. Thus the output represents the arithmetic sum of A and B. If, however, S_0 goes high it is the complement of B, together with an extra signal to the least significant stage, that is fed to the full adders, i.e. the two's complement of B is presented to the adders, and hence the output signals represent A minus B. Suppose we take S_0 low again and this time take S_1 high. The effect of this is to inhibit the propagation of the carry signals between the adder stages. This gives an output pattern which represents the exclusive-OR of A with B. If S_0 is low but S_1 and S_2 are high then one of the adder inputs is inhibited but the other input is fed with the OR of A with B and this goes on to form the output signal. Thus, by using the function-selection lines the same basic adder structure can be made to perform many different operations. The popular SN74181 ALU device, for example, comes in a 16-pin integrated circuit; four of these pins are used for function-control signals, giving the device the capability of performing sixteen different operations.

75

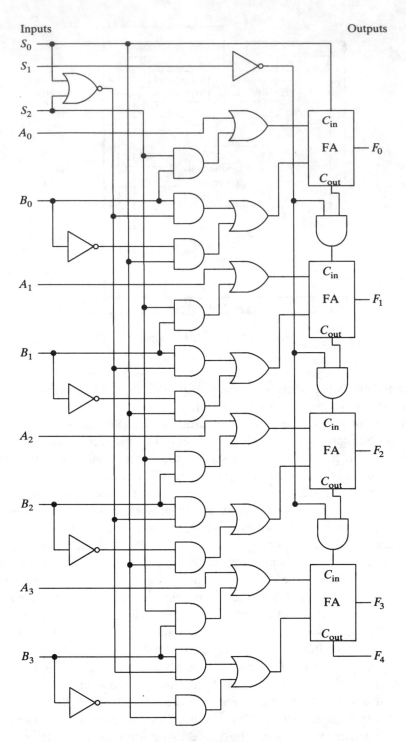

Figure 3.28 (b). Simple ALU structure.

3.6.2 Multiplexors and demultiplexors

The digital multiplexor (MUX) and the complementary device, the demultiplexor (DMUX), were originally designed to facilitate the implementation of time-domain multiplexing (TDM) for the transfer of signals along restricted data paths, where, for cost or other reasons, several signals must share the same transmission medium. In this mode of operation the MUX and DMUX are connected as shown in Figure 3.29 (a). The MUX has two sets of input signals: a set of n control lines and a set of 2^n data lines (there may also be separate timing-control lines). There is usually only one output line. At any one time one, and only one, input line is connected through to the output. Which input line this is, depends upon the code present on the control lines. The DMUX operates in the opposite fashion. Its single data-input line can be connected through to one of several output lines. Which output line is selected is again determined by the code on the control lines. By synchronizing the control codes at the MUX with those at the DMUX, several different signal sources can utilize the same transmission path in a time-sequential manner. Figures 3.29 (b) and (c) show the internal structures of the MUX and DMUX.

One of the major modern day applications of the DMUX is as an address decoder. Consider the truth table for a 1-to-8-line DMUX:

Data input	Control inputs			Outputs							
	A	B	C	0	1	2	3	4	5	6	7
L	x	x	x	H	H	H	H	H	H	H	H
H	0	0	0	L	H	H	H	H	H	H	H
H	0	0	1	H	L	H	H	H	H	H	H
H	0	1	0	H	H	L	H	H	H	H	H
H	0	1	1	H	H	H	L	H	H	H	H
H	1	0	0	H	H	H	H	L	H	H	H
H	1	0	1	H	H	H	H	H	L	H	H
H	1	1	0	H	H	H	H	H	H	L	H
H	1	1	1	H	H	H	H	H	H	H	L

If the control lines are connected to the address bus of a computer system and the data input is held high then the outputs from the DMUX can be used to enable blocks of memory locations. For instance, consider a 16-bit address bus (A_0 to A_{15}). If the control inputs of a 1-to-8 DMUX are connected to the three most significant address lines, the 64-Kbyte memory space will be split into 8-Kbyte blocks, thus:

A	B	C	Addressed area (HEX)
A_{15}	A_{14}	A_{13}	
0	0	0	0000–1FFF
0	0	1	2000–3FFF
0	1	0	4000–5FFF

77

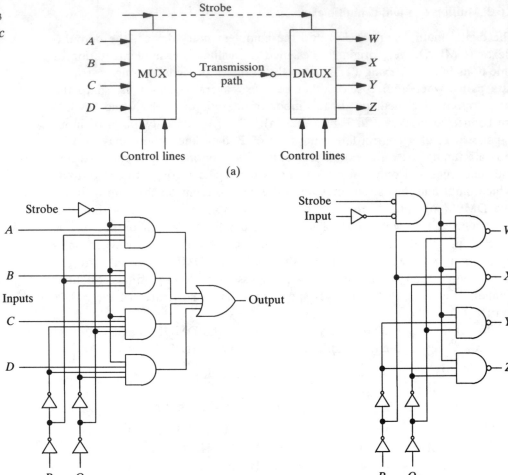

(a)

(b) 4-to-1 multiplexer.

(c) 1-to-4 demultiplexer.

Figure 3.29.

A	B	C	Addressed area (HEX)
A_{15}	A_{14}	A_{13}	
0	1	1	6000–7FFF
1	0	0	8000–9FFF
1	0	1	A000–BFFF
1	1	0	C000–DFFF
1	1	1	E000–FFFF

MUXs are sometimes referred to as universal logic modules (ULM) as they can be used to generate logic functions. For example, a 4-to-1 MUX can generate any combination of three Boolean variables. Consider the function:

$$f = \bar{B}\bar{C} + \bar{A}\bar{B}C + ABC = \bar{B}\bar{C}.1 + \bar{B}C.\bar{A} + B\bar{C}.0 + BC.A.$$

This may be generated by a MUX connected as shown in Figure 3.30. B and C are connected to the control inputs and the various combinations of B and C feed through 1, 0, A or $\bar{A}$, as defined by the equation above.

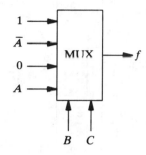

Figure 3.30. MUX function implementation.

The implementation of more complex functions, involving a greater number of variables, can be carried out using larger MUXs. However, the required MUX size increases rapidly with the number of variables and a more usual approach would be to split the complex problem down into parts and to use a cascade of smaller MUXs.

Consider implementing the following function using 4-to-1 MUXs:

$$f = \bar{A}\bar{B}C\bar{D}\bar{E} + \bar{A}BD\bar{E} + A\bar{B}\bar{C}\bar{D}E + A\bar{B}C\bar{D}E + ABCD.$$

Expanding so that each term contains all variables (this is known as the canonical form):

$$f = \bar{A}\bar{B}C\bar{D}\bar{E} + \bar{A}BC\bar{D}\bar{E} + \bar{A}BCD\bar{E} + A\bar{B}\bar{C}\bar{D}E + A\bar{B}C\bar{D}E + ABCD\bar{E} + ABCDE.$$

There are several ways of splitting the function. Suppose, in this case, we use D and E as the end-level control signals. Then

$$f = (\bar{A}\bar{B}C)\bar{D}\bar{E} + (A\bar{B}\bar{C} + A\bar{B}C)\bar{D}E + (\bar{A}BC + \bar{A}BC + ABC)D\bar{E} + (ABC)DE.$$

Opting for B and C as the next-level controls gives:

$$f = ((\bar{A})\bar{B}C)\bar{D}\bar{E} + ((A)\bar{B}\bar{C} + (A)\bar{B}C)\bar{D}E + ((\bar{A})B\bar{C} + BC)D\bar{E} + ((A)BC)DE.$$

This leaves A as the primary input signal and gives the two-level MUX structure of Figure 3.31.

3.6.3 Read-only memories

Read-only memories (ROM) as memory devices are further discussed in Chapter 6. Here we shall only be concerned with their use in implementing Boolean functions. The ROM stores a set of predefined patterns; which of these patterns is presented on the output lines is determined by the signals appearing on the address lines. Hence the ROM may be considered to be storing a truth table. The inputs of the truth table being the address line signals, the outputs from the truth table are the stored contents of the ROM,

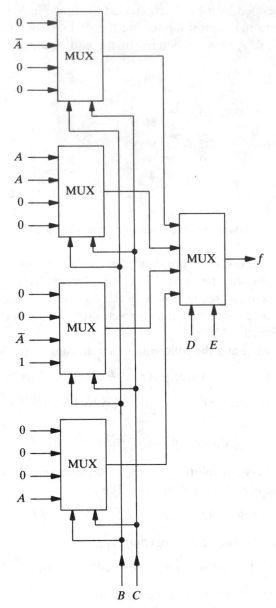

Figure 3.31. Cascaded MUX implementation.

each word in the ROM being equivalent to one row of the truth table. Note that simplification of functions to be implemented using ROMs serves no purpose since the ROM holds the canonical form of the function.

As an example of ROM-based design, consider the requirement for a 4×4 multiplier. In essence the design procedure simply consists of listing required outputs against the equivalent input combination. For the multiplier the (partial) listing is shown in Table 3.2

Implementing the design is then only a matter of storing the defined patterns in the ROM. The multiplier design is well suited to ROM implementation since

Table 3.2. Multiplier listing (partial).

Inputs (address lines)								Outputs (data lines)							
X_3	X_2	X_1	X_0	Y_3	Y_2	Y_1	Y_0	D_7	D_6	D_5	D_4	D_3	D_2	D_1	D_0
0	0	0	0	0	0	0	0	0	0	0	0	0	0	0	0
0	0	0	0	0	0	0	1	0	0	0	0	0	0	0	0
0	0	0	0	0	0	1	0	0	0	0	0	0	0	0	0
0	0	0	0	0	0	1	1	0	0	0	0	0	0	0	0
⋮				⋮				⋮				⋮			
0	0	0	1	0	0	0	0	0	0	0	0	0	0	0	0
0	0	0	1	0	0	0	1	0	0	0	0	0	0	0	1
0	0	0	1	0	0	1	0	0	0	0	0	0	0	1	0
⋮				⋮				⋮				⋮			
0	0	1	0	0	0	0	0	0	0	0	0	0	0	0	0
0	0	1	0	0	0	0	1	0	0	0	0	0	0	1	0
0	0	1	0	0	0	1	0	0	0	0	0	0	1	0	0
⋮				⋮				⋮				⋮			
1	1	1	1	1	1	0	0	1	0	1	1	0	1	0	0
1	1	1	1	1	1	0	1	1	1	0	0	0	0	1	1
1	1	1	1	1	1	1	0	1	1	0	1	0	0	1	0
1	1	1	1	1	1	1	1	1	1	1	0	0	0	0	1

all possible input combinations are exploited. This is not often the case and ROM structures may sometimes appear to be very inefficient. Even the multiplier design becomes cumbersome if the size of the input operands is increased as the size of ROM required increases geometrically with the number of input bits. For larger wordlengths it is common to use a cascade of ROMs. Note that the same technique can be used to produce high-speed division circuits. For this purpose we need not design a new circuit altogether but simply employ a second ROM that will produce the reciprocal of the divisor and whose output is then fed to the inputs of a multiplier ROM.

3.6.4 Programmable logic arrays

Read-only memories are very efficient for the implementation of functions which utilize all, or at least most, of the potential combinations of input variables. However, in practice, this state of affairs rarely holds true and most functions will contain many unused, or "don't care", input conditions. Under these circumstances the use of ROMs becomes increasingly clumsy as less and less of the ROM's capacity is exploited. Note that smaller ROMs could not be used because the ROM size is essentially determined by the number of input variables and not by how many combinations of those variables occur.

If such a situation arises then it is probably better to implement the logic design using a programmable logic array (PLA). The PLA consists of an array of AND gates and an array of OR gates, as shown in Figure 3.32.

The connections within and between these arrays and the connections to input and output pins are made via programmable links, so that the detailed structure of the circuit can be tailored to implement the desired function(s). Figure 3.33 shows the basic arrangement of the links; of course, in a practical PLA, there will be many more gates and many more links than are shown here.

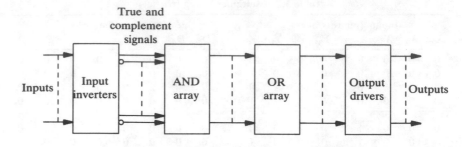

Figure 3.32. PLA: general structure.

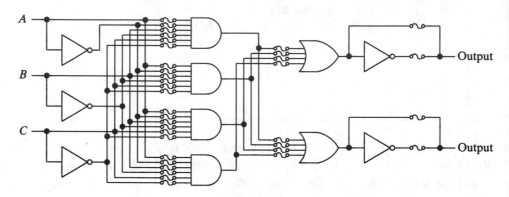

Figure 3.33. PLA: detail.

For some types of PLA the links are made, or left unmade, at the time of manufacture. This is cost effective if many devices, having the same structure, are required. However, it is very expensive to take this route if only a few such devices are required. Under these circumstances a field-programmable logic array (FPLA) is to be preferred. These devices are supplied by the manufacturer with all links intact and the device is dedicated to a particular design, by the user, by means of a programming sequence which disconnects those links that are not required by applying larger voltages across those links than they would experience during normal usage.

When considering implementing functions by PLA it is important to manipulate expressions so as to minimize the number of separate product terms (AND functions) that appear since the number of such terms that are required will determine the size of the PLA. This is in contrast to the ROM implementation technique where the process of minimization was of no consequence.

3.7 Circuit technologies

We will conclude this chapter by looking briefly inside the circuitry used to produce the logic functions we have been using. We will examine five different circuit technologies, three based on the bipolar junction transistor (BJT): transistor–transistor logic (TTL or T^2L), emitter-coupled logic (ECL) and integrated injection logic (I^2L); and two based on the field-effect transistor: n-type metal-oxide semiconductor (nMOS) logic and complementary metal-oxide semiconductor (CMOS) logic.

3.7.1 Transistor–transistor logic

The technology that has dominated digital design for more than a decade is transistor–transistor logic and though its pre-eminent position is now being displaced by other circuits, it is still very widely used.

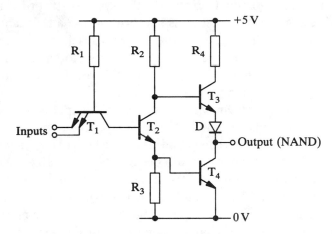

Figure 3.34. TTL NAND gate.

Figure 3.34 shows the structure of a two-input TTL NAND gate (positive logic). Input signals are applied via the multi-emitter transistor T_1. When all the inputs are high current flowing through R_1 passes through the base-collector junction of T_1 to the base of T_2, turning T_2 on. As T_2 is on, current flows in the path R_2, T_2, R_3. The voltage dropped across R_2 lowers the base potential of T_3, keeping it off, whereas the potential formed across R_3 raises the base potential of T_4, keeping it on (diode D ensures that T_3 is off when T_4 is on). T_3 off and T_4 on means that there is a high impedance between the output and the positive rail and a low impedance between the output and ground, giving a low output signal. If any of the inputs is taken low then the current flowing through R_1 is diverted through the low input(s) and is thus no longer available to turn T_2 on. As T_2 is off, there is little voltage dropped across R_2 and the resulting high potential on the base of T_3 turns this transistor on. Similarly, since almost no current flows through R_3, the base of T_4 is low and T_4 is off. T_3 being on and T_4 being off results in a high output signal.

Two important developments that appeared with TTL are the use of multi-emitter transistors—the action of T_1 greatly speeds up the switching of the circuit—and the introduction of the output structure formed by R_4, T_3, D and T_4. This is known as a totem-pole output and also aids high-speed operation since it provides a low-impedance drive for both high and low output conditions, thus lessening the effects of load capacitance on the output line. The use of a totem-pole output does introduce some problems, however. Unfortunately it takes longer for a transistor to turn off than it does for the same transistor to turn on. This means that each time the totem-pole output changes state there is a short period when both T_3 and T_4 are on. This produces a surge of current; it is the purpose of R_4 to limit this current and protect the transistors. However, R_4 cannot be too large or the high-speed drive capability of the output structure will be degraded. The surge of current produces significant noise on the power-supply lines. This noise can cause false switching of other circuits in the system. In order to limit the effect of this self-generated noise, it is necessary to use a large amount of distributed decoupling in TTL circuits, one 0.1-μF capacitor for every two TTL integrated circuits would be typical. The charges stored on the capacitors form local reservoirs which can provide the charge for the current surges without causing the power-supply potential to drop.

The second major problem with the totem-pole output is that it is not possible to connect outputs together. If this is attempted and the two outputs are trying to drive to opposite signal levels then again a large (and in this case no longer transitory) current surge will occur. R_4 is not large enough to limit this current in the long term and the usual result is that one of the output transistors is permanently damaged by junction overheating. There is a considerable demand to connect outputs together in microcomputer systems, where several signal sources may be connected to the same bus line. One option is to dispense with the totem-pole drive in these circumstances and use what is known as an open-collector gate. However this type of output does not have the speed advantage of the totem-pole structure and the more usual approach is to employ gates which have tri-state outputs. These gates do have a totem-pole output structure but with an extra control facility that enables both output transistors to be turned off at the same time: this gives the third output state which is known as a high-impedance output condition.

Figure 3.35 shows a tri-state output structure. It would be quite uncommon for each TTL gate to have its own, separate, enable/disable line for the control of the output. More usually one enable/disable line will be common to all the gates in an integrated circuit package. So long as care is taken that only one tri-state output is enabled at any one time then these gate outputs can be connected together without fear of damage.

3.7.2 Emitter-coupled logic

Emitter-coupled logic is sometimes known as current-mode logic (CML) because of the way in which the circuit operates. The basis of the circuit is an electronic structure known as a long-tailed pair; this same structure forms the core of an operational amplifier. Emitter-coupled logic suffers from several disadvantages compared with other logic families but does have one very significant advantage, that of switching speed. Emitter-coupled logic is the fastest form of logic widely available. Its high speed comes from the fact

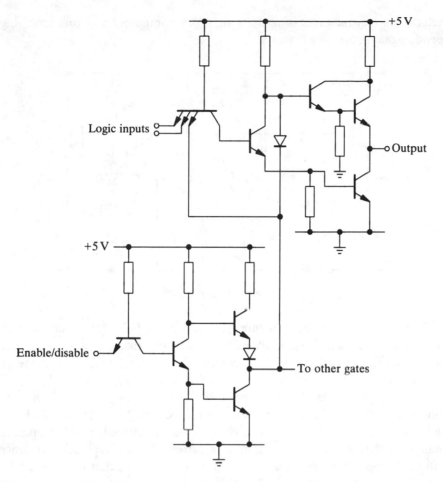

Figure 3.35. Tri-state output structure.

that, unlike TTL, the transistors do not turn on hard but are only switched into what is known as the active region. Since the transistors are not turned on hard there is no build up of stored charge in the base of the transistor which means, in turn, that the transistor can switch off at a much higher speed (note that a special type of TTL, known as Schottky TTL, comes close to this condition but does not offer as high a switching speed as the fastest ECL). The fact that the transistors do not turn on hard does produce detrimental properties, however. Much more power is consumed by a transistor in the active region than by a transistor that is on hard. The extra power required by ECL circuits is expensive to provide and the extra heat generated is even more expensive to remove. The extra heat dissipation also reduces packing density both on the chip and between integrated circuits, further increasing system cost. The fact that the transistors in ECL do not turn fully on also has an effect on logic levels. For ECL the logic swing might be of the order of 0.8 V compared to 2.5 V for TTL. These problems mean that the use of ECL is restricted to those applications where its very high speed of operation can justify the extra costs incurred in its usage. Such applications include the central processing units of super computers and real-time, high-speed, signal-processing systems.

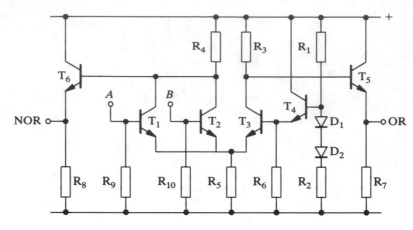

Figure 3.36. ECL gate.

Figure 3.36 shows the structure of a two-input ECL gate, note that dual outputs are available from most ECL devices and that this gate can operate as both an OR circuit and a NOR circuit. The general operation of the gate is to change the main current flow between R_3, T_3, R_5 and R_4, T_1/T_2, R_5, i.e. the current mode changes. With both inputs low T_1 and T_2 are non-conducting, T_3 and T_4 conduct owing to the potential formed by R_1, R_2, D_1 and D_2 (D_1 and D_2 are present to compensate for the effects of temperature changes in the circuit). This results in a current flow through R_3 and hence a potential drop across this resistor. This, in turn, means that the base of T_5 is low. T_5 acts as an emitter follower and hence the OR output is also low. Little current flows through R_4; so the base of T_6 and the NOR output are high. If one, or both, of the inputs goes high then the input transistor(s) turn on (into the active region) and current flows through R_4. This drops the potential on the base of T_6 and hence, via the emitter-follower action of T_6, the NOR output goes low. The input(s) going high also cause an increase in the current flowing through R_5, raising the potential of the common emitter point. Since the base potential of T_3 is fixed, this results in T_3 turning off and the OR output going high.

3.7.3 Integrated injection logic

Integrated injection logic is a relatively recent technology. It is based on bipolar transistors and is intended to challenge the role of the metal-oxide semiconductor technologies in terms of packing density, operating speed and power demand. So far the penetration of I^2L into the market place has not been great, mainly because of the very rapid improvements that have occurred in the competing MOS systems. Currently I^2L is used in some high-speed microprocessors. It is not a technology that is suited to SSI and MSI devices.

The basic I²L gate consists of an n-p-n multi-collector transistor, which operates as an inverter, and a lateral p-n-p transistor, which serves as a current source, see Figure 3.37(a). These elements are usually merged into a single structure as shown in Figure 3.37(b). The single I²L cell thus has one input and several outputs and may be represented schematically as in Figure 3.37(c). However, in terms of seeing what is happening in an I²L structure, perhaps the most useful graphic is that of Figure 3.37(d), which shows the multi-collector transistor as an individual device with the p-n-p transistor being shown in its role as a current source. Note that the I²L circuit contains no resistors; resistors, as compared to transistors, occupy a very large area of silicon. A circuit that does not use resistors will be able to produce much higher on-chip packing densities than is the case for circuits using resistors.

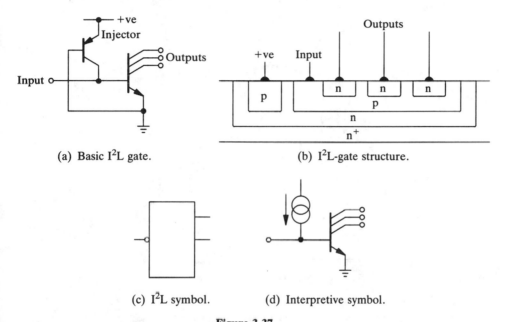

(a) Basic I²L gate.   (b) I²L-gate structure.

(c) I²L symbol.   (d) Interpretive symbol.

Figure 3.37.

The operation of a single cell is very straightforward. If the input signal is low then the current from the current source is sunk in the driving device: it is therefore not available to switch on the multi-collector transistor and all the outputs from this transistor are effectively high (more accurately, from the point of view of the operation of the next level of gating, they will sink no current). If the input line is high then the current from the current source is diverted into the base of the multi-collector transistor, turning it on (thus the output connections can sink currents from the next level of gating).

Logic functions are produced by linking single cells together. Figure 3.38(a) shows how elements could be interconnected to produce both NOR and (via an extra cell which acts as an inverter) OR functions. Figure 3.38(b) shows the production of a NAND function: essentially this is produced by a single I²L element but two extra elements are needed if the input signals are used elsewhere in the system and are required to remain isolated from each other.

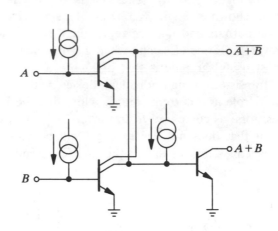

(a) NOR and OR circuits

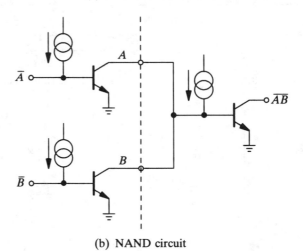

(b) NAND circuit

Figure 3.38.

3.7.4 n-type metal-oxide semiconductor logic

Currently most microprocessors are produced using n-type metal-oxide sem-iconductor technology, though this pre-eminent position is being seriously challenged by the next technology we shall be examining. nMOS logic cir-cuits are very simple and, like I^2L, make no use of resistors. Also, since field-effect transistors (FETs) are used rather than BJTs, the devices are essentially voltage driven rather than current driven. These factors leads to high packing densities and low power consumption and so make nMOS ideally suited to the implementation of complex digital circuits.

Figure 3.39 (a) shows the circuit of a two-input nMOS NOR gate. The upper transistor, T_1, does not take place in any switching operations but acts as

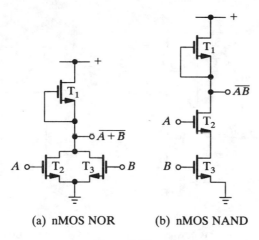

(a) nMOS NOR   (b) nMOS NAND

Figure 3.39.

a load for the two lower transistors (thus eliminating the need for a resistor in this position). When both inputs are low, both T_2 and T_3 are off. There is then a high impedance between the output terminal and ground and the output is pulled high via T_1. If either of the input signals goes high then the transistor associated with that input is turned on and the output is pulled low; most of the supply potential is then dropped across T_1. This yields a NOR function for positive logic. The equivalent positive-logic NAND gate is shown in Figure 3.39(b). Again T_1 does not switch but acts as a load. With both input signals low both T_2 and T_3 are off and the output is pulled high via T_1. Only when both inputs are high and both T_2 and T_3 are on is there a low-impedance path between the output and ground, resulting in a low output signal.

nMOS circuits work over a much wider range of supply voltages than is the case for BJT-based circuits. Transistor-transistor logic, for example, only works reliably with supply voltages in the range 4.5 V to 5.5 V. Many nMOS circuits will operate with a supply voltage between 2 V and 15 V. The output-logic swing from nMOS circuits is close to that of the power supply since nMOS transistors operate as very good switches. nMOS circuits are, however, generally slower that BJT-based circuits since the very high impedances presented by MOS transistors tend to enhance the effects of stray capacitances present in the circuit. Since they are voltage driven, nMOS circuits do consume very little power compared to BJT-based circuits. However, for certain circuit conditions significant amounts of current can still flow in nMOS gates. The final logic family we shall examine was developed from nMOS specifically to produce extremely low power consumption circuits by ensuring low current flow for all gate states.

3.7.5 Complementary metal-oxide semiconductor logic

The structures of complementary metal-oxide semiconductor NAND and NOR gates are shown in Figures 3.40(a) and (b). Consider first the NAND

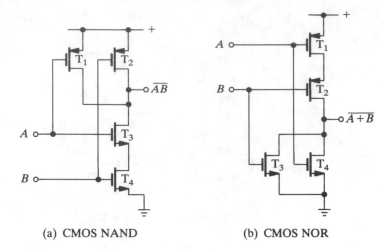

(a) CMOS NAND   (b) CMOS NOR

Figure 3.40.

circuit of Figure 3.40 (a). Each input is connected to a complementary pair of transistors. Input A, for example, is connected to the p-channel device T_1 and the n-channel device T_3. When A is low T_1 is on and T_3 is off; thus the output is high via the T_1 connection. When A is high T_1 is off and T_3 is on: the state of the output will then depend upon B. If B is low T_2 is on and T_4 is off and the output will still be high. However, if B is high T_2 will be off and T_4 will be on. The output will then be low, connected via T_3 and T_4. Notice that no matter what the state of the gate, only one of each pair of complementary transistors is on at any one time. Thus there is never any significant current flow between the supply rails and the power consumption of the circuit is very low.

The CMOS NOR gate operates in a similar manner. The output will only be high when both A and B are low. When A or B are high the output will be low, connected to ground via T_3 or T_4, or both. Again for all circuit conditions one of each pair of complementary transistors is off, leading to a very low power consumption. The power consumption does increase slightly, however, when the gate is changing state when, for a short period, both transistors of a pair are half on. The power consumption of CMOS thus increases with operating frequency.

The very low power consumption of CMOS makes it ideal for applications powered by batteries. All digital watches and calculators are based on CMOS. It is also well suited to situations where the provision of power is difficult such as in remote sensing stations. Early types of CMOS were slow and relatively expensive. However, great strides have been made in the development of faster, cheaper and more compact CMOS circuits. Most microprocessors are available in CMOS versions. Small- and medium-scale integration CMOS devices are produced which have direct pin compatibility with their TTL equivalents. There is little doubt that CMOS will be the dominant technology of the next decade.

3.8 Summary

In this chapter we have examined the basics of Boolean algebra as a means of describing digital systems. The symbols used to record digital structures were also presented together with standard approaches to the design of such systems, using both SSI and MSI devices. Finally, the most popular circuit technologies for the implementation of our designs were considered.

3.9 Further reading

1 *High Speed CMOS Logic Data Manual*, Texas Instruments Ltd., 1984.
2 Hill F. J. and Peterson G. R. (1987). *Digital Systems, Hardware, Organization and Design*, Wiley.
3 Lewin D. (1985). *Design of Logic Systems*, Van Nostrand Reinhold.
4 Mano M. M. (1979). *Digital Logic and Computer Design*, Prentice-Hall.
5 Sedra A. S. and Smith K. C. (1987). *Microelectronic Circuits*, Holt, Rinehart and Wilson.
6 *TTL Data Book*, Texas Instruments Ltd., 1988.

3.10 Problems

3.1 Find the minimal form of the following functions by Boolean manipulation.

(a) $f = \bar{A}CD + A\bar{B}\bar{C} + AB\bar{C} + \bar{A}CD + A\bar{C}D$;   (b) $f = (A+B+C)(\bar{A}+C)(B+\bar{C})$.

3.2 Determine the minimal form of f if

(a) $\bar{f} = A\bar{B}\bar{C} + \bar{A}B + \bar{A}C$;   (b) $f = (A+B+C)A\bar{D}$.

3.3 Prove the following relationships by Boolean algebra. Verify the equalities using perfect induction.

(a) $\bar{A}\bar{B}\bar{C}D + A\bar{B}\bar{C}D + \bar{A}BCD + \bar{A}B\bar{C}D = \bar{A}\bar{B}D + \bar{A}\bar{C}D + \bar{B}\bar{C}D$;

(b) $AB\bar{C} + ABCD + A\bar{B}D + ABC\bar{D} = AB + AD$.

3.4 Show, by Boolean algebra, that

(a) $\bar{A} + \bar{B} = \overline{AB}$;   (b) $AB + A\bar{B}C = AB + AC$.

3.5 Using Boolean algebra find the minimal form of

$$(A+B+C)(\bar{A}+B+C)(\bar{A}+B+\bar{C}).$$

3.6 Minimize the following three-variable functions using Karnaugh maps.

(a) $f = AB + AC + \bar{B}C$;   (b) $f = \overline{ABC + A\bar{B}\bar{C} + \bar{A}B\bar{C} + \bar{A}\bar{B}C}$.

3.7 Minimize, using Karnaugh maps, the following four-variable functions.

(a) $f = (C+\bar{D})(B+C+D)(\bar{A}+D)$;

(b) $f = \bar{A}\bar{B}\bar{C}\bar{D} + ABCD + ABC\bar{D} + \bar{A}\bar{B}CD + \bar{A}BD + \bar{A}BC\bar{D} + A\bar{B}\bar{C}\bar{D} + A\bar{B}C\bar{D} + ABCD$.

3.8 Write down, in the simplest terms, the Boolean functions produced by the circuits shown in Figures P3.8(a) and P3.8(b).

3.9 Repeat problem 3.8 for the circuits shown in Figures P3.9(a) and P3.9(b).

3.10 Using two-input AND and OR gates and inverters draw the logic diagrams representing circuits that will implement the following functions.

(a) $(A+B)(C+D) + \bar{D}(\bar{C}+\bar{A}B)$;   (b) $AB\bar{D} + \bar{D}C + \bar{A}\bar{B}D + \bar{A}\bar{B}C$.

3.11 Using only two-input NAND gates draw the logic diagrams representing circuits that will implement the following functions.

(a) $\overline{\overline{AB}.A.\overline{AB}.B}$;   (b) $\overline{(A+B+\bar{C}).(\overline{AB}+\overline{CD}) + \overline{BCD}}$.

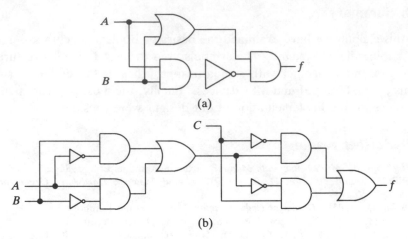

(a)

(b)

Figure P3.8.

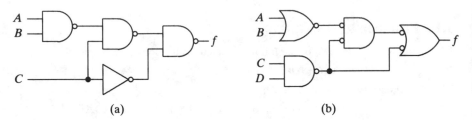

(a)

(b)

Figure P3.9.

3.12 Using only two-input NOR gates draw the logic diagrams representing circuits that will implement the following functions.

(a) $\overline{\overline{A+B}.\overline{B+\overline{C}+D}+\overline{B+C}}$; (b) $(A\overline{C}+\overline{B}\overline{C}+\overline{D}).(A+\overline{B})$.

3.13 Design a logic circuit which will accept four input signals A, B, C and D and produce a 1 output if at least three of the input signals are 1.

3.14 Design a full subtractor circuit. The circuit should accept three input signals: data bits A and B, and a borrow in signal C_{-1}. The circuit should produce two output signals: a difference signal D and a borrow-out signal C. These signals should reflect the result of the arithmetic operation:

$$A-(B+C_{-1}).$$

3.15 Design a circuit that will convert XS3 code to BCD.

3.16 Items placed on a weighing station are also to be sorted into two main height classifications. For this purpose three photoelectric cells are arranged in a vertical pattern above the weighing area. Cell A is the highest sensor, cell B the next highest and cell C the lowest sensor, a further sensor, D, on the scales shows when an item is present. Separate output signals are required each time an object falls into one of the two main height categories, i.e. taller than C and shorter than B or taller than B and shorter than A. These signals should only be produced when the object is present in the weighing area. If an object should appear that is taller than A or shorter than C then a single alarm signal should be activated. This alarm signal should remain until reset via a manually operated switch.

Design the logic circuitry necessary to produce the required output signals.

3.17 A packaging system consists of a motor-driven conveyor belt feeding articles to a carton. A detector on the belt provides a pulse each time an article is deposited in the carton. A second detector gives a continuous output if the carton is empty but not if the carton holds one or more articles. A third detector gives a continuous output when no carton is present. The system is to operate as follows.

The belt is to start running when:

(a) a momentary contact starter button is pressed;

(b) an empty carton is in place.

The belt is to stop running when:

(a) six articles have been counted into the carton;

(b) the carton is removed;

(c) no article is loaded for a period of more than one minute.

A re-settable timer is available which gives a pulse one minute after receiving the last reset signal. Design a system to control the motor drive.

3.18 Design a synchronous counter, using *J-K* flip-flops, that will produce a ten-state code with (5)(4)(2)(1) weightings.

3.19 Using 4-to-1 multiplexors, show how the following function can be implemented.

$$f = ABDE + ABCDE + ABC\bar{D}E + \bar{B}D\bar{E}.$$

3.20 Determine the code that needs to be stored in a ROM if the outputs from the ROM are to correctly illuminate the segments of a seven segment display in response to a BCD input code. The labelling of the display segments is illustrated by Figure P3.20.

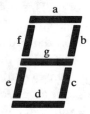

Figure P3.20.

3.21 Based on the simple PLA structure of Figure 3.30, determine which links need to be fused and which links need to be left intact in order to implement the following functions.

    (a) $f_1 = (A+B)C+AB$;   (b) $f_2 = \overline{A+BC} + \overline{BC+AB\bar{C}}$.

4: Architecture of a Typical Microprocessor

4.1 Introduction

This chapter looks at the architecture of a typical microprocessor. The format and types of the machine instructions that a microprocessor obeys are discussed, together with how the instructions specify the location of data. The components of the microprocessor, and how they are connected and coordinated, are described. The architecture of a microprocessor, or any computer processor, is defined by: the set of instructions it can obey, the ways in which the instructions can specify the location of data to be processed, the types and representation of the data, and the format in which the instructions are stored in memory. This definition could be implemented in a number of different ways at the circuit level, but this is transparent to a machine-code program running on the processor. All that would be seen is a difference in performance and cost. The design of a microprocessor architecture is a complex task and can involve thousands of hours of effort. The design engineers have to choose an architecture that can be implemented cost effectively using the current level of technology. It must also meet a clear market requirement and offer a performance advantage over existing products. This has lead to a wide range of architectures, some of which are discussed in Chapter 9. In this chapter the NeMiSyS microprocessor will be used as an example. NeMiSyS is a hypothetical microprocessor that is representative of modern 32-bit architectures such as the Motorola MC68000, although some aspects of its design are very much simplified. The NeMiSyS architecture is also used for the simulator provided with this book.

4.2 Machine instructions

The instructions that a microprocessor obeys are encoded as binary digits in its attached memory system. Each instruction is divided into one or more fields. All instructions have an operation-code field, which defines the instruction's purpose, such as to add or to move data. In addition, the instruction may have one or more operand or address fields, which specify where data is stored. Instruction formats can be classified by the number of operand fields they have. Generally, architectures with two operand instructions are faster than those that have instructions with only one operand. However, the control part of the processor must be more complex and larger in this case. Hence, the choice of the number of operands in the machine instruction format is a design trade-off.

Zero-operand instructions are used when there is no data to be acted upon, or when the operand is implied. For example, the NeMiSyS instruction clear interrupt enable (CLI) has no operands, but clears the interrupt-enable bit in the status register.

One-operand instructions are used when there is only one item of data to be processed: for example, the NeMiSyS instruction INC, which adds one to the contents of its operand. They are also used when there are two data items used by the instruction, but one of them is implied by the operation. The NeMiSyS instruction SSRH has only one explicit operand, but uses the status register as an implied operand.

Two-operand instructions explicitly give two data items. For example, the NeMiSyS instruction ADD, which calculates the sum of its two operands and stores the result in the last one.

Three-or-more-operand instruction formats are not common in microprocessor architectures. However, they are used in RISC machines. For example, the Acorn RISC Machine (ARM) instruction ADD R1,R2,R3.

Most microprocessors use a mixture of instruction formats. However, a microprocessor's architecture is characterized by the maximum number of operand fields in its instruction formats. 8-bit processors generally have one-operand architectures. This is because they are implemented using small scales of integration. Thus, there is limited space on the chip for control circuits and internal registers. The MC6809, for example, has only two general-purpose registers. As there are so few registers, each can have its own set of instructions which imply, from the operation code, the register to be used. This has the advantage that the instruction decoding and execution circuits can be simpler and thus more compact. 16/32-bit processors use very large scale integration, and have more internal registers that need to be addressed as operands. For example, the MC68000 has 16 registers. In this case, having specific instructions for each register would be inefficient; therefore, general-purpose, two-operand instructions are used. These allow the registers to be specified explicitly in the instruction, separately from the operation code. Most of the NeMiSyS instructions have two operands, but zero- and one-operand instructions are used when appropriate. Table 5.2, in Chapter 5, gives a full list of all the NeMiSyS instructions in assembly-language format.

4.2.1 The instruction set

The instruction set of a microprocessor is all the machine instructions that the processor can obey. It defines the operations that the processor must perform. The number of instructions and their complexity governs the size of the decoding and control circuits in the processor. Thus, the instruction set is constrained by the scale of integration that is used to implement the architecture. This is why 8-bit processors generally have less powerful instruction sets than 16/32-bit processors. The tendency toward more complex instruction sets, that this represents, has been reversed by the introduction of reduced-instruction-set computers (RISC).

Instructions can be divided into a number of subsets: assignment, arithmetic/logic, control and conversion. Each of these will be considered in turn.

ASSIGNMENT INSTRUCTIONS

These are instructions that move data around the system. A processor must have instructions that move data between pairs of internal registers, and between these registers and memory. Instructions may also be available for memory-to-memory data transfers. 8-bit processors, like the MC6809, do not have such instructions, and data has to be transferred between memory locations via an internal register. The NeMiSyS processor has a general-purpose two-operand MOVE instruction that handles most of the required assignment operations. Instructions that assign a fixed value to a specific component of the processor, such as the NeMiSyS CLI instruction, are also included in this subset. Input/output (I/O) instructions transfer data values to I/O devices, and are thus assignment instructions. However, many microprocessor do not have any specific I/O instructions, and rely on memory-mapped I/O to access their attached I/O devices. With this method, any instruction that can access memory can be used to control and access an I/O device.

ARITHMETIC/LOGIC INSTRUCTIONS

These are instructions that perform arithmetic (e.g. add) and logic (e.g. AND) operations on data stored in the internal registers and the attached memory of the processor. Most 8-bit processors can only perform arithmetic and logic operations on data stored in internal registers. 16/32-bit processors, such as NeMiSyS, can also perform memory-to-memory arithmetic and logic. Some microprocessors restrict the scope of register-based arithmetic and logic operations to special registers, called accumulators or data registers. Arithmetic is normally limited to integer data, although some microprocessors also support binary-coded decimal data. 16/32-bit processors are beginning to include floating-point hardware. However, in most cases hardware floating-point operations require the use of a mathematics co-processor. These are available as part of the chip families of 16/32-bit processors, such as the Motorola MC68000 series. Special features are provided on these processors to allow the integration of the co-processor into a complete computer system. If a floating-point co-processor is not available, or is too expensive for the application, then the arithmetic has to be performed by software, using the integer facilities of the microprocessor. This is common on most low-cost microprocessor systems.

Arithmetic operations on 8-bit microprocessors do not include multiplication or division. If these are needed, then they have to be provided by software. The built-in hardware multiplication facilities available on 16/32-bit processors are, of course, much faster.

In all computers, data is stored in coded formats, called data types. The most obvious example is the storage of integers as binary numbers. Many architectures support multiple data types. So, for example, a number might be stored in binary integer, binary-coded decimal or floating-point format. Conversion instructions are used to convert one data type to another. However, strict data typing is not present in microprocessor architectures, and what a data item represents is, in many cases, independent of the hardware. Therefore, a wide range of data-conversion instructions are not required. For example, the ASCII code for "a" can be stored in a register. The register will contain the binary representation of the decimal number 61. If this is sent to a printer, then the symbol for "a" will be printed. The register's contents can also be interpreted as an unsigned integer; thus one can be added to it using an INC instruction. The resulting value of decimal 62 can then be interpreted as the character "b".

A less obvious, but important distinction is between addresses and data. An address gives the location of a data item. Memory positions are assumed to be continuous and start at address zero. An effective address is an unsigned integer that is the position of some data in memory. It can be stored in a register and can be arithmetically or logically manipulated. Addresses are also stored in special formats (i.e. addressing modes) in machine instructions. These are converted into effective addresses when an instruction is executed. An instruction that converts an address into an effective address, without referencing its data, is often useful. NeMiSyS has the load-address (LDA) instruction for this purpose.

CONTROL INSTRUCTIONS

The ability to modify the execution order of instructions, as a program is running, is fundamental to the programming of digital computer systems. All processors provide a subset of instructions for this purpose. They are the jump and the conditional-branch instructions. The execution of most instructions alters bits in a program-status register (PSR), which is part of the microprocessor. The bits that are changed depend on such conditions as the sign of a resultant data value. The branch instructions check these bits, and either allow execution of the program to continue with the next instruction in memory, or select an alternative sequence of instructions stored in a different part of memory. The address of this sequence is specified as a field in the branch instruction. In addition, there is always an instruction for performing an unconditional branch or jump, which disregards the bits in the program-status register, and always makes the branch. The branch instructions are used to compile high-level language constructs, such as IF ... THEN ... ELSE or WHILE statements, into machine code. Blocks of instructions that can be shared by different parts of a program are very useful. These are called subroutines, and they are supported by the jump-to-subroutine (JSR) and return-from-subroutine (RTS) instructions.

4.2.2 Addressing modes

Most machine instructions operate on data items stored in registers or memory. The location of these items must be given in the instruction's operand fields. A number of different ways of specifying these locations or addresses are possible. These are called the addressing modes of the microprocessor. Microprocessors offer a range of addressing modes to increase their flexibility, to simplify the writing of assembly language programs and to make implementing high-level language compilers easier. The addressing modes of the NeMiSyS processor are examples of the great variety available on modern microprocessors. In general, the code in an instruction's operand field specifies an addressing mode and a register, and the instruction may be followed by an optional integer constant. The value stored in the register and the constant are used by the processor to calculate an *effective address* for the data, which can then be accessed during the execution of the instruction. NeMiSyS instructions also contain an addressing-unit field which specifies the size of the data accessed as a byte (8 bits), a half word (16 bits) or a full word (32 bits).

ABSOLUTE

In absolute (or memory-direct) addressing mode, the effective address of the data item in memory is given as a 32-bit unsigned integer constant following the instruction. The length of the data at the effective address is specified by the address-unit field of the instruction. The register field of the operand is ignored.

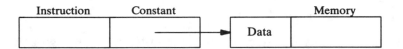

This mode is useful for accessing fixed memory locations, such as memory-mapped I/O devices. It is often used as the standard addressing mode in programs, but this is not recommended. Programs that use this method of addressing for jump instructions, must always be loaded and executed in the same place in memory. While this inflexibility may not be a limitation in the smaller memories of 8-bit processors, it can cause problems with the large address spaces of 16/32-bit microprocessors. Even if the use of absolute addressing for jump instructions is avoided, any data referenced using this addressing mode will still be at a fixed memory location. If these data locations are supposed to be defined within the body of the program, this will clearly cause problems. Indeed, if more than one program is resident in memory, then having fixed-location data can be as great a disadvantage as having fixed-location programs. If subprograms are joined together or linked to form larger programs, the position of a subprogram, relative to the start of the program, can not be predicted in advance. Its load address is unknown.

Hence, absolute addressing for the subprogram's internal variables is inappropriate. The solution to these problems is PC-relative addressing, which is described below.

IMMEDIATE

In immediate addressing mode an unsigned integer constant, following the instruction, is the data item. The register field in the instruction is ignored. The length, and thus the maximum value, of the constant is chosen by the address-unit field of the instruction.

Instruction	Constant
	Data

This mode is useful for accessing constant values in a program. It saves memory space and is faster than absolute addressing, as a second memory access is not required to obtain the data. It cannot be used for destination operands.

REGISTER DIRECT

Register-direct addressing mode specifies that the data is in a register, which is given in the register field of the instruction. There is no constant following the instruction.

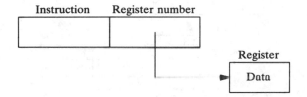

This addressing mode is required to access the internal registers of the processor, in order to perform fast arithmetic and to set up the registers for, say, indexed addressing. Operations on registers are very fast because the processor does not have to access its attached memory system. One of the performance advantages of RISC machines is that these architectures have many internal registers.

REGISTER INDIRECT

The register-indirect addressing mode specifies that the memory location of the data is stored in a register, which is given by the register field of the instruction. We say that the register "points to the data", or "contains the address of the data." There is no constant following the instruction in this addressing mode.

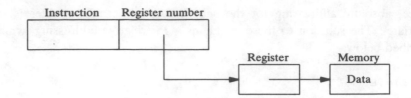

This addressing mode is useful for implementing the pointer data type of high-level languages, such as Pascal. It can be used when a subroutine argument is passed as a memory location. In this way, any operations carried out on the argument will actually change its value outside the subroutine. This is called "parameter passing by reference."

POSTINCREMENT REGISTER INDIRECT

The postincrement register-indirect addressing mode is similar to the register-indirect mode described above. The absolute address of the data is stored in a register as before, but after the data have been accessed the contents of the register are incremented. The size of the increment depends on the addressing unit of the instruction. If the addressing unit is a byte then the register contents are increased by one, if it is a half word by two, or if it is a full word by four. There is no constant following the instruction in this addressing mode.

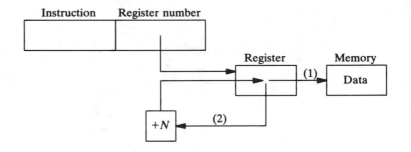

PREDECREMENT REGISTER INDIRECT

The predecrement register-indirect addressing mode is similar to the postincrement mode but, in this case, the register is decremented before the data is accessed in memory. The register is decremented by one, two or four, depending on the addressing unit which can be a byte, a half word or a full word, respectively. There is no constant following the instruction in this addressing mode.

Predecrement, postincrement and simple register-indirect addressing modes are used for similar purposes. However, the first two have the added advantage that tables or arrays of data can be processed item by item, without extra instructions to change the value in the register. A special use for these addressing modes is to support stacks.

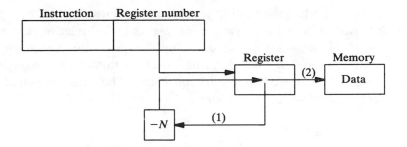

INDEXED

In indexed addressing mode the effective address of the data item is calculated by adding the contents of a register, specified in the operand field, to the value of a full word following the instruction. The value of the full word is constant, but the value stored in the register can be altered while the program is running. The value in the index register is referred to as the offset of the data. This offset is interpreted as a signed integer in two's-complement form. Thus, it can be positive or negative.

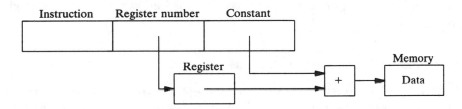

This mode is useful whenever the absolute location of the data is not known until the program is running. As the effective address is calculated using one of the registers, modifying the contents of this will move the expected location of the data. A typical use of indexed addressing is to access a continuous table or array of data items stored in memory. The instruction gives the address of the start of the array, while the offset in the index register selects the data item in the array. Another item can be selected by adding or subtracting multiples of data-item length from the index register.

PC RELATIVE

PC-relative addressing is similar to indexed addressing. In fact, the only difference is that the program counter (PC) is used instead of an index register. The program counter contains the memory location of the next instruction to be executed. By choosing the correct value for the constant part of the address, the location of any data item can be given relative to the next instruction. Thus, wherever the program is loaded in memory, the effective address of the item will be correctly calculated. The location of the data will, in effect, move with the program. This characteristic is called *relocatability*. A particularly important use of relative addressing is in branch instructions,

101

where the effective address is the location of another instruction rather than a data item. NeMiSyS branch instructions use only PC-relative addressing mode. Therefore, to simplify and speed up the processor, the mode and register fields are omitted, and only the offset to the target instruction is given, as a 24-bit field after the operation code. The offset is coded as a signed integer in two's-complement form.

IMPLICIT REFERENCES

Many machine instructions modify the contents of internal processor components, without explicitly referencing them in an operand field. A processor using a one-operand instruction set utilizes this addressing mode in almost all of its instructions. The operation code implies one of the operands, while the other is explicitly stated in the instruction's operand field. A 16/32-bit processor, like NeMiSyS, has many two-operand-format instructions, but nevertheless, uses implicit references in some of its instructions. An example of this is the branch instruction, which implicitly uses the program-status register to determine its operation. Other NeMiSyS instructions include the set-interrupt-enable (SEI) instruction, where again the program-status register is used, and the jump-to-subroutine (JSR) instruction, which uses the stack-pointer register.

4.2.3 Machine instructions and the stack

A stack is a data structure, in which an item taken from the stack will be the one most recently put into the stack. A stack is often called a last-in-first-out (LIFO) queue. Stack insertion and removal are referred to as pushing and popping, respectively. Most microprocessors, including NeMiSyS, use a stack to support subroutine calls and interrupts. Stacks are generally useful for programming; therefore microprocessors have special instructions or addressing modes that support them. Some microprocessors, such as the Z80, have push and pop instructions to access the processor stack. More advanced microprocessors, like the MC68000, use predecrement and postincrement register-indirect addressing modes, with a generalized move instruction, to access the stack. This approach has the advantage that a programmer can use private stacks, as well as the processor stack. When a special register, called the stack pointer, is specified, the processor stack will be accessed, but a private stack can be maintained by using a general-purpose register instead.

4.3 Internal components

The structure of a microprocessor can be divided into two parts for processing and control. The controller decodes machine instructions and generates signals that direct the processing part of the microprocessor. The internal components of the processor are connected by busses, so that data can flow between them. Figure 4.1 shows the general structure of a simplified microprocessor called NeMiSyS.

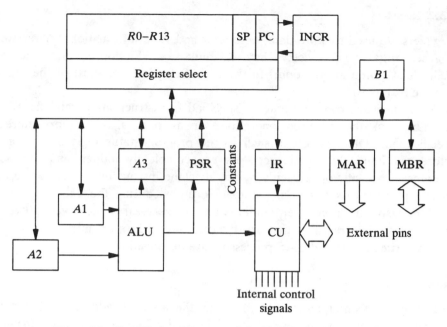

Figure 4.1. Simplified architecture of the NeMiSyS microprocessor.

4.3.1 Arithmetic and logic unit

The arithmetic logic and unit (ALU) is responsible for performing all the numerical and logical calculations needed during the operation of the processor. The arithmetic functions always include the addition and subtraction of signed integers. Many microprocessors have only these two arithmetic operations available, but 16/32-bit processors, such as the MC68000, normally include integer multiply and divide facilities. Most microprocessors confine their arithmetic to binary integer data. Floating-point arithmetic is supported by using a mathematics co-processor, attached to the microprocessor by external busses. Logic operations include operations such as NOT and OR between two data items, and bit-shift operations. The comparison of data items is also performed by the ALU.

The ALU is built using combinational logic, and has two data inputs and a data output. It is designed to cope with the maximum integer that can be stored in the processor's registers. The required data is presented to the ALU inputs and the operation that is to be performed is selected by the control unit. After the ALU has had time to generate the result, this output is stored in a register.

The ALU carries out the arithmetic and logic functions specified in the operation-code field of machine instructions. These functions are applied to data obtained from the effective addresses of operand fields. The ALU may also have to calculate these effective addresses. Hence, a number of the more modern microprocessors have more than one ALU, so that some of these calculations can be performed in parallel, thus speeding up the processor.

103

4.3.2 Registers

Registers are used to store data, addresses and state information. A register is a sequential logic device. They are built from flip-flop circuits, one for each bit of stored information. In the NeMiSyS processor most of the registers are 32 bits wide.

Some registers can be explicitly specified in instruction-operand fields as the target of the required operation, or as part of a memory-address specification. Other registers, such as the program-status register, are only referenced implicitly. Some registers are completely hidden and are not referenced explicitly or implicitly by any machine instructions. They are used by the processor as temporary storage locations during the execution of a machine instruction, and their contents are not necessarily preserved between instructions. The registers that are typically found in most microprocessors, and in particular the NeMiSyS processor, are described below.

PROGRAM COUNTER

The program-counter register (PC) stores the absolute address of the next machine instruction to be executed. This plays a central role in controlling the sequence of machine instructions that the processor executes. Its contents are affected by the branch and jump instructions, and by the interrupt process. The program counter is register $R15$ in the NeMiSyS processor and can be referenced explicitly, for say, PC-relative addressing. In NeMiSyS assembler the name PC is used instead of R15 to clearly indicate that the program counter is being used, rather than a general-purpose register.

INSTRUCTION REGISTER

The instruction register (IR) stores the code of the instruction currently being executed. The operation code is extracted from the IR by the control unit, which determines the sequence of signals necessary to perform the processing required by the instruction. The operand fields are also referenced in order to determine their effective addresses.

BUFFER REGISTERS

These registers interface the microprocessor with its memory system. There are two standard buffer registers: the memory address register (MAR) and the memory buffer register (MBR). The MBR is sometimes known as the memory data register.

Memory address register. This register is connected to the address pins of the microprocessor, each bit of the MAR corresponding to a pin of the address bus. It holds the absolute memory address of the data or instruction to be accessed. The size of the MAR determines the maximum memory address and thus the amount of memory that the processor can directly support. The NeMiSyS processor has a 32-bit MAR. Hence, it can support 4 294 967 296 bytes

(4096 Mbytes or 4 Gbytes) of directly accessible memory, in the form of ROM, RAM and memory-mapped I/O devices. However, the NeMiSyS system that is implemented in the simulator does not use all of this address space. In fact, it provides only approximately 4 Kbytes of memory as RAM and memory-mapped I/O. This is not unusual, because it is both unnecessary and expensive to provide more memory than is required for the software that the system is designed to run. The MAR is not explicitly referenced in any machine instructions, but is used whenever an addressing mode that accesses memory is specified.

Memory buffer register. This register is connected to the data pins of the microprocessor. It stores all data written to, and read from, memory. Its size determines the maximum memory data-transfer unit. The NeMiSyS processor has a 32-bit MBR, and can thus load or store one of its 32-bit registers in a single memory cycle. Some microprocessors, like the MC68000, have a 16-bit MBR and 32-bit general registers. In this case, two memory cycles are required to load a full word (32 bits) into a register. This double read or write cycle for full words is invisible at the machine-instruction level, but it significantly affects the performance of the processor. This difference between internal and external data width, leads to the MC68000 being referred to as a 16/32-bit microprocessor.

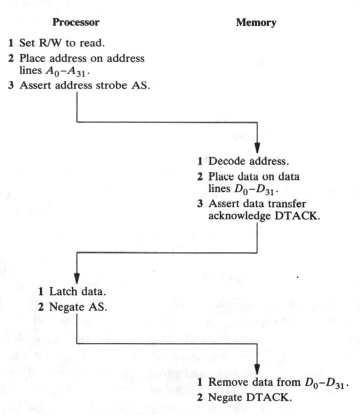

Processor	Memory
1 Set R/W to read.	
2 Place address on address lines $A_0 - A_{31}$.	
3 Assert address strobe AS.	
	1 Decode address.
	2 Place data on data lines $D_0 - D_{31}$.
	3 Assert data transfer acknowledge DTACK.
1 Latch data.	
2 Negate AS.	
	1 Remove data from $D_0 - D_{31}$.
	2 Negate DTACK.

Figure 4.2. NeMiSyS read cycle.

105

Both these registers are used to access memory during the processor's read and write cycles. Figure 4.2 shows a full-word read cycle for the NeMiSyS processor. Having calculated the effective address of an operand, this is placed in the MAR. Signals are then put on the microprocessor's external pins to inform the memory system of the required operation and to co-ordinate the access cycle. This results in the memory system putting the contents of the selected full word onto a data bus, which is connected to the microprocessor. The processor then latches this data bus into the MBR to obtain the required full word of data. Figure 4.3 shows the similar write cycle.

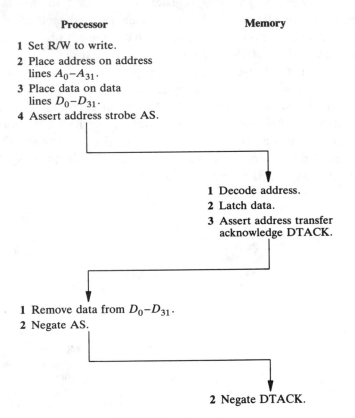

Processor	Memory

1 Set R/W to write.
2 Place address on address lines A_0-A_{31}.
3 Place data on data lines D_0-D_{31}.
4 Assert address strobe AS.

1 Decode address.
2 Latch data.
3 Assert address transfer acknowledge DTACK.

1 Remove data from D_0-D_{31}.
2 Negate AS.

2 Negate DTACK.

Figure 4.3. NeMiSyS write cycle.

PROGRAM-STATUS REGISTER

The program-status register stores the conditional status of the processor. In the NeMiSyS processor, it is 16 bits wide, but not all of the bits are used. Each bit is a Boolean flag indicating the state of one condition. The flags are set during the operation of the processor. They can be used to control conditional branch instructions and processor interrupts, and to manage multiple-word arithmetic. The NeMiSyS status condition flags are as follows.

V (**overflow**): an overflow condition resulted from the last instruction executed.

N (**negative**): a negative integer resulted from the last instruction executed.

Z (**zero**): a zero integer resulted from the last instruction executed.

C (**carry**): a carry was generated by the last instruction executed. This is the result of an arithmetic or shift instruction.

IEN (**interrupt enable**): maskable interrupts are enabled. This bit is switched using the clear-interrupt-enable (CLI) and set-interrupt-enable (SEI) instructions.

BRK (**break**): a break instruction was the last instruction executed. This is a break interrupt, which is non-maskable and internally generated.

UOP (**unknown operation code**): an unknown operation code was found in the last instruction executed. This causes a non-maskable, internally generated interrupt.

IAM (**invalid addressing mode**): an invalid or improper addressing mode specification was found in the last instruction executed. This causes a non-maskable, internally generated interrupt.

STACK POINTER

The stack-pointer register (SP) points to the top of the processor stack, that is, it contains the address of the most recent entry in the stack. In NeMiSyS machine instructions, the stack pointer is referenced as general-purpose register 14, although in assembler the name SP can be used instead of R14. The stack pointer is used implicitly whenever a subroutine is called or an interrupt is serviced, and thus it is critical to the operation of the processor. If an application requires additional stacks, any of the general-purpose registers can be used in the same way as the stack pointer, with predecrement and postincrement register addressing. However, in most cases, separate application stacks are not required and the processor stack can be used. The programmer must be careful when using the processor stack, because any damage or disruption to this stack will normally halt the execution of a program with a serious error. To help control such problems, the MC68000 has two stacks— one for the user mode, that applications programs run under, and one for system mode, that the operating system uses. NeMiSyS is typical of simpler processors and has only one stack.

GENERAL-PURPOSE REGISTERS

These registers are used explicitly to store data and address information. The number of registers varies greatly between microprocessors. Most microprocessors distinguish between data and address registers, often referred to as accumulators and index registers. Data registers are used for arithmetic, while address registers are used for indexing and indirect addressing. This register organization is chosen because it simplifies the internal design of the microprocessor and allows it to be fitted onto less silicon. This is not a design constraint for NeMiSyS because it is a hypothetical microprocessor. Therefore, it has been made more versatile by having only general-purpose registers, all of which can be used for data or addressing. This is similar to the organization of minicomputers such as the DEC PDP/11.

The NeMiSyS microprocessor has 14 general-purpose registers named $R0$ through $R13$. The program counter and the stack pointer can be used in the same way as general purpose registers and are referenced as registers $R14$ and $R15$ when required.

TEMPORARY REGISTERS

Temporary registers or latches are used when data has to be stored during the execution of a machine instruction. This may be required to avoid contention on a shared data path, or while waiting for a subsequent operation. Temporary registers are completely hidden from the user of the microprocessor. They are really a feature of the architecture's implementation rather than its specification.

The NeMiSyS processor has four temporary registers, which are designated $A1$, $A2$, $A3$ and $B1$. Registers $A1$ and $A2$ are connected to the inputs of the ALU, and register $A3$ is able to latch its output. These registers are also used for operand transfer. $A1$ and $A2$ store the values fetched from the instruction operands and $A3$ holds the output value in preparation for delivery to the destination operand. Register $B1$ is used to store data from the other temporary registers when required.

4.3.3 Control unit

The control unit (CU) is at the heart of a microprocessor. It is responsible for controlling the cycle of fetching machine instructions from memory and executing them. This is a complex task and in most microprocessors the control part takes up a considerable proportion of the silicon area of the chip. For example, the control unit occupies 69% of the MC68000 chip area.

The co-ordination of all activity in the processor is performed by the control unit. It is a sequential logic circuit, which steps the processor through a sequence of synchronized operations. This is achieved by sending out a stream of control signals and timed pulses to the components and external pins of the microprocessor. The wires that carry these signals are called control lines or clocking lines. In order to achieve synchronization, both internally and with external components such as memory and I/O devices, an external reference clock pulse is required. This is obtained through one of the microprocessor's external pins.

The registers and ALU are connected by parallel data paths or busses, some of which are shared. During the operation of the processor, data bits move along these busses. As the busses are shared, access to them must be controlled. Only one component at a time should output a signal to a shared bus, and tri-state logic is used to connect the components' outputs to the bus. Tri-state gates can be instructed, by the control unit, to present a high impedance to the bus instead of their normal output signals. This minimizes their effect on the bus. The high-impedance state of tri-state logic is often described as a floating output. Registers input (latch) signals from a bus when instructed by the control unit. The functions of other components, such as the ALU and shift registers, are selected by control-line signals. Two methods are used for building control units: hard-wired and micro-coded.

HARD-WIRED CONTROL UNITS

A hard-wired controller is simply a sequential logic circuit, the states of which correspond to the phases of the instruction execution cycle. Figure 4.4 shows a simplified diagram of how such a controller might be constructed.

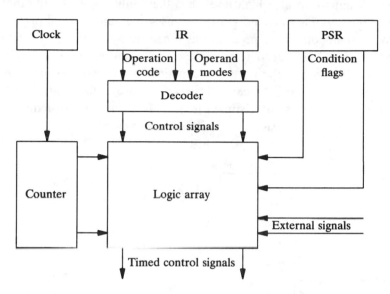

Figure 4.4. Simplified hard-wired controller.

The machine instruction being executed is stored in the instruction register. From here, the instruction is passed to a decoder which generates a pattern of signals. These define what activities are required, but not when they should occur. To complete the execution of the instruction, the operational part of the processor must be stepped through a series of activities. Therefore, a timed sequence of control signals is generated by combining the decoder output with timing signals. The timing is generated by a counter which is, in turn, controlled by a continuous and regular sequence of external clock pulses. As the counter signals are generated, they are combined with the signals from the decoder in an array of simple logic gates. This produces the final timed sequence of control signals, for distribution to the operational components of the processor. The program-status register is connected to the logic array. Thus, depending on the condition flags, alternative sequences of control signals can be generated.

Hard-wired controllers are very compact and fast, but they are very difficult to design; because of this, they have not been used in many microprocessors, the micro-coded controller being preferred. Recently RISC architectures have simplified the control unit by reducing the size and complexity of the instruction set it has to support. This has resulted in control units that can be effectively implemented using hard-wired designs. The improved speed and spare silicon for additional components, such as registers, have produced the performance advantages seen in modern RISC microprocessors.

MICRO-CODED CONTROL UNITS

Machine instructions are sequentially arranged, in the processor's attached memory system, to form a machine-code program. These instructions tell the processor what to do at a high level, but they only imply the internal activities required. A micro-coded controller uses the same stored-program concept to internally direct the processor as it executes a machine instruction.

The information that defines the actions required to execute machine instructions is stored in the form of a sequence of *micro-instructions* or *micro-code* in the controller. Machine instructions are called *macro-instructions* (macro- meaning great) to distinguish them from micro-instructions. Figure 4.5 shows a simplified design for a micro-coded controller.

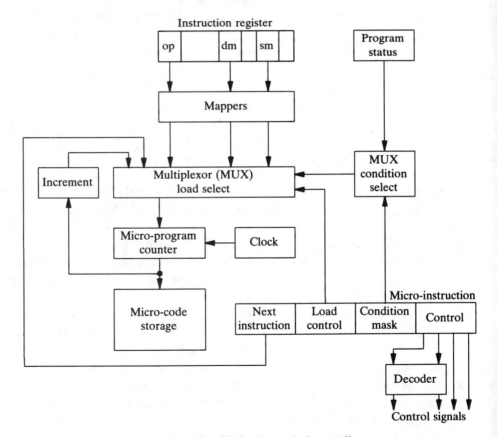

Figure 4.5. Simplified micro-coded controller.

The macro-instruction's operation code is extracted from the instruction register by a mapper, and used to obtain a micro-code address from a mapping table. This address is the start of a sequence of micro-instructions that define the actions required to executed the macro-instruction. The micro-instructions are fetched from micro-storage into the micro-instruction register and decoded to generate a set of control signals. Each micro-instruction defines one phase of the macro-instruction's execution; therefore, a counter is not required to generate timed signals. The controller generates a sequence of timed control signals by executing a sequence of micro-instructions and is synchronized by an external clock signal. A complex decoder is not required for a micro-coded controller because a micro-instruction is designed to be easily decoded.

There are two ways of storing control information in micro-instructions: horizontal or vertical encoding. Horizontal encoding uses a bit to represent each control point in the processor. This is very flexible and requires no decoding, but results in long micro-instructions. An alternative is vertical encoding, where the required combinations of control signals are compressed into a more compact code. This saves space in micro-storage, but decoding circuits are required. Some combinations of signals are not allowed, which leads to a loss of flexibility. In practice, a mixture of the two methods is used to achieve a design compromise.

Micro-instructions contain other fields in addition to control signals. These fields are concerned with flow control in the micro-program executing inside the controller. Figure 4.6 shows the format of the NeMiSyS micro-instruction.

Load control	Next instrn.	Cond. mask	set bit	clear bit	incr.	flag	val	op	S	D	R/W	width
									Control signals			
		PSR			PC	const $\to$ D		ALU	transfer $S \to D$		memory access	

Figure 4.6. NeMiSyS micro-instruction format.

The load-control field indicates how the location of the next micro-instruction to be executed should be derived. It can be the next in micro-storage, or be at the location given in the next instruction field. The choice between these two can be conditional and depend on the contents of the program-status register. If this is the case, then the bits to be tested are selected by a mask in the condition-select field. The most powerful option is to specify that the next micro-instruction address should be derived from mapping tables. The mapping facility uses the operation-code or addressing-mode fields in the instruction register as an index to a table of addresses in micro-storage. These addresses are used by the controller to jump to micro-code subroutines, one for each macro-instruction and addressing mode.

The structure of a micro-program can be understood by considering the data transfers required to execute a macro-instruction, and this will be discussed in the next section.

111

4.4 Instruction execution

The general format of macro-instructions, and how a micro-coded controller can be constructed to execute these, has been discussed. Here, the sequence of events required for macro-instruction execution is considered in more detail. The NeMiSyS processor will be used as an example. A list of NeMiSyS instructions, in assembly-language format, can be found in Table 5.2, but for this discussion the instruction set needs to be defined at its memory-storage level. This is given in Table 4.1. The address formats are given in Figure 4.7.

We could use programs of micro-instructions to describe instruction execution, but this would be difficult to follow. Instead, we are going to use a notation called register-transfer language (RTL). There are many different register-transfer languages but we are going to to use a simple version that is specific to the NeMiSyS microprocessor.

4.4.1 NeMiSyS register-transfer language

This is the register-transfer language that we are going to use to describe the internal operation of the NeMiSyS microprocessor. Here, only the notation of the RTL statements is given. The relation between the actions they describe and the NeMiSyS architecture will be discussed later.

Each component of the processor is denoted by a short code, as follows.

ALU	Arithmetic logic unit	IR	Instruction register
A1,A2	ALU input registers	MAR	Memory address register
A3	ALU output register	MBR	Memory buffer register
B1	Temporary register	PC	Program-counter register
CU	Control unit	PSR	Program-status register
INCR	Incrementer	SP	Stack-pointer register

A transfer of data between these components is written as follows, where the contents of xx is transferred to yy at time or phase 0:

$$t0: [xx] \rightarrow [yy].$$

The brackets [] denote that the components are registers and will store the data until another assignment is made to them. The value stored in the left-hand component is not affected by the data transfer. If a register has fields, as in the case of the instruction register, then these are given as follows:

$$t10: [IR(op)] \rightarrow CU.$$

The fields of the IR that are referenced in this manner are the following.

op	operation code	dm	destination addressing mode
sm	source addressing mode	dr	destination register
sr	source register	bo	branch offset
au	addressing unit		

Table 4.1. NeMiSyS instruction fields.

op: operation codes having 6 bits	
000000 = BREAK	
000010 = ADD	000011 = ADDC
000100 = SUB	000101 = SUBC
000110 = NEG	000111 = NEGC
001000 = INC	001001 = DEC
001010 = AND	001011 = OR
001100 = EOR	001101 = NOT
001110 = LSR	001111 = LSL
010000 = ASR	010001 = ASL
010010 = ROR	010011 = ROL
010100 = MOVE	010101 = LDA
010110 = SSRH	010111 = RSRH
011000 = JMP	011001 = JSR
011010 = CMPU	011011 = CMPS

op: operation codes having 8 bits			
80 = BCC	84 = BCS	88 = BPL	8C = BNE
90 = BEQ	94 = BMI	98 = BVC	9C = BVS
A0 = BLT	A4 = BLE	A8 = BGT	AC = BGE
B0 = BRA		C8 = RTS	CC = RTI
D0 = CLI	D4 = SEI	D8 = CLC	DC = SEC

au: addressing unit

00 $\Rightarrow$ Full word (32 bits).
01 $\Rightarrow$ Byte (8 bits).
10 $\Rightarrow$ Half word (16 bits).
11 $\Rightarrow$ Full word (32 bits).

sm: source addressing mode

0000 $\Rightarrow$ Immediate: byte, half or full word follows.
0001 $\Rightarrow$ Absolute: full word follows.
0010 $\Rightarrow$ Indexed: full word follows.
0011 $\Rightarrow$ Register direct.
0100 $\Rightarrow$ PC relative: full word follows.
0101 $\Rightarrow$ Register indirect.
0110 $\Rightarrow$ Postincrement register indirect.
0111 $\Rightarrow$ Predecrement register indirect.
1xxx $\Rightarrow$ Invalid.

sr: source register number 0...15

dm: destination addressing mode

0000 $\Rightarrow$ Invalid.
0001 $\Rightarrow$ Absolute: full word follows.
0010 $\Rightarrow$ Indexed: full word follows.
0011 $\Rightarrow$ Register direct.
0100 $\Rightarrow$ PC relative: full word follows.
0101 $\Rightarrow$ Register indirect.
0110 $\Rightarrow$ Postincrement register indirect.
0111 $\Rightarrow$ Predecrement register indirect.
1xxx $\Rightarrow$ Invalid.

dr: destination register number 0...15

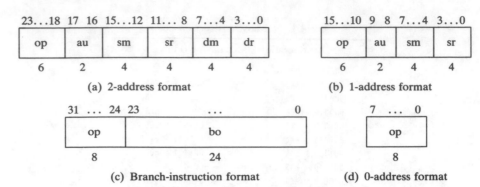

Figure 4.7. NeMiSyS address formats.

The instruction-register fields sr and dr are used to select general-purpose registers when appropriate. No special RTL statements are used to describe this operation because, in the NeMiSyS architecture, the translation of the register field into register-selection signals is performed by dedicated combinational logic. Thus, a control signal on the select-SR line will result in the choice of the correct general-purpose register, without any further intervention from the controller. Such a selection is implied by the following:

t21: [[IR(sr)]] -> [B1],

which specifies that the contents of a general-purpose register, given in the instruction's source-register field, are transferred to register $B1$. Fields are also used to reference PSR flags. The following example, which sets the break flag in the PSR to one, shows how PSR fields are used and also demonstrates how constants can be assigned to registers:

t5: 1 -> [PSR(BRK)].

If a component is combinational, its output will not be retained when its inputs change. Therefore, the output must be assigned to a register before this happens. The way that control signals and inputs to the ALU are specified is shown in the following example:

t2: ADD > ALU([A1],[A2]) -> [A3].

This states that the contents of the input registers $A1$ and $A2$ are processed by the ALU and the result is transferred to the register $A3$. The control signal, which specifies the operation to be performed by the ALU, is given as add in this case, but alternatives are possible, such as shift or subtract. There are no brackets around ALU(...,...) because it is a combinational component. The incrementer is used in a similar way:

t4: 3 > INCR([PC]) -> [PC].

Here, the contents of the program counter are incremented by three, but a different control signal could be used to add other values to the program counter.

Memory access is performed through the memory address and buffer registers. The notation used for a write to memory is:

    t4: [MBR] -> [MS[MAR]],

that is, the contents of the memory buffer register are put into main storage (MS), at an address given by the contents of the memory address register. The notation for a read from memory is similar:

    t12: [MS[MAR]] -> [MBR].

NeMiSyS allows byte, half-word and full-word access to memory. However, for simplicity, no distinction is made. It is assumed that the correct access width, as specified in the instruction register or implied by the activity, is automatically used.

As with micro-code, our RTL must be able to specify conditional execution. The NeMiSyS architecture allows the PSR to be used for this purpose:

    t6: IF [PSR(C)] = 1 JUMP carryset
    t7: 1 -> [PSR(C)].

This indicates that if the condition [PSR(C)] = 1 is true (i.e. the carry flag in the PSR is equal to one), then the next RTL statement is at the label car-ryset. If the condition is false, then the following statement is next. A label is written to the left of a RTL statement, thus:

    carryset t7: [PC] -> [MAR].

The timing flags (e.g. t7) are not labels, and when conditional sequences of RTL statements are specified they can, as above, be duplicated. Only the PSR can be used in conditional RTL statements; therefore, the notation can be shortened without confusion to:

    t6: IF C = 1 JUMP carryset
    t7: 1 -> [PSR(C)]
    carryset t7: [PC] -> [MAR].

This shortened notation without the time flags is used in the NeMiSyS simulator.

If more than one transfer or operation occurs at the same time, then they are separated by a comma, as follows:

    t4: [PC] -> [MAR], 4 > INCR([PC]) -> [PC].

If there is an ordering of activities at a given time, then they are separated with a | symbol:

    t5: [B1] -> [PSR] | JUMP interrupt.

The activities are taken to occur from left to right. Thus [B1] -> [PSR] is performed, and then a jump is taken to the label interrupt. This jump occurs after time *t*5, but before time *t*6. As the examples below show, the only

115

activities that can be ordered in such a way are concerned with flow control within the RTL description.

Finally, it is also possible to use subroutines in RTL, as this example shows:

```
         t3: CALL(getwordatpc)
         t7: [MBR] -> [B1]

getwordatpc t0: [PC] -> [MAR]
         t1: [MS[MAR]] -> [MBR], 4 > INCR([PC]) -> [PC]
         t2: RETURN .
```

The time flags in the subroutine start from t0 because it can be called from a number of locations at different times. The time flag t7, after the call, reflects the three time units needed to perform the subroutine.

The RTL defined above can be used to describe many different data transfers. However, only those that are possible in the NeMiSyS architecture are valid. A glance at Figure 4.1, which is a simplified diagram of the NeMiSyS processor, will show how the internal components are connected, and thus which RTL operations are valid.

4.4.2 The instruction cycle

Machine-code programs are stored in a microprocessor's attached memory system, in either ROM or RAM. In order to run a program, the micropro-cessor must read, or fetch, an instruction from memory, decode it and then execute it. This cycle must be repeated for each instruction in the program that has to be executed, and is called the instruction cycle.

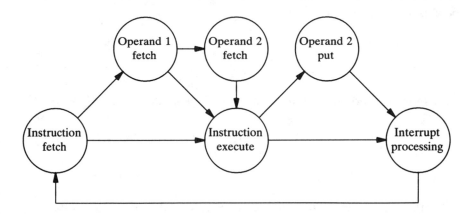

Figure 4.8. Instruction cycle.

Figure 4.8 illustrates the instruction cycle. As this shows, the cycle also involves fetching operands from memory or registers. A register-transfer language, as defined above, will be used to describe the NeMiSyS instruction cycle. The cycle is continuous, but for convenience we will start with the fetch part:

```
fetch t0: [PC] -> [MAR]
      t1: [MS[MAR]] -> [MBR]
      t2: [MBR] -> [IR]
      t3: [IR(op)] -> CU.
```

There, the contents of the program counter are moved into the memory address register. This is the address of the next instruction; thus at $t1$ it is this instruction that is read from memory. The instruction is then transferred to the instruction register. From there, the operation field of the instruction is passed to the controller in order that the instruction can be executed.

In a micro-coded controller, the transfer of the operation field to the controller results in the field being mapped to micro-code, which performs the rest of the cycle. First, if there are any operands they are fetched. There are a number of possible addressing modes; therefore a mapper is used. This selects micro-code routines that calculate effective addresses, and fetch operands into temporary registers, ready for the execution phase of the instruction cycle. At this stage, the length of the instruction is known; therefore the program counter can be incremented. On completion of the execution phase, the result is stored in a temporary register. Finally, the effective address of the destination operand is used to complete the instruction cycle.

The cycle will be considered in more detail by taking the ADD instruction as an example. This is a two-operand instruction. Assuming that the first operand is specified using the immediate addressing mode, and the second using the indexed addressing mode, its execution, after the fetch phase, will proceed as follows:

```
t4: 3 > INCR([PC]) -> [PC] | [IR(sm)] -> CU.
```

There the program-counter contents are incremented by three, because this is the length of an ADD instruction. The source-operand addressing mode is passed to the control unit, which selects the actions required to fetch the first operand. The addressing mode is immediate; thus the following occurs:

```
t5: [PC] -> [MAR]
t6: [MS[MAR]] -> [MBR]
t7: DECODE([IR(au)]) > INCR([PC]) -> [PC]
t8: [MBR] -> [B1].
```

After $t4$, the PC points to the location following the instruction. In this case, it points to the byte, half word or full word that contains the value of the operand. This value is fetched from memory, and put into a temporary register. The program counter is incremented to be ready for the next fetch. The increment is obtained by decoding the IR width field. The destination operand is then decoded by:

```
t9: [IR(dm)] -> CU.
```

The destination-operand addressing mode is indexed; so the result is as follows:

117

```
t10: [PC] -> [MAR]
t11: [MS[MAR]] -> [MBR], 4 > INCR([PC]) -> [PC]
t12: [MBR] -> [A1]
t13: [[IR(dr)]] -> [A2]
t14: ADD > ALU([A1],[A2]) -> [A3]
t15: [A3] -> [MAR]
t16: [MS[MAR]] -> [MBR]
t17: [MBR] -> [A2]
t18: [B1] -> [A1].
```

At $t10$, the PC points to the full word following the instruction. This memory location holds the constant part of operand's effective address. Therefore, it is fetched from memory and the PC is updated. The ALU input registers $A1$ and $A2$ are loaded. The effective address is then calculated and stored in $A3$. This address is transferred to the MAR, in order that the operand value can be fetched from memory, and placed in register $A2$. Finally, the value of the first operand is moved to register $A1$. Registers $A1$ and $A2$ now contain the operand values, and the execution phase is:

```
t19: ADD > ALU([A1],[A2]) -> [A3].
```

When this has finished, the result of the addition is stored in register $A3$, while the address of the destination operand is still held in the MAR. The operand put phase of the instruction cycle can now be started with:

```
t20: [IR(dm)] -> CU.
```

This results in:

```
t21: [A3] -> [MBR]
t22: [MBR] -> [MS[MAR]]
t23 : JUMP fetch,
```

which puts the result of the execution phase into the destination operand. When this is complete, the program counter is pointing to the next instruction and the control unit can proceed with the next instruction cycle.

4.4.3 Jumping and branching

Jump and branch instructions modify the execution sequence of other machine instructions. They do this by changing the contents of the program counter. The jump instruction always branches to the location specified by its single operand. The following RTL sequence describes the execution of a jump (JMP) instruction which uses register-indirect addressing:

```
t4: 2 > INCR([PC]) -> [PC] | [IR(SM)] -> CU
t5: [[IR(sr)]] -> [B1]
t6: [B1] -> [PC] | JUMP fetch.
```

The contents of the general-purpose register specified in the source operand are moved to the program counter. The register *B*1 is used to make the operand fetch compatible with other instructions. The next instruction cycle will begin by fetching the instruction at the address in the program counter. Thus, the program will continue with the instruction specified by the JMP. The program counter is modified by the the incrementer in preparation for the operand fetch, but, in this case, its output is ignored because the addressing mode is register indirect. This simple and elegant method of controlling the execution sequence of machine instructions is used for conditional branch instruction as well. Let us consider the branch on negative (BMI) instruction:

```
      t4: 4 > INCR([PC]) -> [PC] | IF N = 1 JUMP setpc
      t5: JUMP fetch
setpc t6: [IR(bo)] -> [A1]
      t7: [PC] -> [A2]
      t8: ADD > ALU([A1],[A2]) -> [A3]
      t9: [A3] -> [PC] | JUMP fetch.
```

First the PC is incremented to point to the next instruction. The *N* (negative) bit in the PSR is checked. If it is zero, then the branch must not take place and the instruction cycle is completed with the PC pointing to the next instruction. However, if the *N* bit is one, then a new value for the PC is calculated by adding the value in the IR branch offset field (bo) to the PC. This is PC-relative addressing. Branch instructions in the NeMiSyS architecture always use this addressing mode, although unlike other instructions, the offset from the PC is fetched into the IR as an instruction field, and can thus be used without an additional memory fetch.

4.4.4 Calling and returning from subroutines

Calling and returning from subroutines is accomplished with the jump-to-subroutine (JSR) and the return-from-subroutine (RTS) instructions. These operate in a similar way to the JMP instruction, but with the important difference that the stack is involved. First we will look at the JSR instruction. As with the JMP example above, register indirect addressing is assumed:

```
t4 : 2 > INCR([PC]) -> [PC] | [IR(SM)] -> CU
t5 : [[IR(SR)]] -> [B1]
t6 : [SP] -> [A2]
t7 : 4 -> [A1]
t8 : SUB > ALU([A1],[A2]) -> [A3]
t9 : [A3] -> [MAR]
t10: [PC] -> [MBR]
t11: [MBR] -> [MS[MAR]], [A3] -> [SP]
t12: [B1] -> [PC] | JUMP fetch.
```

First the operand, which is the address of the subroutine, is fetched. The stack pointer (SP) contains the address of the top of the stack. Thus, the next free position in the stack is calculated by decrementing the SP. It is

119

decremented by four because the PC is 4 bytes wide. The new SP value is used, by putting it into the memory address register, to write the program-counter (PC) contents into memory. In other words, the PC is pushed onto the stack. The address of the subroutine, which has been stored in register $B1$, is finally put into the PC to complete the instruction. Execution of the subroutine starts on the next instruction cycle. Notice that the PC is incremented at $t4$ in preparation for an operand fetch. However, in this case, it is ignored because register-indirect addressing is being used.

A subroutine is finished by using a RTS instruction. This indicates that the subroutine is complete and that the next machine instruction to be executed is the one following the calling JSR instruction. The processor stack was used by the JSR instruction to store the contents of the program counter. Thus, the stack holds the return address of the subroutine. The processing required to execute a RTS instruction is:

```
t4: [SP] -> [MAR]
t5: [MS[MAR]] -> [MBR], [SP] -> [A1]
t6: 4 -> [A2]
t7: ADD > ALU([A1],[A2]) -> [A3]
t8: [A3] -> [SP]
t9: [MBR] -> [PC] | JUMP fetch.
```

The return address is popped from the stack and put into the program counter, before the next instruction cycle is begun. It is clear that the RTS instruction will pop the top full word off the stack and put it into the program counter, even if this is not the return address. If this happens then the execution of the program will be completely disrupted. Therefore, it is important to ensure that before a subroutine is exited, the stack is returned to its original entry condition.

4.5 Interrupts

A simple program, running on a microprocessor, performs a sequence of actions in a predictable way. It can check for a condition or state at regular intervals and take any necessary action when it occurs. However, this can cause problems if the condition requires immediate action, because the program may not perform the check until it is too late for a satisfactory response. Some conditions, for example the discovery of an invalid instruction in the machine code of the program, can not be monitored by the program itself. A way of handling these situations is to use interrupts. They provide a very powerful method for dealing with unscheduled internal events and interfacing the microprocessor with external devices. Interrupts are particularly useful for devices that require a fast response to requests for service, or that have request timing that is very varied or unpredictable. They allow processors to control real-time systems that require a guaranteed response time.

Interrupts can be internal or external, and masked or non-masked. Internal interrupts are generated by the processor's control unit when it encounters certain instructions or program errors. The NeMiSyS processor, for example, generates an internal interrupt when it executes a BREAK instruction, or when it encounters an invalid instruction. External interrupts are caused by signals on the processor's interrupt-interface lines, which are normally connected to I/O device controllers. The NeMiSyS processor has two interrupt request lines (IREQ and MIREQ) and an interrupt acknowledge line (IACK). Non-maskable interrupts are initiated by MIREQ, while maskable interrupts use IREQ. Maskable interrupts can be disabled by clearing the interrupt-enable bit in the PSR using a CLI instruction. Non-maskable interrupts cannot be disabled. The IACK line is used by the processor to acknowledge that it has accepted a maskable interrupt. Non-maskable interrupts are always accepted and thus do not need an acknowledge line. The IREQ and IACK lines can be used for handshaking with I/O device controllers. NeMiSyS I/O is interfaced with the IREQ and IACK lines. The NeMiSyS MIREQ line is not used. When an interrupt occurs the following events are initiated:

1 the current instruction is completed;

2 the state of the interrupted program is saved in the stack;

3 an interrupt-service routine is executed;

4 the state of the interrupted program is restored;

5 processing resumes at the point in the program where the interrupt occurred.

The interrupt-service routine is specified by putting its address in an interrupt vector before the interrupt occurs. NeMiSyS has two interrupt vectors, which are the full words at address 0000 0000, for maskable interrupts, and at 0000 0004, for non-maskable interrupts. When a maskable interrupt is accepted on NeMiSyS, further maskable interrupts are automatically disabled while the service routine is executing. Looking at the interrupts in more detail, a full description of the process is as follows:

```
        t0 : IF NMI = 1 JUMP nmi
        t1 : IF IEN = 1 JUMP ien
        t2 : JUMP fetch
ien     t2 : IF MI = 1 JUMP mi
        t3 : JUMP fetch
nmi     t1 : 0 -> [PSR(NMI)] | CALL(saveregs)
        t12: 4 -> [MAR]
        t13: [MS[MAR]] -> [MBR]
        t14: [MBR] -> [PC] | JUMP fetch
mi      t3 : 0 -> [PSR(MI)] | CALL(saveregs)
        t14: 0 -> [MAR]
        t15: [MS[MAR]] -> [MBR], 0 -> [PSR(IEN)]
        t16: [MBR] -> [PC] | JUMP fetch
```

121

```
saveregs t0 : [SP] -> [A2]
         t1 : 4 -> [A1]
         t2 : SUB > ALU([A1],[A2]) -> [A3]
         t3 : [A3] -> [MAR]
         t4 : [PC] -> [MBR]
         t5 : [MBR] -> [MS[MAR]], [A3] -> [A2]
         t6 : SUB > ALU([A1],[A2]) -> [A3]
         t7 : [A3] -> [MAR]
         t8 : [PSR] -> [MBR]
         t9 : [MBR] -> [MS[MAR]], [A3]->[SP] | RETURN.
```

This is rather complex; therefore, its operation will be explored by considering the actions performed when the NeMiSyS processor is interrupted on IREQ and maskable interrupts are enabled. The conditional RTL statements are omitted and the CALL(saveregs) is expanded for clarity. Thus, after the current execution cycle is completed we continue with:

```
         t3 : 0 -> [PSR(MI)] | CALL(saveregs)
saveregs t4 : [SP] -> [A2]
         t5 : 4 -> [A1]
         t6 : SUB > ALU([A1],[A2]) -> [A3]
         t7 : [A3] -> [MAR]
         t8 : [PC] -> [MBR]
         t9 : [MBR] -> [MS[MAR]], [A3] -> [A2]
         t10: SUB > ALU([A1],[A2]) -> [A3]
         t11: [A3] -> [MAR]
         t12: [PSR] -> [MBR]
         t13: [MBR] -> [MS[MAR]], [A3]->[SP] | RETURN

         t14: 0 -> [MAR]
         t15: [MS[MAR]] -> [MBR], 0 -> [PSR(IEN)]
         t16: [MBR] -> [PC] | JUMP fetch.
```

The MI (maskable interrupt) flag is reset, the PC and the PSR are saved in the stack and the SP is updated, during $t3$ through to $t13$. The address of the service routine is fetched from memory location zero at time $t14$ to $t15$. Then, at $t16$, the service routine address is put into the PC and the service routine is executed by starting the next instruction cycle. The IEN (interrupt-enable) flag in the PSR is cleared at time $t15$, so that interrupts will be masked during the execution of the the service routine.

At the end of the interrupt-service routine, the return from interrupt instruction (RTI) is used to return to the interrupted program. Provided that the interrupt-service routine restores the general-purpose registers to the state they were in when it was entered and does not damage any of the program's working variables, then the interrupted program will continue unaffected by the interrupt. The execution of RTI is similar to RTS, described above, namely:

```
t4 : [SP] -> [MAR]
t5 : [MS[MAR]] -> [MBR], [SP] -> [A1]
t6 : 4 -> [A2]
t7 : ADD > ALU([A1],[A2]) -> [A3]
t8 : [MBR] -> [PSR]
t9 : [A3] -> [MAR]
t10: [MS[MAR]] -> [MBR], [A3] -> [A1]
t11: ADD > ALU([A1],[A2]) -> [A3]
t12: [A3] -> [SP]
t13: [MBR] -> [PC] | JUMP interrupt.
```

Here, the program-status register and the program counter are popped from the stack before the next instruction cycle is started. Maskable interrupts are enabled, because the IEN flag is restored with the PSR.

4.6 Summary

This chapter has discussed the way in which the digital circuit components described in Chapter 3 can be used to construct a microprocessor. An imaginary microprocessor, called NeMiSyS, was introduced as a representative example. Its instruction set and addressing modes were defined and its internal components were examined. A major component was seen to be the control unit, and a register-transfer language was used to investigate how it managed the instruction cycle. NeMiSyS was then used to explore subroutines and interrupts. The use of NeMiSyS as an example will be continued in the next chapter, where the problems of programming microprocessors in assembly language will be considered.

4.7 Further reading

1 Clements, A. (1985). *The Principles of Computer Hardware,* Oxford University Press.
2 Gorsline, G. W. (1986). *Computer Organization,* Prentice-Hall International.
3 Gosling, J. B. (1982). *Design of Arithmetic Units for Digital Computers,* MacMillan.
4 Protopapas, D. A. (1988). *Microcomputer Hardware Design,* Prentice-Hall International.
5 Stallings, W. (1987). *Computer Organization and Architecture,* MacMillan.
6 Yeun, C. K. (1988). *Essential Concepts of Computer Architecture,* Addison-Wesley.

4.8 Problems

4.1 Figure 4.1 shows a simplified diagram of the NeMiSyS processor. The temporary registers *A*1, *A*2 and *A*3 connect the ALU to the bus.
  (a) Develop this part of the design. The ALU and the registers can be treated as complete components, but show all the control lines and any additional logic gates that are required.
  (b) Connecting devices to the common bus requires special consideration. What type of logic is required in this case? Show, in the design developed above, where this is used.
  (c) Show, in the form of a timing diagram, the sequence of control signals required to perform the RTL operation [A3] -> [A2].

4.2 Using Figure 4.1 as a starting point, show the control lines needed to connect the control unit with the other components of the NeMiSyS processor, in order to direct their operation and data transfer.

4.3 Analyze the instruction set of the NeMiSyS processor on the basis of classification given in Section 4.2.1

4.4 Using NeMiSyS RTL, describe the source-operand fetch phase of the instruction cycle for:
(a) PC-relative addressing;
(b) absolute addressing;
(c) register post increment.

4.5 Using NeMiSyS RTL, describe the instruction cycle for:
(a) a load-effective-address (LDA) instruction with an absolute-mode operand;
(b) a move instruction (MOVE) between a register-indirect operand and a register-predecrement operand.

4.6 Consider the addressing modes available on the NeMiSyS processor. How many memory accesses and ALU operations are required for each mode? Tabulate the results of this analysis and draw conclusions about the relative efficiency of the modes.

4.7 Show, using RTL, the sequence of actions required to service a non-maskable interrupt (NMI) in the NeMiSyS processor.

4.8 Obtain the instruction set of a typical 32-bit micro-processor such as the MC68000 and compare it with the instructions and addressing modes of the NeMiSyS processor. Discuss the usefulness of any differences in the instructions and modes.

4.9 Consider the design of the NeMiSyS processor. Could any improvements be made by increasing the number of busses or other components? Remember that the programmer's view of the processor must remain the same. Draw a simplified diagram of the new design.

4.10 The instruction register (IR) of the NeMiSyS processor is connected to the control unit (CU). The operation of this link requires that the fields of the IR can be selected; show how this can be done. Give a logic diagram to illustrate the design of this part of the processor.

4.11 Give the hexadecimal representation of the NeMiSyS instructions that perform the following operations:
(a) move the full word at memory location 256 to memory location 512;
(b) add the value 7 to the contents of register $R6$;
(c) increment the memory location pointed to by register $R12$.

5: Assembly-language Programming

5.1 Introduction

This chapter will discuss programming microprocessors using an assembly or low-level language. NeMiSyS assembly language will be used as an example. A microprocessor executes programs which are stored in its memory in the form of machine or *object code*. The object code is a sequence of binary digits, containing all the information required to direct the operation of the processor. This code is concise, and relatively easy for the processor to decode and execute. Unfortunately, it is also very tedious, and error prone for a programmer to write. To overcome this problem, the program is written in the form of a text file containing assembly-language *source code*. The source code is translated into object code by a program called an *assembler*. Each statement in the source code is converted into one machine instruction. Mnemonics are used to represent the operation codes, while a special notation is used to represent the different addressing modes available on the target microprocessor. Not all of the lines in a piece of assembly-language source code are converted into machine instructions; assembler directives are used in the source code to instruct the assembler program, for example, to leave some space in the machine code for the storage of data.

Although writing programs in an assembly language is easier than directly producing machine code, it is still a difficult and error-prone language. Unlike high-level languages, there are few rules built into assembly languages. Thus, it is very easy to make mistakes that will go unnoticed until testing is almost complete. Furthermore, there is no structure imposed on the program design by assembly languages, and it is therefore possible to write code that is very difficult to follow, test and maintain, even by the program designer. To successfully develop large assembly-language programs, a programmer must take a disciplined approach to the design task. While a program is being designed, all sorts of apparently clever methods will present themselves to save a little memory, or make the program run just a little faster. However, great care must be exercised at this design stage, because it is possible that these methods will make the program more difficult to test and maintain.

A program written in assembly language is specific to a particular microprocessor. Thus, unlike a high-level language program, it is not very portable. This may not be a problem if the software is always used on the same type of computer. However, if the hardware is radically upgraded, or the software has to run on a range of different computers, then the lack of portability is a severe limitation.

It is clear from the above argument, that writing programs in high-level languages is preferable to the use of assembly languages. Nevertheless, there is still a role for assembly programming. In particular, assembly-language programs come into their own for applications where performance, and the size of a machine-code program, are of overriding importance. The close relationship between the source code and the resulting machine code allows the programmer total control over its form and size. Thus, the design can be fully optimized for a specific application. Some high-level languages do not allow direct access to the computer's hardware or operating-system routines. If these facilities are required and a suitable high-level language is not available, then assembly language will have to be used. However, it may be possible to write interface subroutines in assembler, while the bulk of the program is developed in a high-level language. The study of assembly languages helps us understand computer architecture because of its relationship to the processor's instruction set and the need in assembler to interface directly with the hardware and operating system. It also helps us appreciate the tasks performed by high-level-language compilers.

5.2 Programming model

The programmer does not need to known the detailed internal architecture of the processor being used. A conceptual model of the processor, called the programming model, is sufficient. The programming model gives the objects in the processor that can be referenced through its instruction set and addressing modes. The NeMiSyS programming model consists of the following parts.

5.2.1 General registers

There are 14 general-purpose registers which are designated by R0, R1, ..., R12 and R13 in the NeMiSyS assembly language. These are used to store addresses and data. They are referenced in the operand fields of instructions. Registers R14 and R15 are special registers, but can be referenced as though they were general registers.

5.2.2 Special registers

There are three registers that have a special role in the architecture of the NeMiSyS processor. The stack pointer (SP or R14) and the program counter (PC or R15) can be referenced in the same way as general-purpose registers. The program-status register (PSR) is used by the branch instructions, and is modified by many other instructions. Table 5.1 summarizes the program-status register's flags. Table 5.2 gives the assembly language mnemonics and operands for the NeMiSyS instruction set and shows which instructions affect the PSR. The operation performed by each instruction is briefly described. The bits in the PSR that are affected are marked by an X. There are two instructions, SSRH and RSRH, that allow the programmer to save and restore the contents of the PSR.

Table 5.1. Program-status register.

Bit	Name	Purpose	Bit	Name	Purpose
0	V	Overflow result	4	IEN	Interrupt enable
1	N	Negative result	5	BRK	Break instruction
2	Z	Zero result	6	UOP	Unknown instruction
3	C	Carry generated	7	IAM	Invalid address mode

5.2.3 Data-access width

Most of the NeMiSyS instructions have three versions: a byte, a half word and a full word version. These access different lengths of data (denoted by the .w field following the instruction, see Table 5.2), as follows:

byte	8 bits;
half word	2 bytes, or 16 bits;
full word	2 half words, 4 bytes, or 32 bits.

The effect of access width on the operation of an instruction is best explained by considering the movement of data between registers and memory. If a byte of data is loaded into a register, it goes into the low-order 8 bits of the register. The high-order 24 bits of the register are unchanged (no sign extension). When a byte transfer from a register to memory is performed, the low-order byte of the register is copied into a single byte of memory. Half-word transfers use the low-order 16 bits of a register, and the 16 high-order bits are not affected or used. Full-word transfers involve all 32 bits of a register. The ordering of the bytes in memory, when they are transferred from a register, can cause some confusion. If a register full word is copied into, say, memory address 1000, then the high-order byte will be stored at this address, the next two bytes at addresses 1001 and 1002 respectively, and the low-order byte will be stored at 1003. The process is similar for half-word transfers. It should be noted that this mapping of registers into memory may not be the same on a different microprocessor.

5.3 The NeMiSyS assembler language

5.3.1 Numeric and string values

When numeric and string values have to be specified, they are written as follows.

Decimal	Digits with no prefix (e.g. 24).
Hexadecimal	Digits and the letters A through F, prefixed by $ (e.g. $0A1F).
Binary	The digits 0 and 1, prefixed by % (e.g. %0110001).

Table 5.2. NeMiSyS instruction summary.

Mnemonic	Operands	Operation	V	N	Z	C
ADD.W	S,D	$D := S + D$, with carry	X	X	X	-
ADDC.W	S,D	$D := S + D$, with carry	X	X	X	X
AND.W	S,D	$D := S.D$, bitwise	-	X	X	-
ASL.W	S,D	Arithmetic shift left D for S bits, via carry	-	X	X	X
ASR.W	S,D	Arithmetic shift right D for S bits, via carry	-	X	X	X
BCC	L	Branch carry clear	-	-	-	-
BCS	L	Branch carry set	-	-	-	-
BPL	L	Branch positive	-	-	-	-
BNE	L	Branch not equal	-	-	-	-
BEQ	L	Branch equal	-	-	-	-
BMI	L	Branch minus	-	-	-	-
BVC	L	Branch overflow clear	-	-	-	-
BVS	L	Branch overflow set	-	-	-	-
BLT	L	Branch less than	-	-	-	-
BLE	L	Branch less than or equal	-	-	-	-
BGT	L	Branch greater than	-	-	-	-
BGE	L	Branch greater than or equal	-	-	-	-
BRA	L	Branch always	-	-	-	-
BREAK		Break (goto monitor)	-	-	-	-
CLC		Clear carry flag	-	-	-	0
CLI		Clear interrupt enable	-	-	-	-
CMPU.W	S,D	Compare S and D as unsigned integers	-	X	X	-
CMPS.W	S,D	Compare S and D as signed integers	X	X	X	X
DEC.W	D	$D := D - 1$	-	X	X	-
EOR.W	S,D	$D := S \oplus D$, bitwise	0	X	X	-
INC.W	D	$D := D + 1$	-	X	X	-
JMP	D	Jump to effective address D	-	-	-	-
JSR	D	Jump to subroutine at effective address D	-	-	-	-
LDA	S,D	$D :=$ effective address of S	-	-	-	-
LSL.W	S,D	Logical shift right D for S bits, via carry	-	X	X	X
LSR.W	S,D	Logical shift right D for S bits, via carry	-	X	X	X
MOVE.W	S,D	$D := S$	-	-	-	-
NEG.W	D	$D := -D$, with carry	-	X	X	-
NEGC.W	D	$D := -D$, with carry	-	X	X	X
NOT.W	D	$D := \bar{D}$, bitwise	-	X	X	-
OR.W	S,D	$D := S + D$, bitwise	-	X	X	-
ROL.W	S,D	Rotate D left, via carry, for S bits	-	X	X	X
ROR.W	S,D	Rotate D right, via carry, for S bits	-	X	X	X
RSRH	S	PSR $:= S$ (high-order half word of S ignored)	X	X	X	X
RTI		Return from interrupt	-	-	-	-
RTS		Return from subroutine	-	-	-	-
SEC		Set carry flag	-	-	-	1
SEI		Set interrupt enable	-	-	-	-
SSRH	D	$D :=$ PSR (high-order half word D zero)	-	-	-	-
SUB.W	S,D	$D := D - S$, with carry	X	X	X	-
SUBC.W	S,D	$D := D - S$, with carry	X	X	X	X

String   Any characters enclosed in ' marks (e.g. 'this is a string'). The ASCII code for the characters is used for internal representation.

5.3.2 Assembler directive statements

Assembler directives control the operation of the assembler program. It is important to remember that directives operate at assembly time and do not generate machine instructions. For example, the define-constant (DC) directive provides the program with an initial value in a memory location, but it cannot be used to re-initialize a memory location while the program is running. The NeMiSyS assembler has the following directives.

EQU

The equate directive links a name to a value. This is used to make programs easier to read. For example, to make the bell sound on a printer, it is sent the ASCII code 7. However, rather than use the code directly, it is much better to give it a meaningful name, thus:

```
BELL  EQU 7.
```

Names must be equated to a values before they are used in instruction statements. The equate directives are normally put at the start of the program for easy reference and modification, and so that the names can be used throughout the program.

An equate name can be any sequence of letters, digits and the underscore character that begins with a letter. It must be no more than 16 characters long and it must start in the first column of the line. If an equate name is declared more than once, then the last value defined is used.

DC

The define-constant directive is use to reserve a portion of memory within the program, giving it an initial value. It is qualified by .B, .H or .F, depending on the length of the constant. For example:

```
    DC.B 7          The value 7 in one byte.
*
    DC.B 5,128      The values 5 and 128 in two consecutive bytes.
*
    DC.B 'Hello'    The ASCII codes for the characters of Hello
*                       in five consecutive bytes of memory.
*
    DC.B %01101001  The binary value 01101001 in one byte.
*
    DC.H $02F6      The hexadecimal value 2F6 in one half word or
*                   16 bits.
```

```
*
      DC.F $FFFFFFFF   A full word of binary ones.
*
      DC.F 3,5         Two consecutive full words containing 3 and 5.
```

The constants will be embedded in the machine code where the directives appears in the source code. Define-constant directives can be given labels, so that they can be referenced from instruction statements, thus:

```
                  PROMPT DC.B 'Enter: '.
```

Define-constant directives can be placed anywhere in the source code, but they are normally grouped, for easy reference, at the top or bottom of the program, or with the subroutines that use them.

DS

The define-storage (DS) directive reserves a portion of memory, but does not give it an initial value. It is suffixed with .B, .H or .F to specify the size of the units to be reserved, and is follow by the number of units required, as follows.

```
        DS.B 2     Reserve 2 bytes.
        DS.H 1     Reserve a half word or 2 bytes.
  COUNT DS.F 1     Reserve a full word.
        DS.B $10   Reserve 16 bytes.
```

Define-storage directives, like define-constant directives, can be placed anywhere in a program, but again it is normally best to group them at the beginning or end of programs and subroutines. They are usually given labels, so they can be referenced from instruction statements.

ORG

The origin directive tells the assembler to generate object code for loading into a specified area of memory. It initializes the assembler's internal program counter, which keeps track of where items, such as machine instructions and storage areas, are to be located in memory. If ORG is not used, or it has no address specified, then an address of zero is assumed. It is always good practice to use ORG directives, and one is normally placed at the start of a program, before any instructions or define directives. The ORG directive can be used more than once in a program to create separate segments of code or data in different locations in memory, as follows.

```
           ORG      $100            Data origin.
  BELL     EQU      7
  BUFFSIZE EQU      500
  BUFFER   DS.B     BUFFSIZE        Input buffer.
  ERROR1   DC.B     'Invalid data ' Error message.
           DC.B     'entered'
```

```
            DC.B    BELL            Sound bell on output.
            DC.B    0               End of string mark.
BUFFPNT     DS.H    1               Buffer pointer.
*
            ORG     $400            Program origin.
START       MOVE.H  0,BUFFPNT       Initialize buffer pointer.
```

END

The END directive indicates that there is no more source code to be processed. It should be the last line of the program. Any code following this directive will be ignored. It has no effect on the run-time execution of the program. The BREAK instruction is used to stop a program executing.

Comment lines

Comment lines can be included in the source code by writing an * as the first character on the line. The rest of the line will be ignored by the assembler. For example:

```
        *   this is a comment,
            this is an error.
```

5.3.3 Instruction statements

Each NeMiSyS assembly language statement occupies one line, and has the following format:

```
label instr.w op1,op2   comment.
```

LABEL

A label is a symbolic name for the address of the statement. This name can be used in the operand fields of other statements to reference the labeled instruction or storage location.

A label can be any sequence of letters, digits and the underscore character, that begins with a letter. It must be no more than 16 characters long.

A label is declared by being the first word on a line, and by starting in the first column of that line. A label can be the only word on a line, but this is functionally the same as writing it on the next statement line. This allows more than one label to be associated with an instruction or storage location. A label must be declared only once, but it can be referenced many times. References to a label can occur before or after its declaration in the program.

INSTR

An instruction mnemonic. This is a short code for each of the instructions in the NeMiSyS instruction set. Table 5.2 gives a full list of the instruction and their mnemonics.

131

The instruction mnemonic is the first word on a line, unless it is preceded by a statement label. It must not start in the first column of the line and it must be separated from the statement label, if there is one, by at least one space.

w

The access width of the instruction. Instruction mnemonics are suffixed by a period and an access width which can be B, H or F:

.B   A byte or 8 bits.
.H   A half word, 2 bytes, or 16 bits.
.F   A full word, 2 half words, 4 bytes, or 32 bits.

All mnemonics must be given an access width, apart from the following instructions which have only one access width.

1 SSRH, RSRH, LDA, JMP, JSR and all the branch instructions, which are effectively full-word instructions. However, the SSRH and RSRH instructions implicitly access the PSR, which is only a half word wide. Thus, for SSRH (store status-register half word) high-order zeros are added, and for RSRH (restore status-register half word) the high-order half word of the operand is ignored.

2 CLI, SEI, BREAK, RTS and RTI, which have no operands and thus no access width.

OP1 AND OP2

The operands of the instruction. The number of operands depends on the instruction. Table 5.2 gives the number of operands and their purpose for each instruction.

If there are two operands, then they must be separated by a comma. The operands are specified using an addressing-mode notation, which is described below. There must be at least one space between the operands and the mnemonic or access-width field.

COMMENT

A statement line can be finished with a comment, which must be separated from the previous field by at least one space. The comment is ignored, and can therefore contain any characters. If a label is the only field on a line, there cannot be a comment.

The following are examples of valid NeMiSyS statements:

```
NEXT      ADD.F RO,TOTAL       add count to total
LOOP

          LDA    COUNT(PC),R8
          INC.F  NUMBITS(PC)   increment bit count
          CLI
```

132

```
END_PROG   BREAK                   End of program, return to monitor.
*
IO_STATUS DS.F  1
*          The next statement has two labels, LOOP1 and LOOP2.
LOOP1
LOOP2      NOT.B  IO_STATUS
```

5.3.4 Addressing modes

The addressing modes of the NeMiSyS processor were explained in Chapter 4 and are listed in Table 5.3. Here, the notation used by the NeMiSyS assembly language to specify addressing modes is described.

Table 5.3. Address modes.

Mode	Value	Mode	Value
Absolute	address	Register	Rn
Immediate	#value	Register indirect	(Rn)
Indexed	offset(Rn)	Postincrement register indirect	$(Rn)+$
PC relative	address(PC)	Predecrement register indirect	$-(Rn)$

ABSOLUTE

The value specified is the address of the operand. This can be given in one of the following forms:

1 a numeric value: for example, $0AF5 in

```
                MOVE.F  R5,$0AF5    ;
```

2 an equated name: for example, IOREG in

```
        IOREG EQU    $0AF5

              ...    ...
              ...    ...
              MOVE.F  R5,IOREG   ;
```

3 a label: for example, REGSAVE in

```
              MOVE.F  R6,REGSAVE

              ...    ...
      REGSAVE DS.F.   1           .
```

If absolute addressing is used to reference any DC or DS data items, then the program will always have to be loaded and executed in the same place in memory. This is because the actual memory location of the data will be built into the instruction. The program might be relocated, but the data, which is supposed to be in the body of the program, will not. It is better to use PC-relative addressing instead, and reserve absolute addressing for operands that are truly fixed in memory, such as memory-mapped I/O devices.

IMMEDIATE

With the immediate addressing mode, the value specified is the operand. This mode has the general format #value; for example:

```
INCR EQU    10
BELL EQU    7
     ...    ...
     ...    ...
     MOVE.H #$FFFF,R8      Put $FFFF into register 8.
     ADD.F  #INCR,COUNT    Add 10 to COUNT.
     MOVE.B #BELL,IOREGOUT Put 7 into IOREGOUT.
```

Immediate addressing is used for constants. It cannot be used for output operands. The instruction ADD.F R6,#128 is invalid because the output operand cannot store the result of the addition.

INDEXED

Here the effective address of the operand is calculated by adding the contents of a register to an offset. Its general format is offset(Rn). Some examples are the following.

```
VECTOR DS.F   3                Three-dimensional vector.
       ...    ...
       ...    ...
       LDA    VECTOR,R5        Set vector to (1,1,1).
       MOVE.F #1,0(R5)
       MOVE.F #1,1(R5)
       MOVE.F #1,2(R5)
       ...    ...
       ...    ...
*      Pixel record:
PIXNO  EQU    0
COLOR  EQU    4
XCOORD EQU    5
YCOORD EQU    9
       ...    ...
       ...    ...
       MOVE.F R0,R6           Get pixel-record address.
       MOVE.B NEWCOL,COLOR(R6) Change color.
```

PC-RELATIVE

PC-relative addressing is similar to indexed addressing, that is, the effective address is calculated by adding an offset to the contents of a register, in this case the program counter. The offset is the location of an operand relative to the current instruction. A PC-relative address is specified as label(PC). The assembler calculates the offset and builds it into the instruction. For example:

```
REGSAVE  DS.F   1
STARTVAL DC.F   128
         ...    ...
         ...    ...
         MOVE.F R10,REGSAVE(PC)
         MOVE.F STARTVAL(PC),R10
```

PC-relative addressing should be used whenever possible instead of absolute addressing, because the object code generated is position independent and can be loaded and executed anywhere in memory. It can be used with indexed addressing, as the following example shows.

```
VECTOR DS.F   3                 Three-dimensional vector.
       ...    ...
       ...    ...
       LDA    VECTOR(PC),R5   Set vector to (1,1,1).
       MOVE.F #1,0(R5)
       MOVE.F #1,1(R5)
       MOVE.F #1,2(R5)
```

Compare this with the similar example, given under indexed addressing above, which uses absolute addressing in a LDA instruction.

REGISTER DIRECT

A register-direct address specifies that the operand is the contents of the register. It has the format Rn, and some examples of its use are the following.

```
MOVE.F R1,R5   Copy contents of R1 to R5.
INC.F  R1      Add 1 to R1.
DEC.F  R5      Subtract 1 from R5.
ADD.F  R1,R5   Add R1 to R5.
```

REGISTER INDIRECT

Register indirect addressing takes the contents of the specified register as the effective address of the operand. It has the format (Rn). It can be used as follows:

```
LOW    EQU    $F0
       ...    ...
IOBUFF DS.F   1                 Address of iobuffer.
       ...    ...
       ...    ...
       MOVE.F IOBUFF(PC),R3   Get address of iobuffer from IOBUFF.
       AND.B  #LOW,(R3)       Clear low-order 4 bits of iobuffer.
```

The access width of the instruction applies to the actual operand, rather than the contents of the register. The register itself is always accessed as a full word.

135

POSTINCREMENT REGISTER INDIRECT

This mode is similar to register-indirect addressing, but the register has the access width of the instruction added to its contents after it has been used to address the operand. Its format is $(Rn)+$, and is used as follows:

```
VECTOR DS.F   3                Three-dimensional vector.
       ...    ...
       ...    ...
       LDA    VECTOR(PC),R5    Get vector address.
       MOVE.F #0,(R5)+         Zero all elements.
       MOVE.F #0,(R5)+
       MOVE.F #0,(R5)
```

PREDECREMENT REGISTER INDIRECT

Predecrement register-indirect addressing mode subtracts the access width of the instruction from the specified register before using its contents as the address of the operand. It has the format $-(Rn)$, and can be used, as the following example shows, to step backwards through a list of items:

```
SPACE    EQU    $20
BUFFLEN  EQU    256
BUFFER   DS.B   BUFFLEN
ENDBUFF  EQU    *
*
*        Replace trailing spaces in buffer with zeros:
*           buffer always contains at least one non-space.
*
         LDA     ENDBUFF(PC),R1   Point R1 to end of buffer.
NEXT     CMPU.B  #SPACE,-(R1)     Is next a space?
         BNE     DONE             No: finish.
         MOVE.B  #0,(R1)          Move zero into buffer.
         BRA     NEXT             Continue.
DONE     ...     ...
```

PERMITTED ADDRESSING MODES

In general, any of the addressing modes can be used for the operands of an instruction, but the NeMiSyS microprocessor has some limitations that are reinforced by its assembler:

1 the immediate addressing mode is not allowed for output operands; thus ADD.B R12,#2 is an invalid statement;

2 the input operand of the LDA instruction cannot use the immediate or register-direct addressing modes;

3 JMP and JSR instruction operands can be absolute, register direct, register indirect or PC relative;

4 branch instruction operands must be PC relative: a short notation is used in this case, with just the label being given. Hence, a valid branch instruction is BRA LOOP, rather than BRA LOOP(PC).

5.4 Programming the NeMiSyS microprocessor

In this section, how the NeMiSyS assembly language is used to write programs will be discussed. The general problem of program design will be considered, and then some specific techniques will be studied in detail. Only fairly simple problems will be examined, but this will still provide some insight into writing larger and more complex assembly programs.

5.4.1 Program design

The early stages of writing an assembly program are the same as for high-level language programs. The application must be analyzed and a design developed that meets its requirement specification. There are a number of methods for approaching program design. Some are very formal, but for small assembly programs an intuitively simple method, called stepwise refinement, can be very effective.

Stepwise refinement involves designing a program in stages. Starting with a simple outline, the design is refined in steps until a finished program is obtained, in this case written in assembly language. There is no limit to the number of steps, or how much is changed or refined at each stage. Progress is being made as long as the design moves, in a controlled and understandable way, towards its final form. At the beginning of the process the design is written in a high-level language. Some programmers use a special language, called a program-design language or PDL, but any well-structured language, such as Pascal, will do. The detailed syntax of the chosen language is not critical, as long it is consistent and care is taken with the structure of the program as it develops. An example will make the stepwise refinement method much clearer. Consider a small program to input a line of characters from the keyboard of the NeMiSyS computer, convert it to upper case and then output the result. A first stage might be:

```
begin
      input_line;
      convert_line;
      output_line;
end.
```

A buffer is needed to store the line of data. Therefore, a variable, to be used by these three routines, is declared:

```
VAR BUFFER : array [0..BUFFEND] of byte;
```

Notice that a non-Pascal data type of byte has been used. This is acceptable because it is known that the program is going to be implemented in NeMiSyS assembly language, which supports bytes of data. The input_line activity will have to read characters one at a time—because this is the way that NeMiSyS I/O works—and store them in BUFFER. Input_line should finish when an end-of-line character is read, or when the buffer is full. A refinement for input_line might be the following.

```
begin
    CHCOUNT := 0;
    CH := SPACE;
    while (CH <> END_OF_LINE) and (CHCOUNT <= BUFFEND)
    begin
        get_char(CH);
        store_in_buff(CH,CHCOUNT);
        CHCOUNT := CHCOUNT + 1;
    end
end
```

Two new variables have been used, which are declared as:

```
VAR CH, CHCOUNT : byte;
```

END_OF_LINE, SPACE and BUFFEND are constants and their values can be defined later. So far so good, but let us continue by refining store_in_buff.

```
begin
    CHCOUNT := 0;
    CH := SPACE;
    while (CH <> END_OF_LINE) and (CHCOUNT <= BUFFEND)
    begin
        get_char(CH);
        BUFFER[CHCOUNT] := CH;   { store_in_buff }
        CHCOUNT := CHCOUNT + 1;
    end
end
```

The design should be continued by refining get_char and then using a similar approach to design convert_line and output_line. However, to illustrate the final stages of the method, these steps will be bypassed and instead, most of input_line and the variable declarations will be immediately refined into NeMiSyS assembly code. We obtain the following.

```
END_OF_LINE EQU        $0D          ASCII code for end of line.
SPACE       EQU        $20          ASCII code for space.
BUFFER      DS.B       80           Buffer length 80.
BUFFEND     EQU        79           Buffer end = buffer length - 1.
CH          DS.B       1
            ...        ...
            ...        ...
```

138

```
INPUT_LINE
          MOVE.B      #0,R1           chcount := 0.
          MOVE.B      #SPACE,CH       ch := space.
WHILE     CMPU.B      #END_OF_LINE,CH
          BEQ         ENDWHILE
          CMPU.B      R1,#BUFFEND
          BGT         ENDWHILE
*         Body of while loop.
          get_char(CH)                ** Still to be refined.
          MOVE.B      CH,BUFFER(R1)   ch to buffer.
          INC         R1              chcount := chcount + 1.
*
          BRA         WHILE           Continue while loop.
ENDWHILE
          ... ...
```

Get_char still has not been refined into assembly language. This will be done later, when character input and output are discussed.

In the design shown above, it was possible to directly convert the PDL into assembly language. Unfortunately, this is not always possible, in which case some intuition has to be used to produce the assembly code. This can be difficult if you are new to assembly programming, but will become easier with practice. Developing a program design using stepwise refinement helps by localizing these problems. They can then be solved without worrying about affecting other areas of the design.

5.4.2 The stack

Stacks are data structures that store items on a last-in-first-out basis. Hence, they are often referred to as LIFO structures. The operations of adding data items to a stack and retrieving them are called push and pop (pull), respectively. The NeMiSyS microprocessor has a system stack that is implicitly used during the execution of JMP and RTS instructions, and during interrupt servicing and the execution of a RTI instruction. It can also be accessed explicitly for more general programming activities.

The NeMiSyS stack is located in memory and its address is stored in a register. We say that the register points to the stack. In fact, it points to the last item added, or pushed, to the stack. As a simple example, suppose that all the registers are currently being used by a program, but three more are temporarily needed to perform a calculation. The system stack is used to save the contents of some registers, and restore them after the calculation is finished:

```
          LDA     800,SP     Initialize the stack pointer.
          ...     ...
          MOVE.F  R5,-(SP)   Push registers onto stack.
          MOVE.F  R6,-(SP)
          MOVE.F  R7,-(SP)
```

```
 ...    ...
 ...    ...          Use registers R5, R6 AND R7.
 ...    ...
 MOVE.F  (SP)+,R7    Pop (pull) registers from stack.
 MOVE.F  (SP)+,R6
 MOVE.F  (SP)+,R5
```

Figure 5.1 shows the state of the system stack as this piece of code is being executed. You will notice that the stack grows downward in memory, towards lower addresses, and that the base of the stack is fixed. The stack is used this way because it is compatible with the NeMiSyS processor's method of handling subroutines and interrupts. The stack pointer contains the address of the top of the stack. This value must be loaded at the start of the program if subroutines or interrupts are going to be used, or if the stack is going to be used explicitly, as in the example above. When the system stack is used, the number of push and pop (pull) operations should balance. Particular care must be taken if the stack is being used inside a subroutine or an interrupt-service routine, since any damage will cause unpredictable results when a return instruction (RTS or RTI) is executed.

NeMiSyS programs are not restricted to the system stack. Any general register can be used as a private stack pointer. For example, the following piece of code reverses the order of the characters in a fixed-length string, using a private stack as an intermediate store:

```
STACKBASE  DS.B    50
STACK      EQU     *
*
SLEN       EQU     20
STRING     DS.B    SLEN
           ...     ...
           ...     ...
           LDA     STACK(PC),R0    R0 is the stack pointer.
           ...     ...
           ...     ...
           MOVE.B  #0,R3
           LDA     STRING(PC),R1
MOREPUSH   MOVE.B  (R1)+,-(R0)
           INC.B   R3
           CMPU.B  #SLEN,R3
           BNE     MOREPUSH
*
           MOVE.B  #0,R3
           LDA     STRING(PC),R1
MOREPOP    MOVE.B  (R0)+,(R1)+
           INC.B   R3
           CMPU.B  #SLEN,R3
           BNE     MOREPOP
```

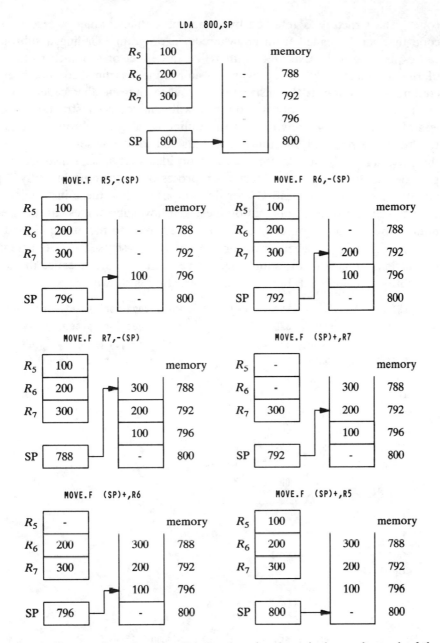

Figure 5.1. Stack operations: the state of the stack and registers is shown after each of the instructions has executed.

5.4.3 Subroutines

When the problem of program design was considered, the design naturally broke up into functional sections, and some of these sections were repeated in different parts of the program. Rather than create programs that are uniformly decomposed to the final level of refinement, it is useful if the hierarchical structure, resulting from stepwise refinement, is retained. This makes the program easier to read, test and modify, and reduces repetitions of the same code. The structure is retained by using subroutines, which are sections of code that can be called from anywhere in a program. Calling a subroutine causes its code to be executed, rather like the effect of a branch to the start of the subroutine. However, when a subroutine has finished, the program continues execution at the instruction following the one that called the subroutine. Execution will always continue with the correct instruction, regardless of where or how many times a subroutine is called. Subroutines allow us to structure a program into functional units and to use these units throughout the program. Subroutines are such a good idea that special instructions are provided to support them. Most microprocessors, including NeMiSyS, provide the instructions JSR and RTS for this purpose. How these instructions are used to call and return from subroutines, and how subroutines can be used to build a structured design, can be shown by returning to the example used earlier in this chapter when program design was discussed. The assembler instructions in most of the subroutines have been omitted, because the design has not been completed, but the overall structure is shown.

```
END_OF_LINE   EQU    $0D              ASCII code for end of line.
SPACE         EQU    $20              ASCII code for space.
BUFFER        DS.B   80               Buffer length 80.
BUFFEND       EQU    79               Buffer end = buffer length - 1.
CH            DS.B   1
*
*             Main body of program.
*
START         JSR    INPUT_LINE(PC)
              JSR    CONVERT_LINE(PC)
              JSR    OUTPUT_LINE(PC)
              BREAK
*
*             Subroutines.
*
INPUT_LINE
*
*             Input line from keyboard.
*
              MOVE.B  #0,R1           chcount := 0.
              MOVE.B  #SPACE,CH       ch := space.
WHILE         CMPU.B  #END_OF_LINE,CH
              BEQ     ENDWHILE
              CMPU.B  R1,#BUFFEND
              BGT     ENDWHILE
*                                     Body of while loop.
              JSR     GET_CHAR(PC)    ch := char.
              MOVE.B  CH,BUFFER(R1)   ch to buffer.
              INC     R1              chcount := chcount + 1.
```

```
*
              BRA      WHILE              Continue while loop.
ENDWHILE      RTS                         Return from subroutine.
*
GET_CHAR
*
*             Read character into CH from keyboard.
*

              ...      ...
              ...      ...
              RTS
*
CONVERT_LINE
*
*             Translate line to uppercase.
*

              ...      ...
              ...      ...
              RTS
*
OUTPUT_LINE
*
*             Output line buffer to screen.
*

              ...      ...
              ...      ...
              RTS
*
              END
```

The JSR instructions call the subroutines. PC-relative addressing mode is used, so that the program is position independent, but other modes could have been used. The subroutines execute until a RTS instruction is reached. Then program execution continues at the statement following the calling JSR instruction. Notice that subroutines are only executed by calling them with a JSR instruction. They must not be embedded in the main body of the program or other subroutines, because the processor will not be able to cope with the unexpected RTS instructions that would be encountered. In the example above, the BREAK instruction at the end of the main program passes control to the NeMiSyS monitor. Thus, the subroutines are only entered via JSR instructions.

When a JSR instruction executes, the processor stores the address of the next instruction on the system stack. At the end of the subroutine, a RTS instruction pops this address off the top of stack to return from the call. If the stack is used inside a subroutine and is not restored to its entry state, the RTS instruction will not pick up the correct return address and the program will fail to run successfully.

It is good practice to write subroutines without side effects. In particular, it is best if a subroutine does not permanently change the contents of any of the registers unless they are used for passing parameters. This is achieved, within the subroutine, by saving the registers in the stack before they are modified and restoring them when the subroutine has finished.

PARAMETER PASSING

The subroutines looked at so far are not general purpose. Subroutines that can accept any data items of the correct type, for both input and output, would be better. The same subroutine can then be used with different variables (e.g. DC and DS directives with different names). These input and output items are called parameters, and there are a number of ways of passing them to a subroutine.

Using registers to hold the parameter values is the simplest method. As an example, take a subroutine, called MAX, that compares two values and returns the largest. The following program uses MAX to select the larger of VAL1 and VAL2, and puts the returned value in VAL3.

```
            MOVE.F   VAL1,R0
            MOVE.F   VAL2,R1
            JSR      MAX
            MOVE.F   R2,VAL3
            ...      ...
            ...      ...
            BREAK
*
MAX         CMPS.F   R0,R1
            BGT      R0_BIG
            MOVE.F   R1,R2
            BRA      END_MAX
R0_BIG      MOVE.F   R0,R2
END_MAX     RTS
            ...      ...
            ...      ...
```

Passing the data items in registers is a form of parameter passing by value. This is satisfactory if the data will fit into a full word, but will not work if the data is, say, a string which is more than four characters long. An alternative is parameter passing by reference. Here, the subroutine is given the parameters' addresses, rather than their data values. This can be illustrated by modifying the above example.

```
            LDA      VAL1,R0
            LDA      VAL2,R1
            LDA      VAL3,R2
            JSR      MAX
            ...      ...
```

144

```
         ...      ...
         BREAK
*
MAX      CMPS.F  (R0),(R1)
         BGT     R0_BIG
         MOVE.F  (R1),(R2)
         BRA     END_MAX
R0_BIG   MOVE.F  (R0),(R2)
END_MAX  RTS
         ...      ...
         ...      ...
         END
```

A disadvantage of these methods is that they can use a lot of registers. Even though NeMiSyS has 14 available registers, it is surprising how these can be used up if there are a lot of subroutines and the registers are being used for other purposes as well as for parameter passing. A variation that reduces the number of register used for parameters is the following.

```
         MOVE.F  VAL1,PARM1
         MOVE.F  VAL2,PARM2
         LDA     PARM1,R0
         BRA     DOIT
PARM1    DS.F    1
PARM2    DS.F    1
PARM3    DS.F    1
DOIT     JSR     MAX
         MOVE.F  PARM3,VAL3
         ...      ...
         ...      ...
         BREAK
*
MAX      CMPS.F  (R0),4(R0)
         BGT     P1_BIG
         MOVE.F  4(R0),8(R0)
         BRA     END_MAX
P1_BIG   MOVE.F  (R0),8(R0)
END_MAX  RTS
         ...      ...
         ...      ...
         END
```

This is rather complex and has the additional disadvantage that it cannot be stored in read-only memory. An alternative, which uses the stack, avoids these problems. The addresses of the parameters are simply pushed on to the stack before the subroutine is called. The MAX example, using parameter passing by reference, will illustrate this method.

```
                    LDA     VAL1,-(SP)    Load parameter addresses onto stack.
                    LDA     VAL2,-(SP)
                    LDA     VAL3,-(SP)
                    JSR     MAX
                    LDA     12(SP),SP     Clear parameters from stack.
                    ...     ...
                    ...     ...
                    BREAK
            *
            MAX     SSRH    -(SP)         Save PSR.
                    MOVE.F  R0,-(SP)      Save registers.
                    MOVE.F  R1,-(SP)
                    MOVE.F  R2,-(SP)
                    MOVE.F  R3,-(SP)
                    LDA     24(SP),R3     Get parameters.
                    MOVE.F  (R3)+,R2      Val3.
                    MOVE.F  (R3)+,R1      Val2.
                    MOVE.F  (R3),R0       Val1.
            *
                    CMPS.F  (R0),(R1)
                    BGT     R0_BIG
                    MOVE.F  (R1),(R2)
                    BRA     END_MAX
            R0_BIG  MOVE.F  (R0),(R2)
            END_MAX MOVE.F  (SP)+,R3      Restore registers.
                    MOVE.F  (SP)+,R2
                    MOVE.F  (SP)+,R1
                    MOVE.F  (SP)+,R0
                    RSRH    (SP)+         Restore PSR.
                    RTS
                    ...     ...
                    ...     ...
                    END
```

Notice that the PSR and all the registers used in the subroutine have been saved and restored. As mentioned before, this is good practice because when the subroutine is called only its parameters will change. It will have no side effects, and the internal working of the subroutine can be ignored when it is used.

5.4.4 I/O operations

The NeMiSyS processor uses memory-mapped data input and output. Like the NeMiSyS processor itself, its input and output support has been invented as a representative example of a real system. Chapter 7 discusses how input/output devices are connected to the processor, but here, how they are used by assembly programs will be described. NeMiSyS uses a single I/O support chip, which is interfaced to a keyboard and a screen, with handshaking to synchronize the operating of the devices and the I/O chip. It is accessed by using memory-mapped registers with the formats and addresses shown in Table 5.4.

Table 5.4. Memory-mapped registers.

Register	Address	Width	Purpose
DBRA	$1000	byte	Port A data-buffer register
DBRB	$1001	byte	Port B data-buffer register
DDRA	$1002	byte	Port A data-direction register
DDRB	$1003	byte	Port B data-direction register
IOSR	$1004	byte	Status register
CNTR	$1005	half word	Counter

We shall consider these in detail.

DDRA AND DDRB

The bits of these registers control the function of the respective data-buffer register bits. 1 specifies that the buffer register bit is input, while 0 indicates that it is to be used for output. In the NeMiSyS computer the buffer registers are connected, via the pins of the I/O support chip, to character devices. Therefore, all the bits should be set the same in each direction register: to one for the keyboard, which is connected to port A, and to zero for the screen, which is connected to port B. Although this is always the same, the programmer must still initialize the ports, by setting the appropriate direction register to the correct value, before any I/O operation is attempted.

DBRA AND DBRB

These registers contain the data bits that are transferred to and from connected devices. In general, their bits can be read or written, depending on the bits of the relevant data-direction register. However, the devices transfer character data in the NeMiSyS computer. Therefore, DBRA should be read as a full byte, while DBRB should only written to in full bytes.

CNTR

This register can be loaded with any half-word integer value, and will proceed to increment every clock cycle until it reaches its maximum value of $FFFF. It will then be set to zero on the next clock cycle, and stay at that value until it is re-initialized. This can be used as a crude timer.

IOSR

The status-register bits are given in Table 5.5. These are either interrupt-enable or ready bits. The latter will be discussed first, leaving interrupt bits

147

Table 5.5. Status-register bits.

Bit	Purpose	Bit	Purpose
0	Port A interrupt enable	4	CNTR interrupt enable
1	Port A ready	5	CNTR ready
2	Port B interrupt enable	6	unused
3	Port B ready	7	unused

until later. Port A is connected to the keyboard. Its ready bit is set to one by the I/O chip to indicate that DBRA has been updated and contains a new character. While this bit is set to one, the value in DBRA will not be changed by the keyboard. It is reset to zero when DBRA or DDRA are accessed by the program. When the ready bit is zero, the keyboard is allowed to put a new character into DBRA. To show how this works in practice, here are the subroutines for reading a character from the keyboard.

```
DBRA      EQU     $1000      Data buffer register A.
DDRA      EQU     $1002      Data direction register A.
IOSR      EQU     $1004      Status register.
RDYA      EQU     $02        Port A ready bit.
*
INIT_GET
*
*         Initialize keyboard get.
*         Call this once at beginning of program.
*
          MOVE.B  #$FF,DDRA  Set up port A for input.
          RTS
*
GET_CHAR
*
*         Get a character from keyboard.
*         Character returned in low-order byte of register R0.
*         Stack pointer must be initialized.
*
          MOVE.F  R1,-(SP)
WAITIN    MOVE.B  IOSR,R1    Wait for keyboard entry.
          AND.B   #RDYA,R1
          BEQ     WAITIN
          MOVE.B  DBRA,R0    Store entered character.
          MOVE.F  (SP)+,R1
          RTS
```

Port B is used by the screen. The ready bit is set to one by the I/O chip to indicate that DBRB has been used by the screen and that more data can be entered by the program. Entering a character into DBRB from the program sets the ready bit to zero. The register should not be changed again until the ready bit returns to one. The subroutines for writing to the screen are as follows.

148

```
        DBRB    EQU    $1001    Data-buffer register B.
        DDRB    EQU    $1003    Data-direction register B.
        IOSR    EQU    $1004    Status register.
        RDYB    EQU    $08      Port B ready bit.
*
INIT_PUT
*
*               Initialize screen character write.
*               Call this once at beginning of program.
*
        MOVE.B  #0,DDRB    Set up port B for output.
        RTS
*
PUT_CHAR
*
*               Put a character on the screen.
*               Character should be placed in low-order
*                 byte of register R0 before PUT_CHAR is called.
*               Stack pointer must be initialized
*
        MOVE.F  R1,-(SP)
WAITOUT MOVE.B  IOSR,R1    Wait until screen ready.
        AND.B   #RDYB,R1
        BEQ     WAITOUT
        MOVE.B  R0,DBRB    Output character.
        MOVE.F  (SP)+,R1
        RTS
```

The counter register CNTR is used in a similar way. Entering a value in CNTR sets its ready bit to zero. The register begins incrementing immediately. When it reaches its maximum value of $FFFF , it is reset to zero and the CNTR ready bit is set to one. If a short pause is needed in a program, the following method can be used.

```
        CNTR      EQU    $1005        Counter register.
        IOSR      EQU    $1004        Status register.
        RDYC      EQU    $20          CNTR ready bit.
        DCYCLES   EQU    $FF00        Start value giving 256 cycles.
                  ...    ...
                  ...    ...
DELAY
          MOVE.H  #DCYCLES,CNTR
CNTR_WAIT MOVE.B  IOSR,R1        Wait until CNTR = 0.
          AND.B   #RDYC,R1
          BEQ     CNTR_WAIT
                  ...    ...
                  ...    ...
```

149

5.4.5 Interrupt-service routines

The routines presented above used a busy-wait method of synchronizing with an I/O device. This is often quite suitable, but it does mean that the processor cannot perform any other task while it is waiting. An alternative method uses interrupts. When interrupt-driven I/O is being used, the program does not have to wait until the I/O device is free or has data for processing. It can proceed with other tasks, which will be temporarily halted (or interrupted) when an I/O device requires servicing. This is accomplished by providing an interrupt-service routine, which is automatically called whenever an interrupt occurs. Before the interrupt-service routine is entered the current instruction is completed; then the contents of the PC and PSR are stored on the stack and interrupts are disabled. The service routine performs any processing that is required, before returning control of the processor to the interrupted program. A RTI instruction is used to return from an interrupt. This restores the PSR and the PC from the stack and re-enables interrupts. The program then continues at the instruction that was going to be executed when the interrupt took place. Restoring the PSR ensures that the condition flags are not altered by the interrupt-service routine. An interrupt can be considered as an unexpected subroutine call, which implies that interrupt-service routines must be very carefully written. They must not affect the status of the interrupted program. The comments, earlier in this chapter, on the benefits of writing subroutines without side effects, must be treated as a rule for interrupt-service routines. The NeMiSyS computer uses maskable interrupts for its I/O devices, and has no external devices connected to its non-maskable interrupt line. To use I/O interrupts on the NeMiSyS processor you must:

1 put the address of the interrupt-service routine in the maskable interrupt vector at address $00000000;

2 enable interrupts on the required devices by setting their interrupt-enable bits in IOSR to one, and set up their data-direction registers (DDRs) appropriately;

3 enable maskable interrupts using the SEI instruction.

After this, if an interrupt occurs from the enabled devices, then the interrupt-service routine will be given control of the processor. Interrupts from an individual device can be stopped at any time by resetting its interrupt-enable bit to zero. All interrupts can be halted by using the SEI instruction. As an example, let us assume that that keyboard and screen I/O is handled using interrupts. The following NeMiSyS assembly code could be used to achieve this.

```
DBRA    EQU    $1000       Data-buffer register A.
DDRA    EQU    $1002       Data-direction register A.
DBRB    EQU    $1001       Data-buffer register B.
DDRB    EQU    $1003       Data-direction register B.
IOSR    EQU    $1004       Status register.
RDYB    EQU    $08         Port B ready bit.
RDYA    EQU    $02         Port A ready bit.
```

```
IENA      EQU     $01             Port A interrupt enable.
IENB      EQU     $04             Port B interrupt enable.
IDIB      EQU     $FB             Port B interrupt disable.
          ...     ...
          ORG     0
          DS.F    2               Vectors.
OBUSY     DS.B    1               Output I/O in progress flag.
          ...     ...
          ORG     $100
START     JSR     INIT_IO
          ...     ...
          ...     ...
          MOVE.B  CHAR,R0
          JSR     PUT_CHAR
          ...     ...
          ...     ...
          JSR     GET_CHAR
          ...     ...
          ...     ...
          BREAK
          ...     ...
          ...     ...
*
INIT_IO
*
*         Initialize I/O interrupt servicing.
*         Call this once at start of program.
*
          LDA     IHANDLE(PC),0   Load interrupt vector.
          MOVE.B  #$FF,DDRA       Set up port A for input.
          MOVE.B  #0,DDRB         Set up port B for output.
          OR.B    #IENA,IOSR      Enable port A interrupts.
          OR.B    #IENB,IOSR      Enable port B interrupts.
          MOVE.B  #0,OBUSY        Mark output not in progress.
          SEI                     Enable maskable interrupts.
          RTS
*
GET_CHAR
*
*         Get a character from keyboard buffer.
*         Character returned in low-order byte of register R0.
*
          ...     ...
          ...     ...
          RTS
*
```

151

```
PUT_CHAR
*
*           Put a character on the screen without waiting.
*           Character should be placed in low-order byte of register R0.
*
            AND.B   #IDIB,IOSR      Disable port B interrupts.
            CMPU.B  #1,OBUSY        Is output in progress?
            BNE     IO_START        No: begin output.
            JSR     PUT_OBUFF       Yes: store character and
            BRA     PCH_FINI        exit.
IO_START    MOVE.B  #1,OBUSY        Mark output in progress.
            MOVE.B  R0,DBRB         Send character to screen.
PCH_FINI    OR.B    #IENB,IOSR      Enable port B interrupts.
            RTS
*
IHANDLE
*
*                                   Interrupt-service routine.
*
            MOVE.F  R0,-(SP)        Save R0.
            MOVE.F  R1,-(SP)        Save R1.
            MOVE.F  IOSR,R1         Interrupt from keys?
            AND.B   #RDYA,R1
            BEQ     NOT_KEY         No: try screen.
            MOVE.B  DBRA,R0         Yes: read character into
            JSR     PUT_IBUFF       input buffer.
NOT_KEY     MOVE.F  IOSR,R1         Interrupt from screen?
            AND.B   #RDYB,R1
            BEQ     NOT_SCRN        No: exit routine.
            JSR     GET_OBUFF       Yes: get character from buffer.
            CMPU.B  #0,R0           Buffer empty?
            BNE     DO_WRITE        No: output character.
            MOVE.B  #0,OBUSY        Yes: mark output not in progress.
            AND.B   #IDIB,IOSR      Disable port B interrupts.
            BRA     NOT_SCRN        Exit routine.
DO_WRITE    MOVE.B  R0,DBRB         Output character.
NOT_SCRN    MOVE.F  (SP)+,R1        Restore R1.
            MOVE.F  (SP)+,R0        Restore R0.
            RTI                     Exit service routine.
            ...     ...
            ...     ...
            END
```

The INIT_IO subroutine initialize the interrupt vector, I/O registers and the flag OBUSY, and enables interrupts. The interrupt service routine is called IHANDLE and uses registers $R0$ and $R1$. These registers might be in use when the interrupt occurs. Therefore, they are saved and restored by the service routine, to prevent any possibility of the program being disrupted. The interrupt routine does not know which device caused the interrupt when it is entered. Hence, it checks or *polls* each of the devices in turn to determine which one should be serviced. It uses the subroutines PUT_IBUFF and GET_OBUFF, and the variable OBUSY to communicate with the rest of the program.

Keyboard input is comparatively simple. When data is input, it is stored in an input buffer, using PUT_IBUFF, until the main program needs it. A subroutine, called GET_CHAR, is used by the main program to retrieve characters from this input buffer when required.

Output to the screen is more complex, and the subroutine PUT_CHAR is given in detail. When output begins, PUT_CHAR moves a character into DBRB and sets the OBUSY flag, which indicates that output is in progress. If more output is requested while this this flag is set, the character is stored in a buffer, using PUT_OBUFF. While PUT_CHAR is running, port B interrupts are disabled to prevent any adverse interaction with the interrupt-service routine.

A screen interrupt is received when an output operation is complete. The service routine knows if the main program has meanwhile called PUT_CHAR by reading the output buffer. If a character is available it is put into DBRB, otherwise the OBUSY flag is cleared to indicate that the buffer is empty and output is no longer in progress.

Interrupts are automatically disabled when the maskable interrupt-service routine is entered. Thus, the service routine cannot be disturbed by another interrupt. If further interrupts must be processed while the service routine is running, then this can be accomplished by using SEI inside the routine. Thus, permitting the service routine to be interrupted. However, this technique is rather difficult to use because all maskable interrupts have to be serviced by the same routine on NeMiSyS, and the result will be, in effect, multiple copies of the interrupt routine running at the same time. Unfortunately, these can interact with each other, giving unpredictable results, unless great care is taken. A safer method is to write the interrupt routine to be as short as possible, in order that very little time is spent with interrupts masked. Any complex processing should be assigned to parts of the program outside the interrupt-service routine.

If non-maskable interrupts are being used, then the programmer has to write a service routine that can be interrupted. The NeMiSyS computer uses non-maskable interrupts to implement its monitor software. Therefore, the non-maskable interrupt vector at address $00000004 should not be altered. (In practice, since the NeMiSyS processor that you will be using is simulated, you will not be able to change this vector, or look at the machine code of the monitor.)

5.5 Summary

This chapter has discussed how assembly-language programs are designed and written. A technique for developing a program design in steps (step-wise refinement) was demonstrated. Building on material presented in Chapter 4, the use of addressing modes, subroutines and interrupts were described. Addressing modes were used in assembly instructions to specify how data stored in registers and memory should be accessed. Subroutines were shown to be a useful facility, and methods for passing parameters to and from them were examined. How interrupts are used from an assembly-language program was explained. The NeMiSyS microprocessor was used throughout this chapter as an example, but similar methods can be used on other microprocessors. Assembly-language programming was shown to be a way of writing very efficient programs. However, it was also demonstrated that assembly languages are difficult to use. Chapter 8 discusses the alternative, i.e. high-level languages.

5.6 Further reading

1 Bennett, A. M. (1987). *68000 Assembly Language Programming*, Prentice-Hall International.
2 Clements, A. (1987). *Microprocessor Systems Design*, PWS.
3 Gorsline, G. W. (1988). *Assembly and Assemblers*, Prentice-Hall International.
4 Levanthal, L. A. (1979). *6502 Assembly Language Programming*, McGraw-Hill.

5.7 Problems

5.1 The input-line routine developed in Section 5.4.1 uses the indexed addressing mode to access a buffer. Rewrite this routine using the register indirect with postincrement addressing mode to write to the buffer.

5.2 Complete the program partially developed in Section 5.4.3 by writing the GET_CHAR, CONVERT_LINE and OUTPUT_LINE subroutines.

5.3 The routine INPUT_LINE in Section 5.4.3 uses a fixed buffer for output. Change it to be more generally useful by passing the buffer as a parameter.

5.4 Design, implement and test assembly-language routines that are the equivalent of the following high-level procedures.

(a) *writestring*(*str*: *string*);

Output *str* on the screen. A string is a sequence of characters terminated by a zero byte.

(b) *writeint*(*int*: *signedint*; *n*: *byte*);

Output *int* as a signed decimal integer on the screen in a field at least *n* characters wide. The sign should not be shown if *int* is positive. The parameter *int* is is a full-word signed binary integer and *n* is a byte unsigned integer.

(c) *writehex*(*hex*: *unsignedint*);

Output *hex* as hexadecimal characters on the screen in a field eight characters wide with leading zeros. The parameter *hex* is a full-word unsigned binary integer.

(d) *writeln*;

Skip the screen to a new line.

5.5 Write a character-input subroutine that checks if a key has been pressed but does not wait for this to occur. If a key has been press then its ASCII code is returned in register *R*0; otherwise zero is returned.

154

5.6 Refine the following PDL subroutine into assembly language.

```
count(i : integer);
    BEGIN
            writeint(i,3);      {see problem 5.4}
            writeln;            {see problem 5.4}
            IF c < 10 THEN
                    count(c+1);
    END
```

Notice that this subroutine calls itself. This is called recursion and is a facility provided by many high-level languages such as Pascal, MODULA-2 and C.

5.7 Given that a string is a sequence of characters terminated by a zero byte, write subroutines that perform the following operations:

(a) return the length of a string;

(b) join two strings;

(c) compare two strings for equality; return 1 if equal and zero otherwise;

(d) search a string for the first occurrence of a given character: return the location of the character if found or zero if not found.

5.8 Write an assembly-language program that calculates and prints the powers of 2 up to 256.

5.9 Write and test an assembly-language subroutine that takes two signed integer input parameters in registers $R0$ and $R1$. It returns [R0] MOD [R1] in register $R2$ and [R0] DIV [R1] in register $R3$.

5.10 Produce assembly-language implementations of the following outline sections of Pascal source code.

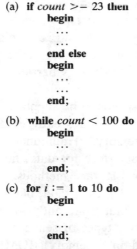

(a) **if** *count* >− 23 **then**
    begin
       ...
       ...
    end else
    begin
       ...
       ...
    end;

(b) **while** *count* < 100 **do**
    begin
       ...
       ...
    end;

(c) **for** $i := 1$ **to** 10 **do**
    begin
       ...
       ...
    end;

5.11 Design, implement and test a sort routine. Its parameters are passed in registers $R0$ and $R1$. They are, respectively, the address of a continuous list of full words containing the numbers to be sorted and the number of items in the list. The list of full words is returned sorted in ascending order.

5.12 Design a program that prints out the message "clock running", while continuously inputting characters from the keyboard and printing them on the screen.

6: Memory Organization

6.1 Introduction

One of the largest parts of any computer system is the memory; not only is the memory required to store the currently executing program, but also the data or variables that may be required or generated during the execution of the program. In modern processor systems more than one program and set of variables can occupy the memory simultaneously, although the processor can only execute one instruction at a time. Hence the need to store all the other instructions making up the program in a memory. It is therefore vital that this important and essential area of any computer is designed correctly to make optimum use of the available memory resources. Consequently this chapter will concentrate on the use and implementation of computer memory.

Computer memory is normally arranged hierarchically with the most expensive semiconductor storage being accessed directly by the central processing unit (CPU) when executing a program and the cheaper magnetic media being used to store programs or data not required immediately by the CPU. Both types of memory store the data as binary digits. Semiconductor memory takes advantage of high and low voltage levels to store data in a latch, while the magnetic media use magnetic poles to indicate the two different states. In both cases an energy barrier exists between the two states so that the data is maintained in the required state; therefore, in order to cause a change of state, energy needs to be supplied. The amount of data that can be stored on the magnetic media greatly exceeds the storage capacities of current semiconductor memory. Traditionally magnetic media have always been used as the secondary store for data; however, this dominance is now being challenged by optical storage methods. Optical methods are not only capable of storing larger volumes of data, but are also a more reliable storage media. These storage techniques are outlined in Section 6.4.2.

Nominally there are two types of semiconductor memory, random-access memory (RAM) and read-only memory (ROM). In both types of memory it is possible to read the binary data from their memory cells. However, data can also be written into the memory cell of a RAM; therefore RAM must be used to store data that is being altered by a program. Of course RAM can also be used to store programs that are being executed. In order to store data in the memory cells of a ROM it is necessary to either build in the required information during manufacture or write in the information using specialized equipment called programmers. This is usually referred to as *programming the device*. A further important difference between RAM and ROM is that if the power to a computer is turned off then normal RAM will lose the stored data: this is termed volatile storage; whereas the ROM will normally retain its data and this is referred to as non-volatile storage. Numerous types of RAM and ROM exist and will be described in the next section.

Both RAM and ROM communicate directly with the CPU, providing immediate access to programs and associated data. A useful feature of this type of memory is the ability to access any one of the memory cells they contain in the same amount of time, independently of the cell's location within the memory array. This feature is known as *random access* and the time taken to access a cell is called the *access time*. The feature of constant access time regardless of position is extremely useful since the CPU does not have to vary its clock speed when reading or writing different memory locations. As both RAM and ROM are random-access memories, some confusion may arise with "RAM". Therefore it may be useful to think of RAM as read-and-modify memory.

The various types of magnetic memories (and optical media) available are accessed from the CPU via devices called *controllers* as their method of operation is more complex. The main types of magnetic memory available are disks, drums and tape; although they may seem dissimilar, certain common features exist. For example, they all use a thin magnetic film coated onto a plastic or similar surface to store the binary data in the form of magnetic flux reversals, the coated materials being referred to as magnetic media. They all employ a means of reading or writing the binary data onto the magnetic medium; however, the shape and form of this read/write head can vary. In the case of disks and tape the magnetic media can normally be removed from the device or drive that is used to access the stored information. Removable media are extremely useful for transferring data between computers or for storing valuable programs and data in a safe place, away from the computer; a feature that is extremely useful and currently cannot be equalled by semiconductor memories for large amounts of data; however, optical storage may provide an alternative.

One drawback of magnetic storage methods is the speed at which the data can be accessed. Normally access times for RAMs and ROMs are expressed in terms of nanoseconds (ns); unfortunately disks have access times of several milliseconds (ms) and tapes need several seconds or even minutes to access the data. The reason for this difference in access speeds is the method used to store the data on the magnetic media. Generally the information on the media is stored in blocks which have to be accessed in a serial fashion. In order to read or write any bit of data it is necessary to scan each bit of the block in sequence until the appropriate section is found; hence it can take quite some time to find the required data. A further difference between disks and tapes is the way in which the blocks are stored on the media. On a disk the blocks are arranged so that they can be accessed in any order, rather like a semiconductor memory. The speed of access to a block of data is dependent upon where the block is in relation to the last block accessed. This is due to the need to move the read/write head physically, by electromagnetic means, from its current block to the desired block. On a tape access is not so straightforward: the blocks are stored sequentially and therefore have to be scanned in sequence until the required block is found, further decreasing the speed at which a particular block can be accessed. Tape drives are one of the slowest media from which data can be accessed; however, for storing important data, they are usually very reliable and robust. Magnetic media will be covered in more detail later in this chapter.

Although accessing data from magnetic media may be slow when compared to semiconductor memories they do have an enormous storage capacity. The cost of disk drives and tape drives can be quite high, due to the high-precision mechanisms required, in comparison to the relatively cheap semiconductor memory. However, their ability to store large amounts of data means that the cost per bit is very small; hence they are used to store the volumes of data associated with even a simple computer. This diversity in roles between semiconductor and magnetic memory also appears in the general terms used to describe their function in a computer. The semiconductor portion of a computer's memory is usually called main or primary memory as it is the main memory used by the CPU. The disk drives and tape systems are usually termed the secondary or auxiliary memory; sometimes it may be referred to as backing store, reflecting their subsidiary role. Generally a compromise is reached in terms of the type and size of memory used to build the computer. Building a computer from relatively expensive semiconductor memory would make it operate very quickly; however, it would be prohibitively expensive for a large amount of memory. Similarly, a tape drive would be relatively cheap; however, the computer would only be able to run very slowly. Figure 6.1 illustrates a typical memory hierarchy as may be found in a moderately sized computer. It is by no means the only configuration that could be used; however, it serves to illustrate what might be expected when studying a real computer. The next sections of this chapter will describe in detail the various types of memory available and how they can be used in a complete microprocessor design.

6.2 Semiconductor memory

Semiconductor memory is generally fabricated on silicon using the same techniques that are used to produce the microprocessor and associated circuits. If it were not for the availability of these cheap and easily used memories then the microprocessor revolution might not have been so dramatic. The introduction to this chapter has indicated that there are two types of semiconductor memory currently available, namely RAM and ROM. This division is quite broad and it is possible to further subdivide these two classes of memory into different, representative, categories. Each of these categories will be discussed in the following sections along with their important characteristics.

Although the internal structure of the memory cells used in RAMs and ROMs are quite different, the macroscopic view of both types is strikingly

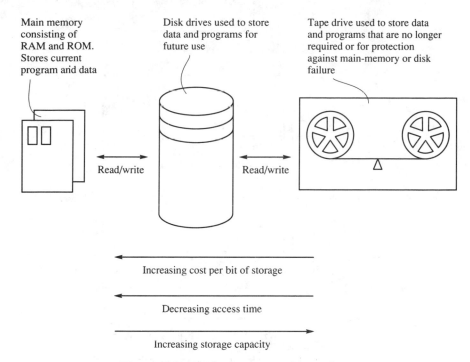

Main memory consisting of RAM and ROM. Stores current program and data

Disk drives used to store data and programs for future use

Tape drive used to store data and programs that are no longer required or for protection against main-memory or disk failure

Read/write   Read/write

Increasing cost per bit of storage

Decreasing access time

Increasing storage capacity

Figure 6.1. Typical memory hierarchy.

similar. The internal architecture of a typical memory, applicable to both RAMs and ROMs, is shown in Figure 6.2. The capacity of a memory is determined by the number of bits it can store or alternatively the number of memory cells it contains. The example memory shown in Figure 6.2 has an n-bit capacity, that is, it can store n bits of data. In order to access any one of these n memory cells it is necessary to supply a unique address to identify the required location within the memory array. In a microprocessor system we usually deal with binary data; therefore the address of the required bit of data must be supplied to the memory chip as a binary number. The number m of address lines required to address anyone of the locations is given by $m = \lceil \log_2 n \rceil$. It is fairly easy to see from this equation that, to make optimum use of the address lines entering the chip, n should be a power of 2, that is, the address will range from 0 to $2^m - 1$. Consequently commercial memories range in size from 32 bits (requiring 5 address lines) up to 4 194 304 bits (requiring 22 address lines) and perhaps beyond. Obviously such large numbers are difficult to remember and write down; therefore a shorthand notation is used to indicate the capacity or size of a memory. This involves using the letters K (Kilo) and M (Mega) to represent the numbers $2^{10} = 1024$ and $2^{20} = 1 048 576$. Thus a 64-Kbit memory would actually contain $64 \times 1024 = 65 536$ bits ($m = 16$) and a 4-Mbit memory would contain $4 \times 1 048 576 = 4 194 304$ bits ($m = 22$). This form of notation can also be used to describe the storage capacities of magnetic media.

159

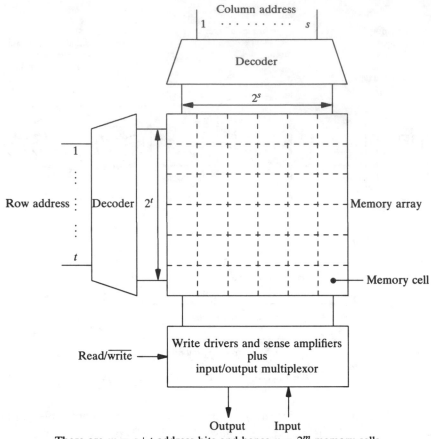

There are $m = s+t$ address bits and hence $n = 2^m$ memory cells.

Figure 6.2. Typical memory-array architecture.

Accessing or decoding the required address is made easier by splitting it into a row and column address for application to the memory array. In the example shown (Figure 6.2) the m address bits are split into t row-address bits and s column-address bits, which greatly simplifies the decoders that are required on the memory chip. Instead of building an m-line-to-2^m-line decoder, only two smaller and simpler t-line-to-2^t-line and s-line-to-2^s-line decoders need be built. If s and t are equal, i.e. m is an even number, then the memory chip can be simplified further as only one type of decoder is required, although two decoders are needed. When a memory chip is accessed the address of the required cell is decoded to activate a single row-and-column line. This in turn activates one of the cells in the memory array. Other signals entering the chip indicate whether the memory cell is to be read from or written to (in a ROM only the read signal is available). If data is read from the cell then the signal is sensed by amplifiers to determine the state of the cell, either a logic 1 or 0, before it is placed on the data outputs of the chip. If a write operation is indicated then write drivers would force the selected cell to take up the state indicated by the input data. From this description it is possible to identify certain specific lines that are associated

160

with a RAM: these are address lines, data input/output lines and control signals to determine whether a read or write operation is required.

6.2.1 Random-access memory (RAM)

Random-access memory (RAM) is found in almost every computer system, large or small, for storing not only programs but data associated with the program and subroutine linkages via a stack. Currently there are two common types of RAM, namely *static* and *dynamic*. The differences arise from the way the memory cell, for storing the data, is constructed. Static memory is far easier to use from an engineering viewpoint; however, due to the size of their memory cell, dynamic memories are the densest types currently available.

STATIC

The structure of a typical static nMOS memory-bit cell is shown in Figure 6.3.

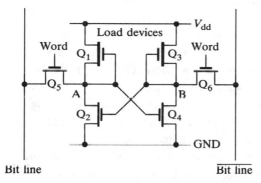

Figure 6.3. A static nMOS memory cell.

The cell consists of six field-effect transistors (FET) (Q_1 to Q_6). Two of the transistor pairs Q_1–Q_2 and Q_3–Q_4 form a cross-coupled inverter pair which acts as a latch to store the data. Transistors Q_5 and Q_6 are used to access the latch for the purpose of reading or writing data. The FETs used in the memory cell can be considered to be simple switches controlled by the gate input. The operation of this cell is quite straightforward. If transistor Q_2 is turned on by point B being at logic 1 then Q_3 will also be turned on, sustaining this state. By similar reasoning, point A will be at logic 0. As long as no external influences are applied to the memory cell it will retain this state as long as power is maintained to the chip. The operation of the cell is then as follows.

Reading from the cell. In order to read the cell the two gate lines labelled "word" are taken to logic 1, turning on the two FETs Q_5 and Q_6. This is achieved by decoding the correct address to access this cell. Once Q_5 and Q_6 are turned on the two lines "bit line" and "$\overline{\text{bit line}}$" take on the states of points A and B, that is 0 and 1, respectively, in this example. These bit lines are connected to sense amplifiers which determine, for the sake of argument, that this cell contains a logic 0. Once the data has been obtained from the cell the value it contains appears on the external data pins of the memory chip for use by the CPU. If A and B had been 1 and 0, respectively, then the cell would have contained a logic 1, in this example. Thus this simple cell can retain binary information.

Writing to the cell. The first stages of the write operation are very similar to those of the read operation. The address of the required cell is decoded and the required "word" lines driven to logic 1, thus turning on the two transistors Q_5 and Q_6. However, this time, instead of the two bit lines taking up the states of A and B, they are driven by amplifiers, controlled from the external data pins of the chip, to take up the correct state to store the data, either a 0 or 1. Therefore the values of A and B are overridden and the memory cell is forced to take up the new value. Obviously if the value of the data to be written to the memory cell is the same as the value already present then no change takes place.

As can be seen from the above functional description the operation of a typical static memory cell is quite straightforward. A real device may be made up from thousands of these cells with the associated decoders and amplifiers also present. The pinout for the 6264 CMOS static RAM chip, shown schematically in Figure 6.4(a), includes 13 address lines giving 8 K (2^{13}) of addressable locations. There are also 8 data lines, which means that each address location is associated with 8 bits; this is usually referred to as an address word. The 8 bits of data are called a byte: therefore this memory is capable of storing 8 Kbytes of data or $8 K \times 8$ bits of data. Arranging the data from the chip in this fashion makes it easier to interface the CPU to the memory, as most CPUs either have an 8-bit, 16-bit (2-byte) or 32-bit (4-byte) data bus. One further point to note about the data lines is that they are bidirectional, that is, data can be written and read from these lines depending upon the state of the control lines. The control lines are also responsible for setting the data lines to a high-impedance state, where they are neither inputs or outputs but merely float. This high-impedance state allows multiple memory chips to be connected to the same data bus. On this particular memory chip there are 4 control lines performing the functions shown in Figure 6.4(b). The write-enable ($\overline{WE}$) line determines whether a read or write operation is being performed on the memory. The two lines $\overline{CS}_1$ and CS_2 are called chip-select lines and are used when more than one memory chip is connected to the address- and data-bus lines (see Section 6.3.1). The output-enable ($\overline{OE}$) line is used to control the state of the data lines, active or floating. The internal architecture of this static RAM chip is shown in Figure 6.5. As can be seen from this diagram, the arrangement is very similar to the general case discussed earlier and shown in Figure 6.2, with row and column decoders and a memory array.

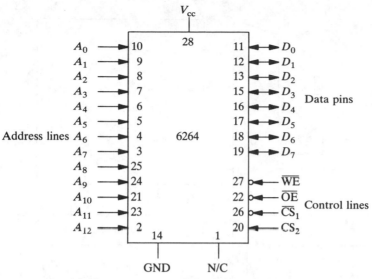

(a) Pin arrangement of the 6264 static RAM.

$\overline{WE}$	$\overline{OE}$	$\overline{CS}_1$	CS_2	Mode	Data pins
x	x	1	x	Not selected	High impedance
x	x	x	0		
1	1	0	1	Output disable	High impedance
1	0	0	1	Read	D_{out}
0	x	0	1	Write	D_{in}

(b) Truth table for the operation of the 6264 control lines.

Figure 6.4. The 6264 8-K × 8-bit static CMOS RAM.

When designing a computer system that uses memory it is necessary to have data characterizing the performance of each memory chip in order to make an informed choice, as well as to perform worst-case timing analysis to make sure the design will work in all cases. The data is usually supplied by manufacturers in the form of a data sheet. The data sheet lists all the various physical parameters of the device from operating-temperature range to the maximum and minimum voltages at which the device can be used. One particularly useful part of the data sheet is that containing the timing diagrams or waveforms that illustrate, graphically, the inter-relationships of the signals needed to drive the memory device from the CPU. The memory chip does not in reality operate in zero time: it takes several nanoseconds for signals to propagate around the chip and addresses to be decoded before any data appears at the output of the chips. A much simplified set of timing diagrams is shown in Figure 6.6 for the 6264 static memory chip.

The timing diagram in Figure 6.6(a) shows the relationships between the various signals for a read cycle. The main cycle takes place between points X and Y on the diagram. The time between points X and Y is called the read-cycle time and is the minimum time required to access the data and prepare for the next cycle, in this case 100 ns. The use of the two parallel lines

163

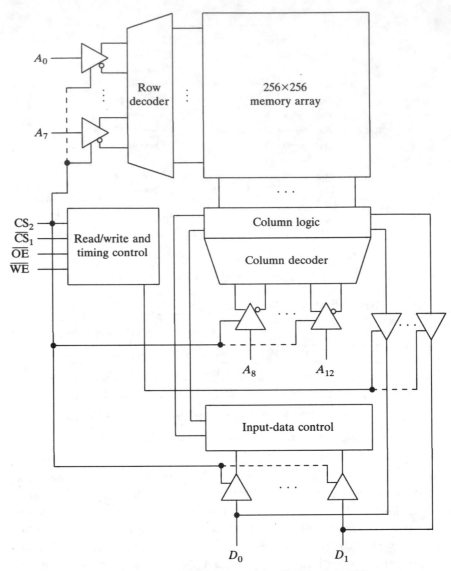

Figure 6.5. Internal architecture of the 6264 static RAM.

between X and Y to represent the address lines is a conventional shorthand to make the diagram easier to understand. The lines indicate that several associated lines are changing simultaneously to form a stable state, the value of which is irrelevant to the actual representation. Since real signals do not instantaneously change state, sloping lines are used to indicate that changes may be taking place: this is shown at the points X and Y for the address bus. The signals $\overline{CS}_1$, CS_2 and $\overline{OE}$ are shown changing to their required state, the shaded areas indicating that the state is unknown. However, they must settle to the required values within the read cycle, when the address data must also be stable. A short time after $\overline{OE}$ becomes stable (point C) the output data drivers are enabled. This is indicated by the data-out lines taking on a similar

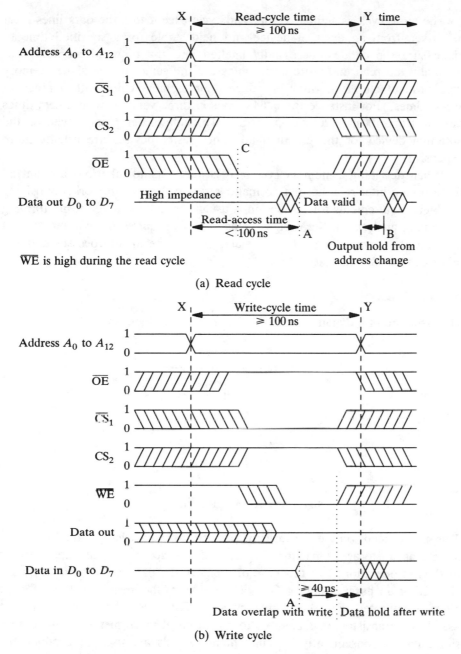

(a) Read cycle

(b) Write cycle

Figure 6.6. Timing waveforms for the 6264 static RAM.

parallel-line representation. However, the data outputs can usually be enabled more quickly than the data can be accessed from the memory array; therefore the data does not become valid until after the *read-access time* (from X to A) has elapsed. Once the data is valid it will remain so for a short period after the end point (Y) of the read cycle. This is due to the propagation delays in the system: this time period is called the *output-hold time*.

165

The write cycle is similar to the read cycle except that the data lines must be driven from an external device and held stable for a specified time to make sure the data is successfully latched into the required memory cell. Normally the read and write cycle times are the same for the same memory type. Manufacturers normally produce their memory chips with a range of access times. For instance the 6264 comes in three versions with access times ranging from 100 ns, 120 ns and 150 ns. This allows designers to choose the optimum device for the job in hand, since faster devices are usually more expensive.

When designing a memory system it is necessary to take into account all these timing inter-relationships to make sure the design will work. Satisfying the logic requirements for the design is usually trivial compared to performing the timing analysis on the design. This analysis requires the combination of not only the memory data sheet but also that of the microprocessor and any other chips in the interface.

DYNAMIC MEMORY

The structure of a dynamic memory cell is shown in Figure 6.7.

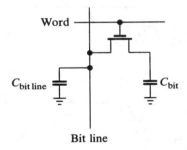

Figure 6.7. A dynamic memory cell.

This circuit needs only one transistor per cell. The transistor is dual purpose: it acts as a storage capacitor C_{bit} and allows access to the capacitor for read/write operations. In this type of memory the data is represented by the charged or uncharged state of C_{bit}. The design of the memory cell itself is far more complicated than the equivalent static RAM cell; however, because it uses fewer transistors it is possible to put more of these memory cells onto a single chip. Consequently dynamic memory cells are used to produce the densest memory chips available.

The operation of the dynamic RAM (DRAM) cell is quite simple. However, as discussed later, this hides one of the significant design problems with DRAM. To read from the DRAM cell the word line is decoded to take it high. If capacitor C_{bit} is charged then the charge will be shared with the bit line capacitance $C_{bit line}$. An amplifier on the end of the bit line then senses this charge to generate the valid output data in the form of a logic 1 or 0. Similarly, if C_{bit} is uncharged then, when enabled, there will be no charge redistribution and a different logic state would be represented. To write data to the cell simply involves charging or discharging the bit capacitance when it is enabled.

From the above description it is quite clear that unless the charge on C_{bit} is restored to its proper value after a read operation the charge would soon reach a state that would be invalid for the stored bit. Similarly, as the capacitor used to store the charge is not perfect it slowly discharges so that the cell will eventually *forget* the information. DRAM manufacturers usually allow about 2 ms for this to occur. The first problem is easily solved by simply writing the read data back to the accessed cell in order to restore the charge. In order to stop the DRAM from *forgetting* the information it is necessary to periodically refresh the data, preferably before 2 ms from the last read. The refresh is easily achieved by simply reading every bit cell in the memory array in under the 2-ms limit. This is not as difficult as it may appear. The array can be read on a row-by-row basis so that several cells are refreshed concurrently in fewer read operations than by simply reading every cell. This is greatly simplified by each column in the array having its own sense-and-write amplifiers. Although the operation of a DRAM is considerably more complex than its static RAM counterpart, there are special chips available that make interfacing them to a CPU fairly straightforward.

The pins on the 4116 dynamic RAM chip are shown in Figure 6.8. The 4116 is a 16-K × 1-bit device: from the previous discussion this should require 14 (2^{14}) address lines; however, the device only uses 7 address lines. Therefore to access the memory correctly it is necessary to multiplex the address lines, that is, lines A_0 to A_6 also act as address lines A_7 to A_{14}. This address-bus multiplexing is achieved internally by means of the $\overline{RAS}$ (row-address-select) and $\overline{CAS}$ (column-address-select) control lines.

The timing diagram for a read access is shown in Figure 6.9. This time access to the RAM is not straightforward, as demonstrated by the following description of a complete access cycle. Part of the address called the row address is applied first and must be stable before $\overline{RAS}$ goes low and latches the data into the row decoder (see Figure 6.8(b)). The controlling device, i.e. the CPU, must then apply the next part of the address, called the column address, which is then latched when stable by $\overline{CAS}$ going low. At this point the complete address has been input to the RAM chip. The state of the $\overline{Write}$ line then determines whether a read or write operation is to take place. When a write operation occurs the data to be written must be held stable while the $\overline{Write}$ line is low, also the $\overline{Write}$ line must go low before the $\overline{CAS}$ line is taken low in order to keep the internal timing correct. As can be seen from the diagram, the cycle time and the access time are not equal, as is the case with static RAM, due to the fact that certain internal housekeeping must occur before the next access can take place. The 4116 has a random read- or write-cycle time of 375 ns but an access time of 150 ns.

Refreshing the DRAM is simply a matter of applying a row address followed by the $\overline{RAS}$ signal without any input data or taking $\overline{CAS}$ low, as shown in Figure 6.10; therefore only the rows are addressed. The values in the 64 bits of each row are sensed and rewritten in order to refresh the data. If

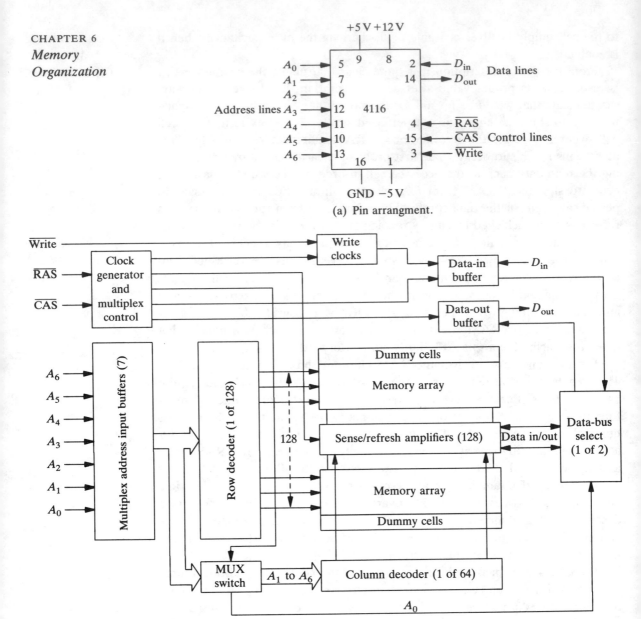

(a) Pin arrangment.

(b) Internal architecture of the dynamic RAM.

Figure 6.8. The 4116 16-K × 1-bit dynamic RAM.

each column address is accessed in turn then only 128 operations are required in order to completely refresh the full 16-Kbit memory. If each cycle takes 375 ns then in 48 μs (128×375 ns) the complete RAM will be refreshed: this is only 2.4% of the allowed 2 ms between refreshes. Thus there is still sufficient time for normal read and write accesses from the CPU. If each cell had to be accessed individually, it would take 6.2 ms to refresh the entire RAM; this is well outside the 2-ms limit.

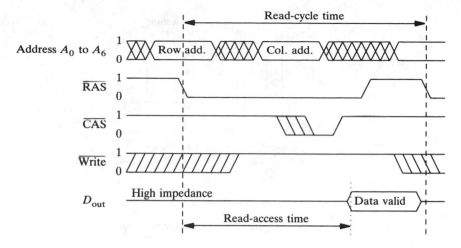

Figure 6.9. Timing waveforms for the read cycle of the 4116 dynamic RAM.

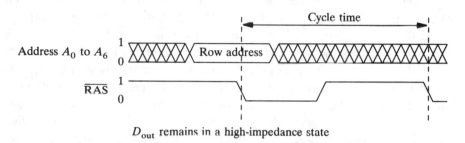

D_{out} remains in a high-impedance state

Figure 6.10. $\overline{RAS}$ only refresh-cycle timing diagram.

As DRAMs rely on charge storage to indicate binary 1 or 0 they are very susceptible to external influences altering the correct state to a faulty or erroneous state. One major cause of this problem are alpha-particle strikes from within the silicon substrate discharging a particular cell. These are the so called soft errors, as the cell is not damaged: only the information in the cell is lost. Therefore to counteract these problem DRAMs are usually used in computer system such that the data they hold is encoded using a Hamming code (see Chapter 2). This code is special in that it can detect and correct single-bit errors that may occur in the information word. The penalty for using the Hamming code is that extra bits have to be stored along with the information. If 8 bits (1 byte) is to be stored then 5 extra bits are needed, whereas 32 bits (4 bytes) require 7 bits. All the information bits plus the code bits go through an error-detection/correction circuit before reaching the CPU; this scheme is illustrated for an 8-bit data word in Figure 6.11. Using this scheme allows the CPU to continue reading the data even if a soft error occurs. It also allows the CPU to continue if a permanent fault occurs with a DRAM chip; however, this time the fault can be signalled and the system easily repaired before any catastrophic failure occurs. DRAMs also suffer from the problem of causing noisy supply lines due to the large currents they

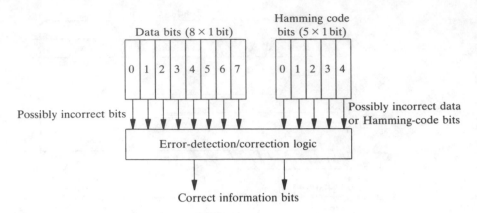

Figure 6.11. Error detection/correction using a Hamming code.

draw when turned on. This can cause glitches on the power supply which may effect other chips on the same board. In order to minimize this problem capacitors must be placed close to the DRAMs between all the power and ground lines: these are called decoupling capacitors. If the noise is not suitably attenuated then the DRAMs can exhibit intermittent faulty behavior.

The description of access to DRAM data has been greatly simplified; however, it illustrates the complex nature of such a device. To aid the designer in interfacing DRAMs to the CPU many manufacturers have designed controllers which sit between the CPU and the DRAMs. The controllers correctly multiplex the address with the correct sequencing of $\overline{\text{RAS}}$, $\overline{\text{CAS}}$ and $\overline{\text{Write}}$. A typical DRAM controller is shown in Figure 6.12. They are also capable of performing the refresh cycles by occasionally denying the CPU access to the RAM. The DRAM controller correctly sequences the refresh operation so that no cells are inadvertently omitted. The controller is designed to allow the CPU access to more than one bank of DRAM chips; therefore the address lines from the DRAM controller are given a high driving capability so that the system performance is not degraded by using a large number of RAM chips. This situation arises due to the fact that DRAMs are normally 1 bit wide and hence 8 chips are needed to store a single byte. Most CPUs need to use a DRAM controller to efficiently handle this type of memory, although the controller is not essential to use DRAMs. The Z80 CPU has a built-in refresh register and can use DRAMs without any great problems (see Chapter 9).

6.2.2 Read-only memory (ROM)

Read-only memory is very useful in the design of a computer because of its non-volatile nature. When a computer is switched on the CPU must find some coherent instructions to execute so that it can begin to perform its tasks. The easiest way to achieve this is to use a ROM. The ROM contains program instructions and perhaps data that allow the CPU to initialize various input/output devices and generally sort out the numerous tasks it

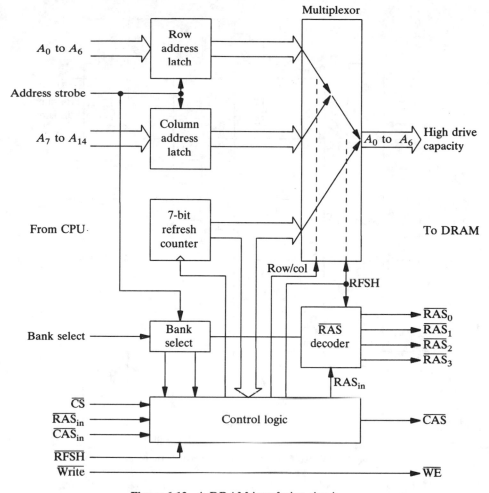

Figure 6.12. A DRAM interfacing circuit.

needs to perform. The ROM may contain the complete program that the CPU is going to execute; for instance, in small computers the ROM can sometimes contain a BASIC interpreter or the control program for a domestic appliance. Alternatively, in larger, more complex computers, the ROM may contain the instructions to tell the CPU how to access the programs it should be executing; these programs could be located on magnetic media, such as a floppy disk. The ROM would be programmed with the instructions for reading the disk and storing the program from the disk in RAM. The program read into RAM could then be executed in order to obtain more programs from the disk and read them into RAM. This sequence of events is usually called bootstrapping or simply the *boot* operation, the analogy being that the computer is pulling itself up by its bootstraps.

The organizational details of a ROM are very similar to a RAM: it has address lines, data-output lines and a read line. However, the memory cells used to store the binary data in the ROM are quite different. A very much

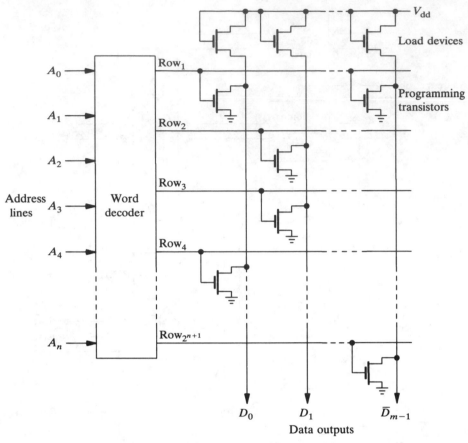

Address lines A_0, A_1, A_2, A_3, A_4, A_n — Word decoder — Row_1, Row_2, Row_3, Row_4, $\text{Row}_{2^{n+1}}$

V_{dd} — Load devices — Programming transistors

D_0 D_1 $\bar{D}_{m-1}$

Data outputs

(a) MOS ROM containing 2^{n+1} words by m bits.

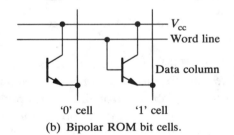

V_{cc}
Word line
Data column

'0' cell   '1' cell

(b) Bipolar ROM bit cells.

Figure 6.13. Typical layout of a ROM.

simplified schematic diagram of a MOS ROM is shown in Figure 6.13 (a). The transistors at the crosspoints of the data columns and word rows are used to represent the stored data. When the incoming address is decoded the selected-word line goes high, turning on any transistors that are connected to the line. The transistors that are turned on will pull the data columns down to a logic zero, all the other data-column lines will remain at their normal logic 1 state. Thus by placing a transistor at a crosspoint it is possible to program a 0 into that location. Using current silicon processing technology, this scheme is very simple to implement and extremely dense ROMs of this type can be produced. Unfortunately this type of ROM can only be programmed when it is actually manufactured. Once programmed it is impossible to alter. Also, unless a large number of ROMs are required, it can be very expensive for each individual device. This method of programming a ROM is called *mask programming*. A bipolar ROM is organized along the same lines as a MOS ROM. However, in this case programming is achieved by either connecting or not connecting the base of the bipolar transistors to the word line, as shown in Figure 6.13 (b).

PROGRAMMABLE READ-ONLY MEMORY (PROM)

The mask-programmed ROMs are not ideal when a program is being developed, nor are they useful when only a small number of computers with the same data or program will be built. This problem can be solved by the use of programmable ROMs (PROMs). The structure of the PROM memory cell is shown in Figure 6.14.

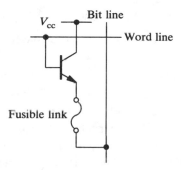

Figure 6.14. Programmable ROM cell.

As can be seen from the diagram the structure of the cell is not dissimilar to the cell shown in Figure 6.13 (b), except for the inclusion of the fusible link. The fusible link can be selectively *blown* for every cell in the ROM. If the link is present then the cell is used to represent a logic 1, since selecting the transistor will take the bit line up to V_{cc}. If the link has been blown then the cell represents a logic 0. The links are blown by using a special piece of equipment, called a PROM programmer, which is capable of reading in data that represents the bits to be stored. When programming takes place the links to be blown are selected and a larger than normal current is passed down the emitter of the transistor: this causes the polysilicon or nichrome fuse link to open circuit. If the PROM is programmed incorrectly then there is no second chance: once programmed these devices cannot be reprogrammed.

ERASABLE PROGRAMMABLE READ-ONLY MEMORY (EPROM)

The PROM described in the last section is useful when only a few dedicated ROMs are required. However, when a new computer or program is being

173

developed, minor alterations may require the ROM to be reprogrammed several times during the prototype stage. A reusable ROM would therefore be a great advantage: an EPROM gives just this flexibility. The EPROM can be reused several hundred times without degrading its performance; also the program it contains will remain intact for a very long time, provided it is carefully protected. Sometimes an EPROM is used to permanently store the correct version of a program in low-volume computer products.

The EPROM's functionality is achieved by its very versatile memory cell, as shown in Figure 6.15.

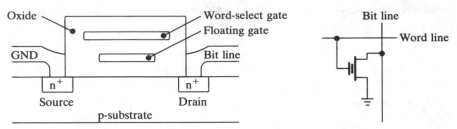

(a) Physical construction of the EPROM cell.   (b) Logic diagram for the EPROM cell.

Figure 6.15. EPROM cell.

This form of cell is called a floating-gate transistor or stacked-gate cell. In this transistor the n-channel of the MOSFET has two polysilicon gates. The lower gate, or floating gate, is left unconnected. The upper gate is connected to the word line as in the ROM structure discussed earlier and is used to turn the channel on, since it is equivalent to a normal n-channel MOSFET. With the word line at logic 1 and the floating gate uncharged, the transistor is turned on taking the bit line low. The low on the bit line represents a logic 1 on the output.

Programming the cell involves charging the floating gate, rather like the plate of a capacitor. The charge is placed on the floating gate by applying a higher than normal voltage, 12 V to 25 V, depending on the technology of the EPROM, to the drain and the select gate. This establishes a channel between the source and drain. The high electric field between the source and drain causes the electrons travelling through the channel to be accelerated. Some of the accelerated electrons gain enough energy to penetrate into the gate oxide layer. In this region they come under the influence of the positive voltage on the word-line gate; a number of the electrons then become trapped on the floating gate, giving it a negative potential. This process is self-limiting as the electrons on the gate prevent further electrons entering the oxide. When the programming voltage is removed from the cell it will be left with a negative charge on the floating gate. The negative charge has the effect of inhibiting the gate from turning on when a positive voltage is applied to the word-select gate. The cell is therefore programmed to produce a logic 0 state at the output.

The cell can be erased by illuminating it with ultra-violet light with a
wavelength of 253.7 nm. The light gives the trapped electrons on the floating
gate enough energy to allow them to return through the oxide layer, effectively discharging the floating gate. The erasure is accomplished through a
quartz window in the EPROM package. The whole EPROM can be erased
in one operation lasting approximately 10–30 minutes. It is worth noting that
both sunlight and fluorescent lights contain ultra-violet light and could erase
the EPROM over a period of time. Consequently the quartz window should
be covered during normal use. The window also allows normal light to enter,
possibly causing photo currents to flow in some parts of the circuit which
could cause problems when accessing the device.

The internal arrangement of the 2716 (2-K × 8-bit) EPROM is shown in
Figure 6.16.

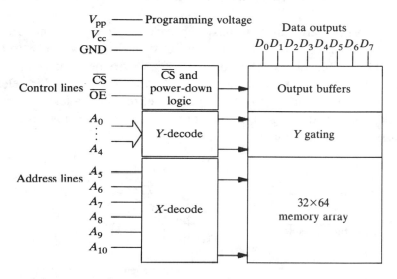

Figure 6.16. Internal architecture of the 2716 2-K × 8-bit EPROM.

The memory array is arranged as an 8-bit wide, 32-column by 64-row array.
There are 11 address lines giving the 2-Kbit address space. The $\overline{OE}$ and $\overline{CS}$
signals are similar to those found on the RAMs. The timing diagram for a
read operation is also very similar to that of the static RAM. It has a 5-V
operating voltage and a 25-V programming voltage. Access time is 450 ns
with a power dissipation of 550 mW when being accessed and a dissipation of
161 mW when not enabled. To program the entire 2-Kbyte array takes
approximately 2 minutes, as the $\overline{CE}$ line needs to be pulsed for 50 ms on
every programming step. During the programming step the V_{pp} pin must be
taken to the required 25 V. Manufacturers usually give precise instructions
on how their devices are to be programmed. Failure to follow these steps
could result in destruction of the device. For this reason it is recommended
that a commercially available EPROM programmer is used to program the
device. The programmers are normally capable of dealing with a wide range

of devices from different manufacturers. Currently EPROMs containing 4 Mbits (512 K × 8 bits) are available commercially—larger sizes will no doubt follow.

To completely erase and reprogram an EPROM would take about 25 minutes; also, specialized hardware for the programming and erase operations is required. If only one bit needs to be altered then quite an array of equipment is needed. For some operations this route can be a disadvantage. One possible solution to this problem is to use EEPROM which can be reprogrammed by simply applying an external 10 V to 15 V on a programming pin. The chip can thus be reprogrammed without removing the device from the circuit.

The transistor structure that makes up part of the array of the EEPROM is shown in Figure 6.17 (a). The structure of the cell is very similar to that of an EPROM. One difference is the use of an extremely thin oxide region between the floating gate and the drain. As the oxide is so thin, electrons can tunnel through and charge the floating gate, provided a high-enough electric field exists in this region. Using this scheme it is possible to charge or discharge the floating gate.

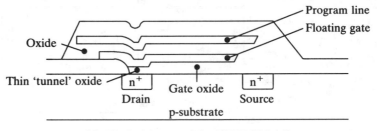

(a) Physical layout of the EEPROM cell.

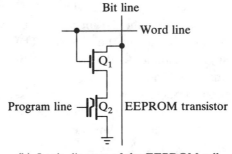

(b) Logic diagram of the EEPROM cell.

Figure 6.17. EEPROM cell.

A memory cell using the specialized transistor structure is shown in Figure 6.17 (b). The cell is erased by applying 20 V to the program line and the word line. The bit line is then taken to ground. This arrangement holds the drain of the floating-gate transistor Q_2 to ground; thus electrons can tunnel through the thin oxide and charge the floating gate in about 9 ms. Writing data into the cells is very similar. If a logic 0 is to be written into the cell then the bit line is taken to 20 V; for a logic 1 the bit line is taken to 0 V. The program line is then taken to 0 V with the word lines taken high. The floating gates where a 0 is to be stored are discharged by the electrons tunnelling back to the drain. Again this operation takes about 9 ms. These times are very much faster than the 25 minutes required to erase and program the EPROM.

Reading the cell is achieved by taking both the row and program lines high. If the cell has a charged floating gate a logic 1 output is produced. Uncharged gate cells produce the logic 0 output. Read-access times for EEPROMS are similar to those of standard RAM devices; however, the number of write cycles is limited.

6.3 Semiconductor memory-system design

So far only the building blocks of a microprocessor main memory have been discussed. To produce a useful memory structure these *building blocks* must be assembled in the correct manner. Consideration needs to be given to the amount of memory required, whether RAM or ROM will be needed, which type of RAM and ROM should be used and the optimum method of interfacing all the memory to the microprocessor. The design engineer therefore has a complex task in choosing the best solution. The complete system must be fast, use as few chips as possible and be able to fit onto the specified printed circuit board. All these criteria can have an effect on the final cost of the product and how it competes with rival products.

6.3.1 Address decoding

The amount of memory available on a single chip has greatly increased in recent years, aiding the design engineer in his task. The growth in memory size is quite staggering considering that the first custom-built memory chips contained about 1024 bits (1 Kbit) of memory. Current devices can store 4 Mbits of data in a single chip, an increase of 4000% in approximately 12 years. 16-Mbyte chips will soon be sampled!

Most 8-bit microprocessors only have 16 address lines and can therefore address 64 K (2^{16}) locations. The modern 16/32-bit processors can address up to 4 Gbits of information (2^{32}), which is far larger than any currently available memory chip. The 64-K memory address space of the 8-bit microprocessor could arguably be occupied by a single modern RAM chip; however, this does not make allowances for the need for ROM to hold a program or the need for external peripherals, such as disk drives. Similarly the 16/32-bit processors would require quite a number of even the largest current RAM chips to completely fill the available address range, which is obviously impractical. Therefore a scheme is needed to allow the address lines or bus from the microprocessor to be decoded in such a way that different memory chips can be selected within the address space.

To illustrate the problem faced by a designer, consider the following example. A design engineer has been told to build a microprocessor system with $4\,K \times 8$ of EPROM and $4\,K \times 8$ of RAM, using a microprocessor that only has 16 address lines. From the previous discussion, the chosen microprocessor is capable of addressing $64\,K$ locations, which is more than adequate for the designer's purpose, as the memory to be used requires only 12 address lines ($2^{12} = 4\,K$). Therefore the design shown in Figure 6.18 is produced.

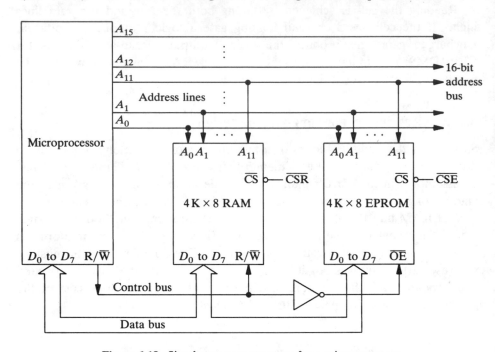

Figure 6.18. Simple memory structure for a microprocessor.

If both $\overline{CSE}$ and $\overline{CSR}$ are taken low when the CPU tries to read from either of the 4-Kbyte memories by using the address line A_0 to A_{11}, then both the EPROM and RAM will place data onto the data-bus lines D_0 to D_7, causing a conflict between the data. The microprocessor would therefore read in garbage. To distinguish between the two memories the chip-select lines, $\overline{CSE}$ and $\overline{CSR}$, need to be used. If either chip select is high then the outputs of that device will enter their high-impedance state and float. This situation will effectively relinquish control of the data bus to the memory that is being accessed and valid data will be read, or written, by the microprocessor.

The state of the chip-select lines can be determined by the unused address lines A_{12} to A_{15}. A relationship between these remaining address lines needs to be derived such that both chip-select lines are never low simultaneously. The operation of deriving the relationship is termed *address decoding*. There are many strategies that can be used for address decoding, some of which will be examined in the following sections.

The simplest form of address decoding is to use all the address lines to specify each physical memory location uniquely. This is rather like using a single memory chip for the whole address space. In the example case, full address decoding, as this scheme is sometimes called, can be obtained by implementing logic functions that act as follows.

A_{15}	A_{14}	A_{13}	A_{12}	$\overline{CSE}$	$\overline{CSR}$
0	0	0	0	0	1
0	0	0	1	1	0
0	0	1	0	1	1
⋮	⋮	⋮	⋮	⋮	⋮
1	1	1	1	1	1

Using this scheme, each physical memory location is uniquely identified (for RAM in 1000–1FFF and no data conflicts can occur. The hardware required to perform this decoding is shown in Figure 6.19(a).

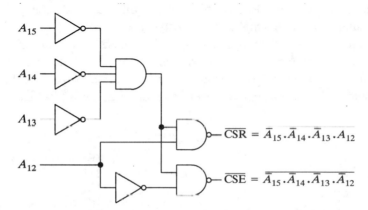

$$\overline{CSR} = \overline{\overline{A}_{15} \cdot \overline{A}_{14} \cdot \overline{A}_{13} \cdot A_{12}}$$

$$\overline{CSE} = \overline{\overline{A}_{15} \cdot \overline{A}_{14} \cdot \overline{A}_{13} \cdot \overline{A}_{12}}$$

(a) Logic diagram for the full decoding scheme.

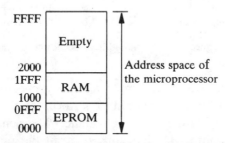

(b) Memory map produced by the decoder in (a).

Figure 6.19. Full address decoding scheme.

If the above inputs are applied to the circuit then the required chip-select signals are generated. When any form of decoding is used, the designer and software programmers who will use the board must know the addresses of the various memory components. It is no good writing a program that will store the data generated into a ROM. A simple and very effective means of representing where the memory is situated in the memory space of the micro-processor is shown in Figure 6.19 (b). The information shown in this figure is called the *memory map* of the microprocessor and gives details of what addresses are required to access the various memory types. In this example the EPROM occupies the addresses from 0000 to 0FFF, the RAM occupies the addresses from 1000 to 1FFF, and from address 2000 to FFFF there is no memory. The memory addresses have been specified in hexadecimal nota-tion; representing the addresses in binary form would be tedious for 16 bits and even worse for 32 bits. The use of hexadecimal allows an easy conver-sion to the binary bit pattern while allowing a concise and easily understood representation of the address. From this memory map it is easy to see the arrangement of the microprocessor memory. If the microprocessor wishes to access address 0AE2 then this is clearly going to be in the EPROM at its own internal address AE2; whereas the address 1AE2 will be inside the RAM at address AE2. Therefore there is no conflict of interests between the two memory devices. The example chosen to illustrate this technique is very sim-ple and, in practice, would probably not use full address decoding.

An alternative and simpler method that avoids conflicts with this memory arrangement is partial decoding, so called because only some of the unused address lines are required. For example, in order to decode the memory space for the previous two memories, address line A_{15} could be attached directly to $\overline{\text{CSE}}$ and through an inverter to $\overline{\text{CSR}}$, as shown in Figure 6.20 (a).

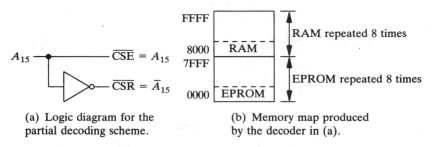

(a) Logic diagram for the partial decoding scheme.

(b) Memory map produced by the decoder in (a).

Figure 6.20. Partial address decoding scheme.

This arrangement is extremely simple and very cheap to implement, since only one gate, an inverter, is required to perform the decoding, as opposed to 7 gates using the full address decoding scheme. The penalty paid for this sim-plicity is that the entire 64 K locations of available microprocessor address space is now occupied by two 4-Kbyte blocks of memory, as shown by the memory map of Figure 6.20 (b). Although both memory devices are only 4 Kbytes in size, they both occupy 32 Kbytes of address space. This means that both devices are effectively repeated 8 times in their respective halves of the address space. Therefore addresses 0000, 1000, 2000, 3000, 4000, 5000, 6000 and 7000 would all refer to the same location, the first address of the EPROM. This situation applies to all the locations in the EPROM. A simi-lar situation exists for the RAM chip.

Partial address decoding is useful for small dedicated systems where size and component cost are considerations; for example, in dedicated controllers that will never require expansion. However, partial decoding does have its drawbacks. Memory expansion can be difficult when such a scheme is used; also incorrect memory writes could accidently corrupt RAM data unintentionally.

The third scheme to be used for address decoding is essentially a hybrid of the previous two schemes. This method involves the division, or decoding, of the memory space into fairly small manageable units. With a 64-K address space, 4-Kbyte or 8-Kbyte blocks are not unreasonable. Generally the largest memory device to be used will determine the size of the block. In the simple example above a 4-Kbyte block would be chosen. Each of the blocks of memory could then be subdivided into smaller blocks by either the use of partial decoding or full decoding as required for the particular application. The increasing use of microprocessors with a large number of address lines makes the use of all but block decoding prohibitively expensive due to the large number of address lines that need to be decoded. However, for future designs memory-chip sizes will increase, easing the problem slightly.

Various techniques can be used to implement the previous decoding schemes. The simplest method of producing a decoder involves using random logic in the form of AND, NAND, OR and NOR gates. The schemes shown in Figures 6.19 and 6.20 are essentially random-logic decoders. These types of decoders are very rarely used to decode complex memory maps, as they are very inflexible and a large number of gates (chips) would be required. One advantage of using random logic is that it can be made to operate at high speed if fast gates are used.

The most likely schemes to be used for decoding will involve the use of n-line-to-2^n-line decoders or PROM circuits. Increasingly, programmable array logic (PAL) (see Chapter 3) is also being used as a cheap and simple means of producing complex decoders; however, special software and programmers are required to fully utilize their potential. The use of n-line-to-2^n-line decoders is eased by the production of single-chip versions which can decode 4 lines to 16 lines (74LS154) or 3 lines to 8 lines (74LS138). These decoders are similar to those found inside memories, in that a particular binary input pattern will only activate a single output line. For example a 4-line-to-16-line decoder with four binary inputs and sixteen binary outputs would decode the input 0001 to produce a single logic 0 on line 1 with all other outputs remaining at logic 1. As the output is taken to logic 0 it is easy to interface directly to the chip-select line of a typical memory device. Figure 6.21 illustrates the use of a 3-line-to-8-line decoder in producing a block-decoded structure with 8-Kbyte blocks. The top 3 address lines from, in the example case, an 8-bit microprocessor are taken to the inputs of the decoder and each output is connected to the chip-select lines of the memory: two memories are shown in the example. The unused lines can be left unconnected for future expansion, which could be important in some systems. The 74LS138 has three enable lines which can be used to disable its activity and drive all of the outputs to logic 1 regardless of the binary input pattern.

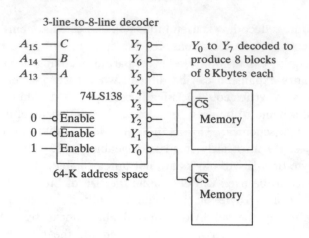

3-line-to-8-line decoder

A_{15} — C Y_7 — Y_0 to Y_7 decoded to
A_{14} — B Y_6 — produce 8 blocks
A_{13} — A Y_5 — of 8 Kbytes each

74LS138 Y_4
           Y_3 — $\overline{CS}$
0 — Enable Y_2 —   Memory
0 — Enable Y_1 —
1 — Enable Y_0 —

64-K address space $\overline{CS}$
Memory

Figure 6.21. Memory decoding using a 3-line-to-8-line decoder.

These enable lines can further assist the designer by allowing other control signals or even another decoder to control the decoding operation.

It should now be apparent that the decode operation is simply an exercise in Boolean algebra in trying to determine Boolean functions that realize the required decode function to derive the correct chip-select signals. A Boolean function can be represented by a truth table. If this truth table, which is simply a collection of zero and ones, is stored in a PROM then the same decoding effect can be achieved without having to solve the Boolean function. As the previous section has shown, a PROM has address lines that can be used to input the variables of the function and data lines which can be used to produce the outputs of the truth table. A complication arises if there are a large number (m) of address lines to decode, since it will be necessary to produce bit patterns for all the 2^m words in the PROM. As m increases, the number of PROM words will increase exponentially. However, this is outweighted by the extremely versatile nature of the PROM and the ability to produce blocks of differing sizes, rather than the fixed block sizes of the n-line-to-2^n-line decoders. The PROM technique is simply a more verbose method of implementing the decoding logic. Using special software it is possible to minimize the decoding equations so that they may fit in a PAL. This technique is much better than the PROM as well as being easier to maintain. Changing the memory map now simply involves reprogramming the PAL. The same technologies used in PROM design can be found in PAL designs, EEPROM becoming the more dominant. PROMs have been chosen here for simplicity of explanation. All the techniques discussed with respect to PROMs can be applied to PALs for memory decoding. PALs can also be used to implement logic other than address decoders.

6.3.2 Examples of memory-decoding schemes

To illustrate how the above techniques can be used in a practical situation, consider the hypothetical memory map of Figure 6.22 (a). The memory map

Address

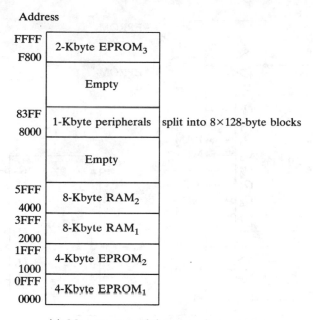

FFFF F800	2-Kbyte EPROM$_3$
	Empty
83FF 8000	1-Kbyte peripherals
	Empty
5FFF 4000	8-Kbyte RAM$_2$
3FFF 2000	8-Kbyte RAM$_1$
1FFF 1000	4-Kbyte EPROM$_2$
0FFF 0000	4-Kbyte EPROM$_1$

split into 8×128-byte blocks

(a) Memory map of the example computer.

Device	Address lines															
	15	14	13	12	11	10	9	8	7	6	5	4	3	2	1	0
EPROM$_1$	0	0	0	0	0	0	0	0	0	0	0	0	0	0	0	0
	0	0	0	0	1	1	1	1	1	1	1	1	1	1	1	1
EPROM$_2$	0	0	0	1	0	0	0	0	0	0	0	0	0	0	0	0
	0	0	0	1	1	1	1	1	1	1	1	1	1	1	1	1
RAM$_1$	0	0	1	0	0	0	0	0	0	0	0	0	0	0	0	0
	0	0	1	1	1	1	1	1	1	1	1	1	1	1	1	1
RAM$_2$	0	1	0	0	0	0	0	0	0	0	0	0	0	0	0	0
	0	1	0	1	1	1	1	1	1	1	1	1	1	1	1	1
Peripherals	1	0	0	0	0	0	0	0	0	0	0	0	0	0	0	0
	1	0	0	0	0	0	1	1	1	1	1	1	1	1	1	1
EPROM$_3$	1	1	1	1	1	0	0	0	0	0	0	0	0	0	0	0
	1	1	1	1	1	1	1	1	1	1	1	1	1	1	1	1

(b) Address bits used in the memory map.

Figure 6.22. Typical memory map and address table.

is very tightly defined and is quite complex. The memory map shows that no memory repeats are allowed. This means that full address decoding must be used. The table in Figure 6.22 (b) gives the same information in a slightly more useful form for the design of a memory decoder. In order to illustrate the different methods that have been discussed, two approaches to the design of the memory decoder will be taken. The first approach will use block decoding with some random-logic decoding. The second method will use a PROM to perform the main decoding operation.

The largest memory block to be fitted into the memory map is 8 Kbytes. Therefore the memory map will be sectioned into 8 × 8-Kbyte blocks using the

183

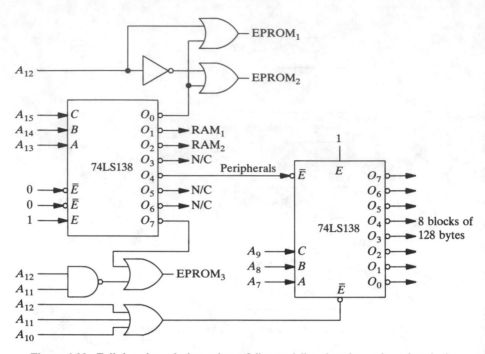

Figure 6.23. Full decoder solution using a 3-line-to-8-line decoder and random logic.

address lines A_{15}, A_{14} and A_{13} applied to a 3-line-to-8-line decoder, in this case a 74LS138, as shown in Figure 6.23. Within these blocks a finer-resolution decoding scheme can be used, where necessary, to obtain the required decoding level. The two EPROMS at the low-address range can be separated by the use of address A_{12} in the true and inverted form; as can be seen from the table in Figure 6.22 (b), they both occupy the first 8-Kbyte block. These lines are ORed with the O_0 output of the 74LS138. As RAM$_1$ and RAM$_2$ are 8 Kbytes in size, no further decoding is required for this section. The next 8-Kbyte block is ignored. The peripheral addresses are decoded into the next 8-Kbyte block. In order to get the required resolution on the 8 further devices that reside in this 8-Kbyte block, another 74LS138 decoder is used. Output O_4 of the 74LS138 is used to enable the second-level 74LS138 decoder along with the OR of address bits A_{12}, A_{11} and A_{10}, which more fully specify this block of memory. The next three address lines (A_9, A_8 and A_7) are then used to address the 8 peripheral devices through the 74LS138 decoder.

The next two 8-Kbyte blocks are again ignored. The final 8-Kbyte block has to be accessed in the last 2 Kbytes of its range. In order to achieve this, address lines A_{12} and A_{11} are ORed with O_7 from the first level of decoding to produce the required signal for EPROM$_3$. This decoding scheme requires quite a number of chips in order to decode the address range fully. This example is quite simple, with real examples producing more complex results, especially if more than 16 address lines are available from the microprocessor.

The second method of decoding the memory map uses a 256×8-bit PROM such as the 38S22. The bit table in Figure 6.24 (a) lists the values that must be

Hex	A_5	A_4	A_3	A_2	A_1	A_0	O_7	O_6	O_5	O_4	O_3	O_2	O_1	O_0
			Address range							PROM output bits				
00	0	0	0	0	0	0	1	1	1	1	1	1	1	0
03	0	0	0	0	1	1	1	1	1	1	1	1	1	0
04	0	0	0	1	0	0	1	1	1	1	1	1	0	1
07	0	0	0	1	1	1	1	1	1	1	1	1	0	1
08	0	0	1	0	0	0	1	1	1	1	1	0	1	1
0F	0	0	1	1	1	1	1	1	1	1	1	0	1	1
10	0	1	0	0	0	0	1	1	1	1	0	1	1	1
17	0	1	0	1	1	1	1	1	1	1	0	1	1	1
18	0	1	1	0	0	0	1	0	1	1	1	1	1	1
1F	0	1	1	1	1	1	1	0	1	1	1	1	1	1
20	1	0	0	0	0	0	1	1	1	0	1	1	1	1
21	1	0	0	0	0	1	0	1	1	1	1	1	1	1
3D	1	1	1	1	0	1	0	1	1	1	1	1	1	1
3E	1	1	1	1	1	0	1	1	0	1	1	1	1	1
3F	1	1	1	1	1	1	1	1	0	1	1	1	1	1

(a) Contents of the PROM required for decoding the memory map

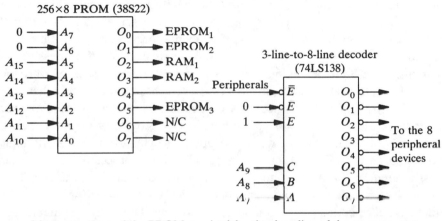

(b) Connections of the PROM required for the decoding of the memory map

Figure 6.24. PROM decoding technique for the example memory map.

stored in each location of the PROM. Rather than list all 256 address locations and every associated value, this table simply indicates the range of addresses that should contain the value shown. The PROM chosen to perform the decoding has 8 address lines; however, only six of these are used in this example. The reason for only using six address lines is that there are only 8 data outputs and thus only a maximum of eight different devices can be distinguished. Treating the peripherals as one device in our example means that six devices must be distinguished; however, splitting the peripheral block into eight individual devices requires thirteen control lines, which cannot be handled by a single PROM. Therefore, in order to optimize performance, the PROM will decode the memory in 1-Kbyte blocks, the smallest unit present (peripherals considered as 1 block). Using 6 lines (2^6) it is possible to decode 64×1-Kbyte blocks. The results of this exercise are shown in Figure 6.24 (a). In this example the two unused data outputs have been decoded to produce 0 outputs when the empty memory is accessed. This information could be used by the microprocessor to indicate that it has accesseed non-valid addresses. The final decoder for this memory map is shown in Figure 6.24 (b). The peripheral addresses unfortunately still have to be decoded by a 74LS138. However, the number of chips have still been significantly reduced from the previous example. This example has also shown a possible benefit of using a PROM in that uncommitted outputs can be used to signal other information to the microprocessor; also the processor memory map can be very easily changed by reprogramming the PROM with different values.

6.3.3 Memory-management schemes

Memory management refers to the way in which the address lines eminating from the processor are altered in order to make better use of the available physical memory. The major hurdle to understanding the use of memory management is the fundamental differences between physical addresses and logical addresses. The logical address refers to the actual addresses that the microprocessor can access by means of its own address lines. An 8-bit microprocessor has a 16-bit address bus; therefore it is capable of addressing 64 K locations. The 8-bit processor thus has a logical address space of 64 K locations. The physical address space represents the locations that have real memory chips which can be accessed for reading or writing data. For example, if the 8-bit processor was configured with 32 Kbytes of memory then the processor's physical memory would occupy 32 Kbytes even though its logical address space is 64 K locations. It is possible for a processor to have more physical memory locations than logical memory locations: cost is usually the limiting factor. Memory management thus deals with the mapping of logical addresses specified from the processor onto the available physical memory.

Memory management is rarely used on small processor systems. Generally the techniques to be discussed will normally only be found on larger machines. The main reasons for this is that memory management not only involves designing specialized hardware but also the programs that run on the machine need to be just as specialized. Hence the design of such a computer system involves a team of hardware and software engineers to produce a machine that can use to the full the extra facilities offered by the memory-management scheme.

BANK MEMORY SELECTION

The first memory-management technique to be discussed deals with the problem of how to add more main memory to a processor, increasing the physical memory beyond the logical memory addressable from the processor itself. This involves the mapping of a small logical address space onto a larger physical address space. This problem plagues designers of systems using the popular 8-bit processors. Nearly all 8-bit systems only have a 64-K logical address space, which was adequate when memory chips were only 2 Kbytes to 4 Kbytes in size. However, it is now possible to use DRAMs which would occupy the entire logical address space of the 8-bit microprocessors. This problem does not affect the more modern processors with a 20-bit or 32-bit address bus, but in the future this situation could change. Large memories are useful in that they assist software writers in producing programs which are more efficient and can run at increased speeds.

The problem of a small logical address space is one of mapping or selecting parts of the larger physical memory to access data or run programs. Figure 6.25 shows the problem that is faced with this arrangement. Essentially there are only 16 address bits ($2^{16} = 64\,K$) to access the $1\,M$ ($2^{20} = 1\,M$) of available physical memory which requires 20 bits; therefore 4 extra bits must be found in order to allow accesses to any part of the physical memory. Two possible solutions will be presented to this dilemma; however, both solutions require special software to complement the new hardware.

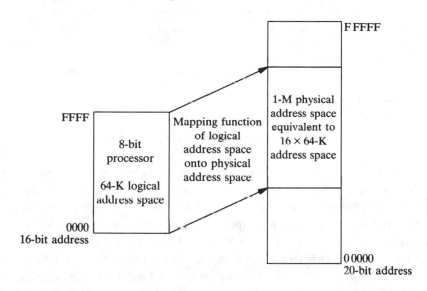

Figure 6.25. Mapping a small logical address space onto a large physical address space.

The first solution involves partitioning the physical memory into discrete blocks. In this case 32 blocks or banks, as they are more commonly called, of memory would seem to be reasonable. Extra hardware could then be used to access any one of the 32 banks of 32-Kbyte blocks of memory. Figure 6.26 shows one possible method of achieving this objective. This is not the only possible solution but it serves to illustrate the bank-selection technique. Using this scheme it is necessary to allow the processor access to some memory which does not change its position. Therefore a 32-Kbyte block of memory, labelled *base* RAM, is fixed in the memory map and decoded by address line A_{15} to occupy the first 32 K locations of the logical address space. The latch used to select one of the 32 other RAMs must also fit into the

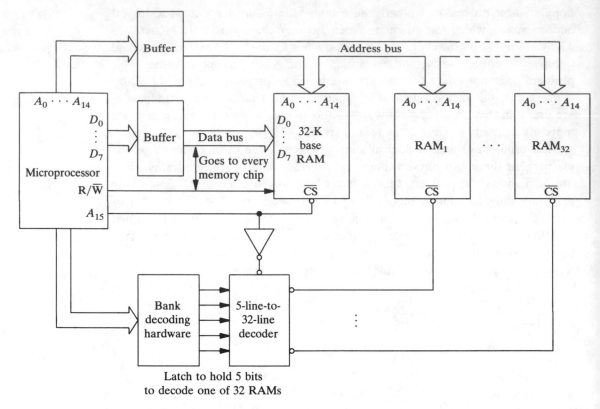

Figure 6.26. Bank selection using a decoder.

memory map at a fixed location in this range. Selection of a memory bank involves writing a 5-bit code into the latch to select the required chip. For example, if a program running in *base* RAM needs to access data stored in RAM_2, for example, then it simply writes the binary value 00010 into the decoding latch to select the correct RAM. Access to this RAM is then as if it were occupying the higher 32 Kbyte blocks of memory: $\overline{A}_{15}$ is used to decode the selected RAM addresses. Usually specialized software is required to ease the transition between banks of memory. Problems arise when a program is running in one of the selectable RAMs and needs to access data in a different selectable RAM. Also, if a program running in one selectable RAM has to jump to another part of the program in another selectable RAM then difficulties arise. In the first case the data for the program and the program itself cannot work correctly. In the second case the jump instruction needs to be able to access the decode latch. These problems are not insurmountable; however, they greatly increase the complexity of the software used in this system.

Bank switching of memory really becomes effective when large data tables are being used by the software, for example in word processors. Alternatively, if ROMs containing specialized programs replace the RAM chips, then these programs can be almost instantly accessed without requiring access to the secondary store.

The alternative approach to bank switching involves dividing the address bus into two groups of bits. One set of bits is used as normal to address the memory. The other set of bits is expanded by additional hardware to the required full number of bits for access to the complete physical memory. For example, in the above case the 16 address bits could be split into a group of 4 bits and a group of 12 bits. The 12 bits of the address come directly from the lower portion of the processors address bus: thus the memory can be accessed in blocks of 4 Kbytes. The other 4 bits come from the high-order address lines. These 4 bits are put through a special piece of hardware called a mapping table, which, in this example, is a 16-location memory and 8 bits wide. The 16 locations are addressed by the 4 bits with the 8 output bits forming the rest of the address for the physical memory. Therefore by storing values in this mapping table it is possible to access any location in the physical memory. Simply changing the values in the mapping table allows any one of sixteen 4-Kbyte blocks to be accessed, that is, the complete 64-K logical address space of the processor can be mapped to any of the physical memory locations. Figure 6.27 shows the above example using the mapping-table method. Obviously the number of allowed input and output bits can be varied to suit the required memory organization.

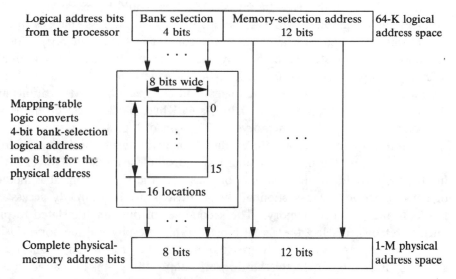

Figure 6.27. Bank selection using a translation table.

The mapping table will translate the logical address into the physical address totally independently of the running software. However, other software, usually called the operating system, must take care of altering the addresses in the mapping table so that the correct locations in the physical memory are accessed. This technique is so useful that special chips, e.g. 74LS610, exist to perform this translation operation without the need for large numbers of extra discrete chips. This system is superior to the former method since the memory boundaries are software configurable and not fixed, as in the previous method. This makes it far easier to deal with the problems previously discussed for data access. However, jump instructions to memory locations not currently mapped into the logical address still require attention.

VIRTUAL-MEMORY MANAGEMENT

Mapping a large logical address space onto a small physical address space is a serious problem with modern processors having a large number of address lines. For example, the MC68000 has a possible logical address space of 16 Mbytes or about 24 address lines. This would be quite expensive to populate with even the largest memory chips available; however, this is now changing. The method used to deal with this mapping problem is usually called virtual-memory management. The techniques used for virtual-memory management also enable a processor to run more than one program concurrently, although only one instruction at a time can be executed. Multitasking, as it is sometimes called, is extremely useful and can appear to make a simple processor seem as if it is executing many different programs in parallel.

Virtual-memory management is more complex than the bank-selection management techniques discussed previously. The address translation has to take in the logical address and, as before, produce a physical address as before; however, in this case the physical address could be exceeded and this problem has to be dealt with. As before the logical address space is broken into blocks, called pages when used in the context of virtual memory. Blocks are normally associated with the physical memory whereas pages normally refer to the logical address space; however, a page is usually equal in size to the block. So, for example, if the logical address space has 2 M locations and the physical address space 512 K, then if a 64-Kbyte block is chosen there will be 8 blocks in the physical memory and 32 pages of logical addresses.

Several different methods are available for dividing up the logical address in order to access the physical memory. The simplest method involves dividing the logical address into two sections, as with the translation-table method of bank selection. One section of the address is used to directly access the blocks in the physical memory. The second set of bits are translated to produce fewer bits to select the actual blocks within the physical memory. Using the above example, the hardware to perform the translation might be as shown in Figure 6.28. In this hardware example the 21-bit logical address is broken into a 16-bit offset address used to access locations within each of the physical memory blocks. The remaining logical addresses are passed through the translation table to produce 3 bits to access any of the 8 blocks in the memory. This arrangement is very similar to Figure 6.27.

A difficult problem now arises in that only 8 pages are allowed in the physical memory at one time; however, the processor is capable of accessing 32 pages. Consequently, each page cannot be uniquely identified. To overcome this problem an extra bit is included in the translation table to indicate whether or not the particular page is present in memory, the "present" bit in Figure 6.28. If this bit indicates the page is present then the translation can

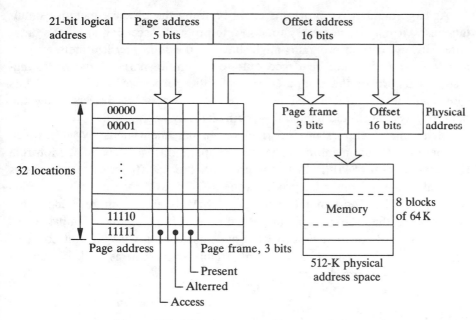

Figure 6.28. Virtual-memory address translation.

proceed; however, if this bit indicates the page is not present then the page-frame address is useless. If this situation arises when a page is not present in the memory then the operating system must take steps to obtain the required page from elsewhere, for example a hard disk, and write it in to an available page slot. The bits for the page frame can then be updated and the memory access can continue.

Access to an available page is perfectly satisfactory until all the blocks of physical memory are used up. When this situation arises some algorithm must be used to decide which page can be written back to the disk and the new page written into the physical memory. This processing is usually known as paging or swapping. Most computers reserve areas of their hard-disk drives for swapping pages in and out from the physical memory.

The simplest algorithm for deciding which page to swap out is to use a *first-in-first-out* (FIFO) policy, that is, the oldest page is chosen to be swapped out of the physical memory for replacement by the new page. An alternative approach is called the *least-recently-used* (LRU) algorithm. This method needs to keep track of how long it is since each page has been accessed. The rationale behind this scheme is that a heavily used page should remain in memory since there is a high probability that it will be required again.

To further speed up the swapping operation it may not be necessary to write the page being replaced back onto the disk if it has not been altered, e.g. if it is a program page. To this end another bit is usually included in the page table to indicate whether or not a write has taken place to a particular page, indicating when a page can be overwritten without the need to be saved to the disk. This information is represented by the "altered" bit in Figure 6.28.

191

As large numbers of programs could be using the physical memory simultaneously, it may be necessary to stop one program accessing another's data. Alternatively, certain programs might have to reside in physical memory permanently and must not be paged out; for example, parts of the operating-system software of the computer. This is the purpose of the "access" bits shown in the page table of Figure 6.28. The access bits indicate how the operating system should handle this particular page.

Many manufacturers produce memory-management units (MMU) which can perform virtual-memory management operations: for example, Motorola produces the MC68451 PMMU for use with the 680X0 series of processors and National Semiconductor produces the 32082 MMU for use with the 32000 series of CPUs. Most of the commercial MMUs are far more complicated and versatile than the example presented here. However, the example serves as quite a gentle introduction to what appears to be a very difficult concept. Some commercial systems allow even variable page sizes as well as allowing different levels of translation.

CACHE MEMORY

Computer designers are always looking for methods of speeding up a system. Several studies of computer programs has shown that accesses to data and program code tend, on average, to be confined to small portions of memory at any one time. This phenomenon is known as *locality of reference*. Also a number of microprocessors can run at speeds in excess of the cycle times of most RAMs; consequently the processor is being slowed down by accessing memory. Both of these observations led to the concept of a cache memory. This memory is unlike the main memory in that it is usually very small and of a comparable speed to the microprocessor. Cache memory is quite expensive. However, only a small amount is required; therefore the cost is not as excessive as populating the whole main memory with fast RAMs.

The idea behind cache memory is that the active portion of a program or data table is stored in this fast memory. When the CPU makes an access to memory the cache is examined first to see if the required data is present. If the data is present then the access is completed with no further action. However, if the data is not present then the main memory is accessed, as before. Obviously, the hope is that data will be found in the cache and thus speed up memory accesses. When data is found in the cache memory this is termed a hit. With most computers a hit rate of 90% can be expected, that is, 9 in every 10 accesses to the cache will be successful. If the computer has a cache with an access time of 90 ns and main memory with an access time of 350 ns then the average memory access time, given a 90% hit rate, will be 116 ns, or less than half the main-memory access time for very much less cost than populating the main memory with expensive memory ICs. A typical cache arrangement is shown in Figure 6.29 (a).

Several different techniques for organizing a cache memory are shown in Figure 6.29. The first scheme, shown in Figure 6.29 (b), uses associative

192

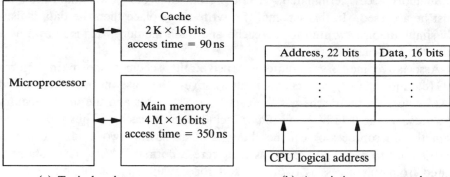

(a) Typical cache arrangement.   (b) Associative memory cache.

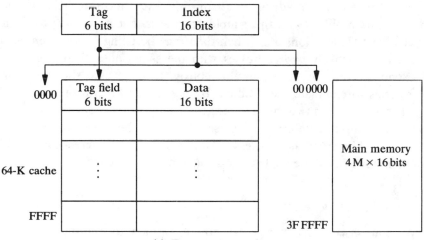

(c) Tag-memory cache.

	Tag field 6 bits	Data 16 bits	Tag field 6 bits	Data 16 bits
0000				
	⋮	⋮	⋮	⋮

(d) Multiple tag-memory cache.

Figure 6.29. Cache memory.

memory. This type of memory is very special in that if a value is applied to its input it will indicate if the value is present in the memory, producing the data associated with the value, if successful. Therefore operation as a cache memory is very simple: the address to be accessed is applied to the associative memory and the main memory. If the address is present in the cache memory then the data associated with the address can be read or written and the memory access terminated. If not present then the slower main memory must be accessed. In this scheme, if a write takes place then the data is usually simultaneously written to the cache and main memory: this is known as a write through cache.

Associative memory is quite expensive; the schemes shown in Figures 6.29(c) and 6.29(d) are cheaper alternatives that use normal high-speed RAM chips. In these schemes the cache has 2^m words and the main memory 2^n words. A tag field of $n-m$ bits is included in the cache. Thus the address from the microprocessor can be thought of as having two fields. The first field is used to access the cache as if it were a normal RAM. The data generated from this access is then checked for equality to the tag-bit field. If these are equal then the data can be used. If unsuccessful, main memory must be accessed for a read, or data can be written into the cache and main memory. The problem with this simple scheme is that if accesses to memory are occuring 64 K locations apart then data may be being overwritten when it is needed. Effectively two addresses with the same index field but different tags cannot occupy the cache simultaneously. The scheme shown in Figure 6.29(d) tries to eliminate the problem by paralleling several memories. The example shown allows two index values to occupy the cache with different tag fields. Associative memory can be used to check the tag fields for equality with the tag bits from the address. The cache can be increased by adding extra bits; however, the complexity also increases.

6.3.4 Direct memory access

The previous discussions have shown how main memory can interact with the other types of memory available, namely disk drives. Although disk drives are comparatively slow when accessing data, they can transmit the resulting bit stream to the microprocessor at very high speed. For example, a floppy disk is capable of passing 250 Kbits per second to the CPU, or over 32 Kbytes per second. A hard disk is even faster at 1 Mbit per second, or 128 Kbytes per second. The normal 8-bit CPUs are just capable of reading this data from a floppy disk and writing it to RAM. However, even the fastest modern processor has difficulty coping with the flow of data from a hard disk. A processor would spend a lot of its valuable time simply reading and writing data to disks, which is a waste of processor time. A better solution would be to off-load this heavy but simple data-transfer task to a dumb piece of hardware. The technique chosen to quickly transfer the data from a high-speed device is called direct memory access (DMA) (see Chapter 7). The DMA technique involves transferring data from an external device, such as a disk but not limited to a disk, directly into the processor's memory without the intervention of the processor; in fact, the processor could be disabled during the transfer. Two schemes exist for the direct transfer of data:

1 the processor is disabled while the entire block of data is transferred to memory;

2 the DMA operation takes place while the processor is not accessing memory: this method is called cycle stealing since it uses memory cycles that the CPU does not require.

The DMA chip, as it is called, is quite complex functionally and often intricate to program correctly. A typical set up for a DMA chip in a computer system is shown in Figure 6.30.

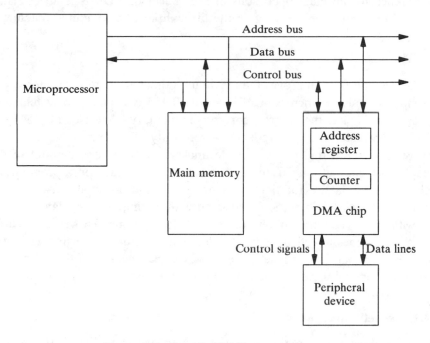

Figure 6.30. A typical DMA arrangement.

The DMA chip is connected to the peripheral from which data is read or written via a data bus and control lines, the control lines are used to indicate when data is to be read or written. More than one peripheral can be supported by most DMA chips. The DMA chip is then connected onto the buses from the processor like the memory chips, the DMA chip looks to the processor like another memory device. However, the DMA chip is clever enough to take over the address, data and control buses when it needs to input or output data.

When the processor needs to use the DMA chip it usually indicates, by writing to an internal register within the DMA chip, how much data is to be obtained and the address in main memory where it should be read from or written to. The DMA device will then gain control of the buses by requesting, via hardware lines, the processor to relinquish control of the buses: only one bus master is allowed at a time in order to avoid conflicts. Once control has been obtained the data transfer takes place. The address register is used to access the main memory at the correct location. After each byte is written to or read from the memory the address register is incremented and the data count is

195

decremented. This sequence of events repeats until the data count reaches zero. Control of the buses is then relinquished by the DMA chip and the processor can continue. An alternative method involves the DMA chip either only transferring a single byte every time bus control is gained or waiting for a period in the cycle of the processor when it is not accessing main memory and transferring a byte at a time. Both of these methods, although slower, cause very little interference to the main operations of the processor. DMA is an extremely powerful technique for speeding up the throughput of a computer system.

6.4 Mass storage

As RAM technology improves the maximum possible size of main memory for a computer is always increasing. However, RAM is volatile, so switching the power off will delete the program currently in memory. The use of ROM would solve the problem but would reduce the versatility of the computer. A form of storage is required that is non-volatile and can store a large amount of data cheaply, as a computer needs to have access to a large number of programs and their associated data. Traditionally magnetic methods have been used for storing the volumes of information associated with a computer. Today this is still true with the development of disk drives and tape systems. However, the use of optical techniques to reliably store large amounts of data is beginning to challenge the dominance of magnetic media. The final sections of this chapter will look at the optical methods available for storing data.

6.4.1 Magnetic storage

The method used to store data on a dynamic magnetic medium is the same for nearly every variation. The magnetic material, normally iron oxide, is coated in a very thin film onto a non-magnetic substrate. Hard disks use an aluminum substrate, whereas floppy disks and tape use a flexible plastic substrate, e.g. mylar. The digital data is written onto or read from the surface magnetic coating using a read/write head. The magnetic head consists of a highly permeable split-ring core wound with a fine wire coil. The data is written to the magnetic surface by inducing a current in the coil. The current produces a magnetic flux which generates a field in the gap of the core: this in turn magnetizes the surface of the medium in one of two directions depending upon which way the magnetizing current was flowing. The tape will now remain magnetized until rewritten. Figure 6.31 (a) shows this arrangement along with data stored on the tape. If this tape is passed under the head then at the flux reversals, where the rate of change of the magnetic field is the greatest, a voltage will be produced at the output of the coil. Figure 6.31 (b) shows the voltage variations that will be produced if the tape in 6.31 (a) is passed under the head. This is the general scheme used in most magnetic storage devices. The writing and reading processes are analog in nature therefore for efficient use of the storage media the digital data must be encoded in some form before storing and decoded to retrieve the stored data. This process of encoding and decoding involves

196

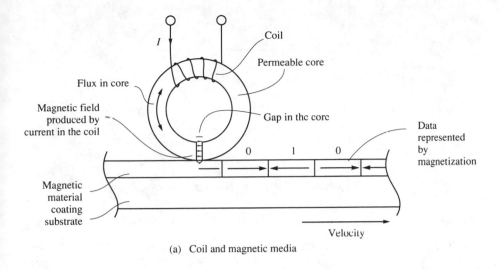

(a) Coil and magnetic media

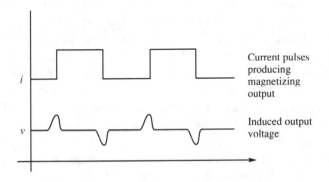

(b) Current and voltage waveforms in the coil

Figure 6.31. Constructional details of a magnetic head.

complex analog amplifier design techniques in order to minimize the effects of external noise and the problems of missing pieces of iron oxide on the medium. There are many different ways in which 0 and 1 can be represented by the magnetized regions on the tape. To illustrate these techniques several different storage methods will be studied along with their associated problems.

The technique chosen to encode the digital data must satisfy several different conflicting criteria. At one end of the scale the data must be stored reliably, as the data is valuable and should not be corrupted. At the other end of the scale the density of stored data must be maximized, to make efficient use of the media. The data density can be increased by reducing the distance, d, between flux reversals as shown in Figure 6.32. However, as d is reduced the voltage pulses produced when reading back the data begin to interfere with one another, due to their trailing and leading edges, causing the pulse height to decrease. The pulse will be more difficult to sense, which in turn will make reliable decoding of the data more difficult; this phenomenon is known as *pulse overcrowding*.

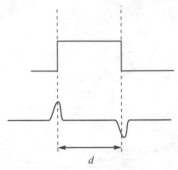

Figure 6.32. Pulse-overcrowding problem.

DIGITAL ENCODING TECHNIQUES FOR MAGNETIC STORAGE

The simplest encoding technique for storing binary data is called return-to-zero (RZ). The RZ technique, illustrated in Figure 6.33 (a), involves applying a current pulse in one direction to write a logic 1 and a current pulse in the other direction to write a 0. Between writing data the current returns to zero. (The voltage induced in the coil during the read operation is also shown in the figure.) As there is no current flowing in the coil between writing each bit, problems can arise when overwriting old data. If the data pulses are not accurately aligned with the old flux patterns then corruption of the data can occur, as old data pulses may occur between the new pulses. Therefore the tape must be erased before being used again. To retrieve the data from the tape an extra clock track must be accurately recorded alongside the data to indicate when the middle of a bit cell has been reached and the data is valid. This is called a non-self-clocked scheme and can be quite a disadvantage. Generally encoding methods which also include the clock in the same data stream are preferable. These encoding techniques are known as *self-clocking*.

The problem of correctly erasing old data can be solved by always maintaining a current in the read head. This encoding method is called return-to-bias (RB) and is illustrated in Figure 6.33 (b). In this scheme a current pulse only occurs when a 1 is to be recorded: at all other times the current flows in the opposite direction. Therefore 0s are recorded as if no data were being written. Obviously, using this scheme it is not necessary to accurately align the new and old data since the old data is effectively erased when the new data is written to the tape. However, this technique still requires a clock track to be recorded alongside the data.

The non-return-to-zero encoding (NRZ), illustrated in Figure 6.33 (c), does not require old data to be erased from the tape. In this method the current reversals only occur when a $1 \rightarrow 0$ or $0 \rightarrow 1$ transition takes place in the data stream to be encoded; therefore a series of consecutive 0s or 1s will

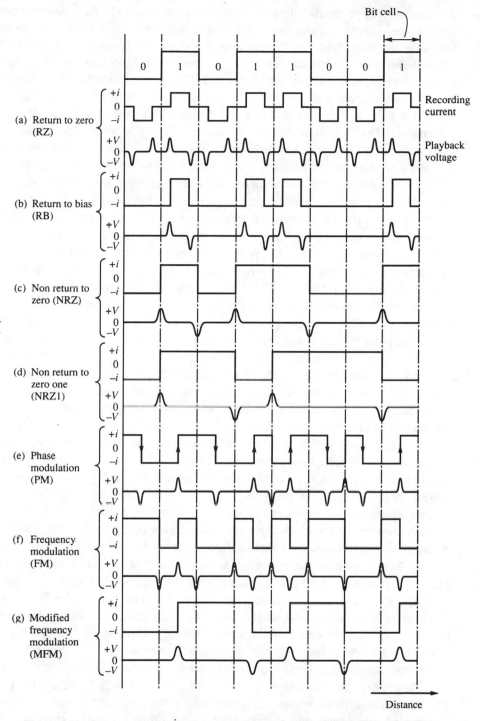

Figure 6.33. Some examples of magnetic-media encoding and decoding methods.

produce no flux reversals. Since the magnetic coating is always being magnet-ized in one direction or the other old data is automatically erased. The prob-lem with the basic NRZ method is that if a single bit is misread then all the subsequent bits are also misread. An improved NRZ method is called non-return-to-zero-one (NRZ1): see Figure 6.33(d). With this scheme transitions in the head current only occur when a 1 is to be written; no flux changes occur for a 0. This means that a 1 is indicated by either a negative or posi-tive pulse. The 0 is indicated by no pulse. Thus if a single bit is misread only one bit is affected. Unfortunately both the NRZ and NRZ1 methods are non-self-clocking and still require a clock track to be recorded. However, NRZ techniques are capable of dense data storage since a maximum of only one flux transition per bit cell is required, which is the minimum achievable.

Most tape drives use phase or Manchester encoding, illustrated in Figure 6.33(e), sometimes referred to as phase or bi-phase modulation (PM). The data is encoded as follows. A 1 is represented by a positive flux transition in the middle of the bit cell. The 0 is similarly represented by a negative transi-tion in the middle of the bit cell. Therefore a positive read voltage indicates a 1 and a negative read voltage indicates a 0. The frequency of the pulses produced varies between the input-data frequency for a sequence of ones and zeros and twice the input-data rate for a stream of ones or zeros. This encod-ing method greatly simplifies the design of the read amplifiers. There is always a transition in the middle of a bit cell. Therefore it is possible to gen-erate a clock signal from the data: hence this scheme is self clocking. How-ever, because of the number of flux transitions per bit cell, pulse crowding occurs at a lower density than for NRZ.

The phase-modulation method is difficult to decode. The frequency-modulation (FM) technique is easier to decode and yet has all the advantages of the phase-modulation method. The FM method is widely used on single-density floppy-disk drives because of its simplicity and therefore cheapness. In this scheme a flux transition always occurs on the boundary of a bit cell, making recovery of the clock from the data stream relatively straight forward. A logic 1 is represented by a flux change in the middle of a bit cell, while a 0 has no flux change in the bit cell. Therefore a 1 is indicated by a voltage pulse in the middle of a bit cell while a zero has no pulse, as shown in Figure 6.33(f). The efficiency of this scheme is not very great due to the fact that a clock pulse is recorded with every data bit. This problem can be overcome by the modified frequency-modulation (MFM) method shown in Figure 6.33(g). This encoding method is usually referred to as double-density recording, since for a given bit density on the medium MFM is capable of recording twice the amount of data as FM. MFM works by eliminating all but the essential clock pulses. The essential clock pulses only occur between streams of more than one 0 data bit. At all other times the clock can be recovered from the data itself. The main drawback of MFM is that decoding the data can be quite difficult, although chips do exist to perform this task. The next few sections will look at the various types of media available for storing data.

The surface of the drum is normally divided into tracks as shown in Figure 6.34.

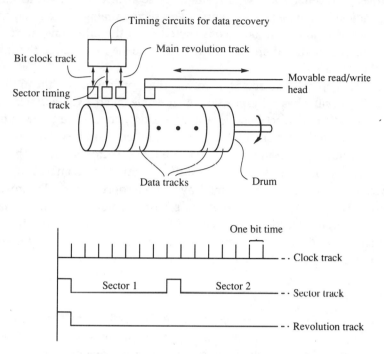

Figure 6.34. Physical layout of the drum.

The tracks are further subdivided into sectors which can hold a fixed number of data bits. For example, there may be 64 tracks on the drum with 256 sectors and 512 bits per sector; therefore the drum can store up to 1 Mbyte (64×256×512 bytes) of information. To speed up data-transfer operations a complete sector is read or written in one operation. A drum may be accessed by a single head on a movable arm (scanning head) or, to speed up access to the drum, each track may have its own head, allowing simultaneous access to all the tracks (fixed head).

The nature of the stored data means that sectors have to be accessed serially as the data passes under the head, although each track can be accessed in any order. The access time to the data in each track is determined by the rotational speed of the drum. If the drum is rotating at 3600 rev min^{-1} then the worst-case access time, also called the *latency time*, will occur when the required data has just passed under the head. One complete revolution of the drum is needed, taking 16.6 ms. This only applies to the fixed-head drums; a movable-head drum will also require time to find the selected track, called the *seek time*.

A drum is addressed by specifying the track number and the sector number within the track. In the example case, 6 bits are needed to address the track and 8 bits to specify the sector within the track: in all 14 bits are needed to address any 512-bit sector. To allow data to be accessed correctly, several timing tracks are permanently recorded on the drum. The timing tracks are used to indicate when each bit occurs and where the sectors occur in the track. Another track with a single pulse recorded is used as an index mark to indicate when sector 1 occurs. The sector timing track is superfluous since this information could be reconstructed from the bit timing track. However, this increases the complexity of the logic needed to decode the information on the drum.

Access to a particular sector involves the application of the 6-bit track address. This will physically position a movable head over a track or electronically select the correct head in a fixed-head system. The 8 lower bits are then compared with a count of the number of sector pulses from the drum. When the sector pulse counter equals the applied sector address, the data is either written from or read to a serial shift register. The contents of the shift register can in turn be written or read by the microprocessor. The revolution pulse from the drum is used to reset the sector counter. The bit timing track is used to correctly decode or encode the binary data. The encoding method used to store the data on the drum is usually NRZ1 as it is very dense and the timing tracks are already available.

HARD-DISK STORAGE

A disk storage system is organized on very similar lines to the drum storage discussed in the previous section. The disks themselves consist of a rigid circular aluminum sheet coated with a thin layer of magnetic material; both sides of the disk can be used to record data. The layout of the data is in the form of tracks arranged concentrically from the central spindle. Each track is then divided into sectors. Again the minimum quantity of information that can be transferred in one operation is a sector. A typical hard-disk arrangement is shown Figure 6.35 (a).

The read/write head is moved radially across the disk by a precision actuator. The head does not touch the disk surface, as this would quickly destroy both the disk surface and the head, but floats a few microns above the surface. The head is aerodynamically shaped so that when in the laminar air flow just above the spinning disk it will tend to float. Light springs on the head arm resist the upward movement of the head to maintain the correct gap. Modern head designs are very efficient and must contain grooves to spill some of the air. The separation between the head and disk is so small that even the grease from a fingerprint can destroy the airflow. This may cause the head to hit or crash into the disk, damaging the disk surface with the consequential loss of data: hence head crashes are to be avoided. The air gap is so crucial to the operation of the system that many hard-disk systems are operated in very carefully controlled environmental conditions.

Hard-disk drives are relatively expensive items due to the very complex and precise engineering involved in their construction. The amount of storage can be increased by stacking several disks on one spindle. Each disk

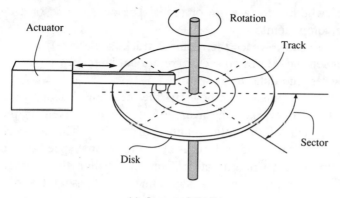

(a) Layout of the disk.

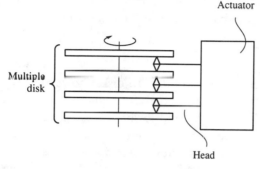

(b) Disk-drive construction

Figure 6.35. Physical layout of a hard disk.

surface is given its own head, except the outer two surfaces, in order to improve access time. This arrangement is shown in Figure 6.35 (b). The disk is maintained rotating at a constant speed at all times. When the disks are stationary an area of the disk surface must be provided where the heads can land in safety without causing any damage, this operation is usually called *head parking*. Occasionally a head is provided for all tracks, as in the fixed-head drum arrangement; however, this is very expensive.

Normally hard-disk drives are permanently associated with a computer and cannot be removed. The disk drive contains a controller which interfaces the disk to the computer, possibly using DMA techniques. The controller is responsible for taking the addresses of the sectors from the computer and driving the heads to the required location, whereupon a read or write operation can take place. Disk drives are sometimes called random-access devices because of the ability to step the head to any sector on the disk. However, this is not strictly true as the data must be read or written sequentially in the sector when it moves under the head. Hard disks are more correctly referred to as semi-random-access devices.

Hard-disk capacity ranges from about 5 Mbytes to over 1 Gbyte for large computers. As the discussion about semiconductor memories has shown, not only is the storage capacity important but also the time required to access the stored data, whether for a write or read. The access time of a disk is made up of three components:

1 time for the head to reach the required track (seek time);
2 time to reach the addressed sector (rotational latency);
3 time to access the data in the sector.

The seek time is dependent on how fast the mechanical actuators can step the head from track to track. The average seek time is given by the step time multiplied by half of the total number of tracks. The average rotational latency is given by half the reciprocal of the rotational speed of the disk. For example if the disk is rotating at $3600\,\mathrm{rev\,min^{-1}}$ then the average rotational latency will be $8.3\,\mathrm{ms}$. Given a step time of $0.5\,\mathrm{ms}$ and a disk with 200 tracks, the average seek time will be $50\,\mathrm{ms}$. Obviously the worst-case seek time would be $100\,\mathrm{ms}$. The access time to the data is commonly ignored because it is so short in comparison to the other times.

FLOPPY DISKS

Floppy disks are very similar to the hard disks; however, they are normally of lower mechanical quality and are therefore much cheaper. Also, floppy disks are usually of smaller capacity and lower access speeds than hard disk. The medium was produced, initially, by IBM as a semi-random-access device that could be used on cheaper computer systems where the cost of the then hard-disk systems could not be justified. Currently the Winchester hard disk with capacities of 5–70 Mbytes is competing very successfully with floppy-disk drives, with many being fitted as standard to personal computers (many single-user workstations now have hard disks of greater than 300-Mbytes capacity). However, this is not to dismiss the floppy disk. The floppy disk uses many standard formats and is extremely versatile for transferring data between computers. Indeed, this book comes with a floppy disk for transferring a program between machines.

The original floppy disk was 8 inches in size; however, sizes of $5\frac{1}{4}''$, $3\frac{1}{2}''$ and $3''$ are now available. Although the sizes may vary, their construction is very similar. All types contain a flexible disk which is coated with a magnetic material. The disk is then surrounded by a semi-flexible or rigid plastic cover. The disk itself has a hole in the center which is clamped by the spindle in the drive mechanism. The medium can be removed from the drive and thus many different disks can be used in the same drive, although not simultaneously. There is a slot cut in the cover to allow the drive heads access to the magnetic surface. The heads in this case make contact with the medium; hence the lifetime of the disk is limited, especially if one portion of the disk is continually accessed. Both sides of the disk can be used. Unlike the hard disks, a floppy disk is only spun up to speed when it is to be accessed: at all other times the disk is stationary. For this reason several seconds need to elapse before reading the disk so that it can come up to the correct speed, usually $300\,\mathrm{rev\,min^{-1}}$.

The data is stored on the floppy disk as tracks and sectors, as in the case of the hard disk. However, because of the low precision of the device the data density cannot be as high as that on a hard disk. Ranges from 80 Kbytes up to about 1.44 Mbytes are possible on a $3\frac{1}{2}''$ disk; however, it is possible to store over 2.7 Mbytes of data on a modified $5\frac{1}{4}''$ disk. This capacity is small when compared to a hard disk; however, it is an ideal size for the machines that will use the disks, as many only have a main memory of less than 1 Mbyte. Normally either FM, single-density or MFM, double-density, recording is used to write the data on the disk.

Construction of a typical $5\frac{1}{4}''$ floppy disk is shown in Figure 6.36.

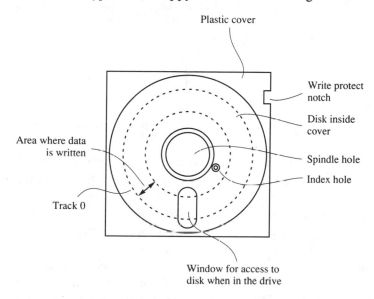

Figure 6.36. Construction of a typical $5\frac{1}{4}''$ floppy disk.

The index hole shown in Figure 6.36 is detected by an opto-sensor to indicate the beginning of a track. All other information about the sectors is written onto the disk by a formatting sequence before the disk is used for data: this type of disk is called soft sectored. Another disk type, called hard sectored, has several index holes equally spaced around the disk where the sectors occur. As no information about the sectors is stored on the disk itself, more data can be placed on this type of floppy disk. However, soft-sectored formats appear to be the more popular choice for most computers since they are more reliable.

The IBM 3740 format, shown in Figure 6.37, specifies all the information required to identify tracks and sectors along with synchronization data. The format divides each sector into four fields: the ID record, ID gap, data-field record and data gap. The address mark used to start the record consists of an all-1 byte. The following information indicates the track and sector address of the data record. An ID gap follows: this is simply a buffer zone which allows the controller time to verify that the address is correct and to switch to write mode if required. The data address mark is another all-1 byte

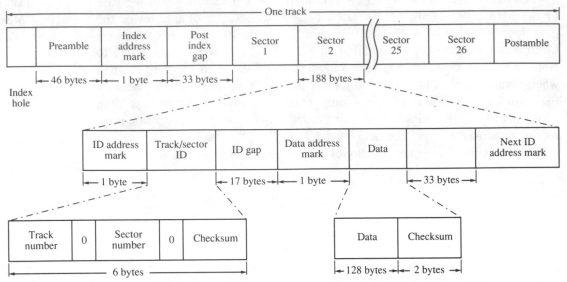

Figure 6.37. The format of an IBM 3740 soft-sectored track.

to indicate the start of the data record. The data record is 128 bytes long plus 2 checksum or cyclic redundancy (CRC) bytes. The checksum bytes are used to check that the data contains no errors. The checksum is simply the addition of all the data bytes over 16 bits, ignoring any carry outs that may occur. Thus if the data is re-read without errors then the recalculated checksum should equal the read checksum. The CRC is more complicated in formation and is therefore more powerful, detecting a wider range of errors. It is also more easily calculated from a stream of serial data, unlike the checksum. The data stream is divided by a special number using hardware long division. The remainder after the division is then recorded as the CRC bytes. On re-reading, the data plus the CRC bits are put through the same divider. If the remainder is zero at the end of the operation then no error occurred, whereas a non-zero remainder indicates an error has occurred. When data is written to a disk sector it can be immediately re-read to check it contains no errors. If errors are present then it can be re-written and rechecked. If several write/verify attempts are unsuccessful then this sector can be marked as bad and another sector used for the data. A buffer of 33 bytes after the data record allows the head to prepare for the next sector by switching to read mode if the previous sector involved writing data.

A typical $5\frac{1}{4}''$-disk format may have 40 tracks per side with 8 sectors per track, with each sector containing 512 bytes. With a density of 48 tracks per inch this format is capable of storing 320 K of double-density usable data, the rest of the storage being used for the address marks, etc. This is a fairly old format, with modern drives capable of using 135 tracks per inch (tpi) on a $3\frac{1}{2}''$ disk giving 640 Kbytes of formatted storage. Since the size of the sectors decrease towards the center of the disk, the data density must increase on progressing towards the inner tracks; this also complicates the design of the electronic reading circuits.

Currently there are three main magnetic tape storage techniques: cassette, cartridge and reel to reel. The primary use of tape is to store large amounts of data at a very low cost per bit. The tape consists of a plastic backing with a magnetic iron-oxide coating on one side similar to a disk. The heads used to read and write tapes are of almost identical construction to those used for disks. The data is stored using similar methods to those used on the disk. However, access to the data is very much slower than a disk since the data is arranged linearly along the tape, whereas the disk is arranged in a two-dimensional form, making access far easier.

The function of a tape transport or drive is to move the tape past the read/write head at a constant velocity. It is also necessary to search for particular blocks of data; this means that the tape must be stopped, started and rewound quickly. As the reels of tape have a high inertia, the mechanisms for driving the tape need to be complex if the tape is not to be broken. The highest-quality drives usually decouple the actual reel drives (dispensing the tape and winding on the tape) from the tape movement past the heads.

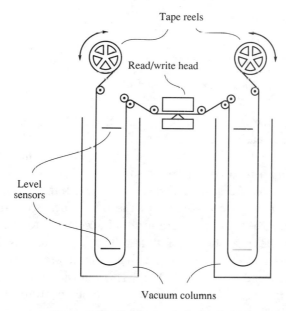

Figure 6.38. Vacuum-column tape drive.

The tape is pulled past the read/write head by the friction between a capstan (or tube) rotating at a constant speed pressing the tape against an idler wheel. Removing the idler wheel removes the drive to the tape; brake pads may also be applied to the tape to stop it. Thus the reels do not drive the tape directly. To stop and start the tape effectively, without breaking, there must be a small amount of buffer tape. One method of buffering the tape is shown in Figure 6.38. In this scheme loops of tape are sucked by a vacuum into two columns. Sensors in the column maintain the loops at a constant length by controlling the speed of the motors driving the reels. The small loops of tape allow almost instantaneous stopping and starting of the tape when the idler is applied and removed, respectively. Although the reels themselves cannot be stopped and started so quickly, the tape buffered in the columns allows them momentarily to continue to payout and take up tape.

Vacuum tape mechanisms are rarely used now because they are difficult to maintain reliably; they are also noisy. Modern tape drives use tension springs to perform the same operation as the vacuum column. The springs allow the tape to remain in tension even when it is stopped or started quickly, as shown in Figure 6.39.

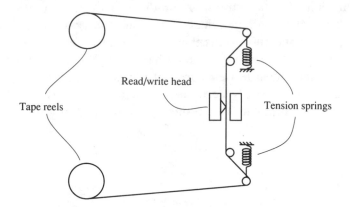

Figure 6.39. Spring-tensioned tape drive.

The data on the tape is usually stored on 9 parallel tracks with a read/write head for each track. Eight of the tracks are used to store the data and the ninth stores the odd parity of the eight data bits, therefore errors can be detected during a read operation, as shown in Figure 6.40 (a). The data is stored in blocks as on the disk. However, each block must have a gap, called the *inter-block gap* (Figure 6.40 (b)), to the next block so that the tape has time to stop after a read/write operation and accelerate to the correct speed before reaching the next block of data. The tapes usually have physical marks at the start and end of the tape: these marks are called "beginning of tape" (BOT) and "end of tape" (EOT). This allows the tape drive to know its position on the tape. The tape density is usually given in terms of bits per inch (bpi), e.g. 1600 bpi is a common format.

The major use of tape drives on modern computers is to back up data on non-removable hard disks, so that if a head crash does occur the data can be restored to a position before the catastrophe. The major cost of a tape drive goes in the mechanics and electronics required to stop and start the tape in the inter-block gap. As tapes are mainly used for backing up large quantities of data in one operation, the need for such precise stopping and starting is not critical. This has brought about the introduction of streaming tape drives which do not possess the complex start/stop mechanism and as a consequence are a great deal cheaper and smaller in size. If a streaming tape must be

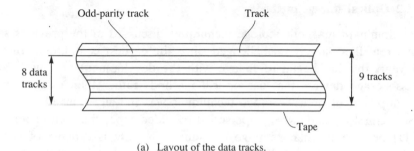

(a) Layout of the data tracks.

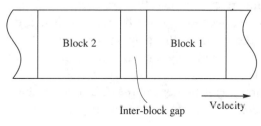

(b)   Data blocks on the tape.

Figure 6.40. Magnetic tape-storage details.

stopped then it has to be rewound past the inter-block gap so that it can gain enough speed to read the next block. As the tape is mainly designed to read and write large amounts, or streams, of data without stopping, this complex maneuver is not a drawback. The problem can be overcome by using cheap semiconductor memory as the buffer. Thus the computer only reads and writes to RAM, which is accessed by the tape drive, making block-access operations very simple. Another format that is becoming popular is the $\frac{1}{4}''$ cartridge tape which can store up to about 60 Mbytes of data on a tape no bigger than a normal pocket calculator. The drive is about the same size as a $5\frac{1}{4}''$ floppy disk drive. This format is often used to back up the Winchester disk used in workstation computers.

As disk-storage capabilities on workstations and personal computers increase rapidly there is a need to be able to backup all this data in case of mishap; therefore a cheap and convenient tape storage system is becoming a vital component of these machines. A German company has developed a tape system which combines helical scanning and digital audio-tape (DAT) technology to store over 1.2 Gbytes of data on a 60-meter DAT cassette. The use of helical scanning increases the effective length of the tape to over 11,280 meters. The DAT cassette is relatively cheap in comparison to other magnetic-tape media and is widely available. The main use of the system is to allow automatic backups of important computer data without the need for an operator to remember to backup the system, thus easing the difficulties of inexperienced users. This storage scheme has been proposed as an international standard which means that several manufacturers will be producing the equipment.

6.4.2 Optical storage methods

The dominant magnetic storage techniques discussed in the previous sections have remained largely unchallenged since the earliest computers. In the last 17 years this has begun to change with the development of a new range of mass-storage devices that use optical methods rather than magnetism. The development of optical storage techniques was initially spurred on by video-disc technology and, when superseded by video tape, the audio compact disc (CD) became the main storage medium. The CD is capable of storing 74 minutes of digitally encoded audio information on one side of a special 12-cm disc. The digital data is protected by an error-correcting code to help prevent serious audio imperfections due to scratches on the surface of the disc. Improvement in disc technology has brought with it a dramatic reduction in the cost of both the medium and playing hardware. Unfortunately, like normal LPs, the CD is a read-only medium, the recorded data being determined during manufacture.

The high storage capabilities of the CD readily lends itself to the mass storage of computer data. The arrangement used to read the CD is shown in Figure 6.41.

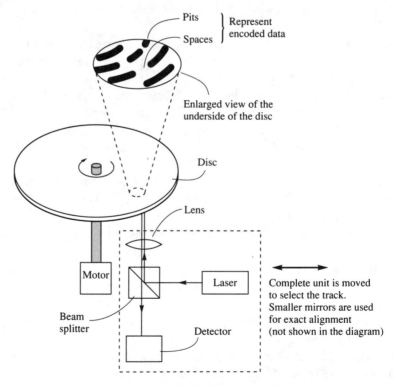

Figure 6.41. Optical storage methods.

This is very similar to the hard-disk arrangement. However, the organization of the data on the disk is very different. The data is stored as a single spiral path, each revolution of which is called a "track" (note this "track" is not the same as tracks on magnetic media). The tracks are further subdivided into sectors which have a unique identifying "address." The address, in this case, is usually specified as a time, rather than sector number and track number. Therefore a sector address is given as the minute of play, the second within the minute and the sector number within that second of play. In one second it is possible to read 75 sectors. This addressing scheme is a direct result of the CD's audio origins. Also, the CD uses a constant-linear-velocity (CLV) reading (and recording) scheme; therefore the relative speed of the disk head and disk surface remain constant. The CD drive maintains the CLV by changing the rotational speed of the disc (varying from 200 to 500 rev min^{-1} from inner to outer tracks) as the head moves from track to track, unlike magnetic disks whose rotational speed remains constant, which is known as constant angular velocity (CAV). The use of CLV results in sectors which have the same linear dimensions: hence more sectors can be stored on the outer tracks of the disc (this compares to a fixed number of sectors/track for most magnetic media). Using the CLV method of storing the data means that much more information can be contained on a disc; also, as the bit-packing density remains constant, the decoding electronics are much easier to design.

The data bits on the disc are represented by the reflectivity of the disc's surface. The low-power laser beam is focussed by the lens to a small point (Figure 6.41) on the surface of the disc. The light is then either reflected back and detected or scattered by pits on the surface and therefore not detected. The presence or absence of a bit does not directly indicate a 0 or 1 since the data is in fact heavily encoded; therefore the resulting read signal must be decoded to recover the original information. The encoding scheme endeavors to produce a sequence of pits and spaces that are of a very narrow range of lengths in order to simplify the reading electronics. The codes used to correct the data coming from the disc are far more robust than those found on an equivalent audio CD. A good code is necessary for data storage since if a single bit is lost then this could cause the computer to crash with possibly catastrophic results; however, a missing bit on an audio disc would go unnoticed by the listener. The code chosen allows the disc to return bit-error rates that are comparable with the best magnetic storage media. Even with extensive encoding to protect the data it is still possible to store about 500 Mbytes of data on a single CD.

Accessing a particular sector on the disc is quite straightforward. The first stage involves moving the laser spot, by physically moving the laser, to the approximate vicinity of the desired track containing the sector to be read. The addressing information from the hit track is then read and the number of tracks in error is calculated. The second stage is then initiated by moving a low-inertia reflecting surface in the required direction to cause the laser spot to move the calculated number of tracks: the sector can then be read. The alignment of the laser spot on the required track is very complex, due to the spiral nature of the stored data and warping of the disc, involving an intricate control algorithm. Although the average seek time can be about 500 ms, track-to-track access times can be lower than 1 ms, which compares favorably with many hard disks.

Optical discs range in size from 500 Mbytes to several giga bytes. If more data storage is required (as is usually the case) then a "jukebox" or "auto-changer" system can be used. This system may contain several discs that represent a complete database; however, only one or two of these discs can be loaded into the drive for reading, the other discs being automatically swapped into the drives under software control. Currently the optical-disc market is split into three distinct groups:

1 read-only discs, called CD-ROMs;

2 write-once, read-many (WORM) discs;

3 alterable discs.

CD-ROMS

The CD-ROMs can only be read from and as such cannot be classified as true mass storage. These discs are produced in exactly the same way a audio CDs: hence they are relatively cheap to mass produce. Many manufacturers now produce drives to read CD-ROMs that interface directly to most popular computers.

The main use of CD-ROMs is to distribute textual material in machine-readable form. A large number of publishers now sell dictionaries, reference books and other texts on CD-ROMs for use with other software packages, such as word processors. Another growing field is interactive CD-ROMs where, in conjunction with special software, the computer can be used for training purposes, all the visual information needed being stored on the CD-ROM and called up in the required sequence by the software.

WRITE-ONCE, READ-MANY (WORM) DISCS

The WORM drive has been around since 1978 when Philips first demonstrated a 12″ optical data disc. These discs store about 1 Gbyte of data as a series of pits on one side of the disc, just like the CD. However, in this case data can be written to the disc by using a high-power laser to burn a pit in a thin layer of plastic-coated tellurium. The same laser, at a reduced power, can then be used to read back the stored data; thus the drive can perform a read-and-write operation.

The WORM discs have an exceptionally high data capacity, fast access times (comparable with many Winchester drives) and naturally form an archival store of any file changes. When a file is altered it must be written to a new section of the disc, thereby leaving the old copy on the disc as a record of the change. Alternatively, only the changed section of the file might be written back to the disc, thus optimizing disc usage. The WORM disc is very useful in applications where a record of any transactions or file changes are needed for security purposes.

One further application of WORM drives is to distribute large corporate databases to other sites. Normally this may be done using magnetic tapes or discs, both of which do not travel well. The information could also be sent by modem on a telephone line; however, this is very expensive for megabytes of data. The WORM drives can offer a very cheap, robust and reliable method of transferring large volumes of data.

ALTERABLE OPTICAL DISCS

The alterable (or erasable) optical disc, unlike the WORM disc, can be written to many times, acting very like a Winchester disk. Several approaches have been taken to this problem, the most promising of which uses a magneto-optical (MO) technique. The initial discs were only capable of storing, a not very impressive, 40 Mbytes (most Winchester drives are much larger); however, this figure has now been taken to well over 650 Mbytes, making them very competitive with many Winchester drives.

The MO technique involves using a recording method which aligns magnetic domains in a thin metallic layer when the laser beam is focussed on that area. The magnetic reversals on the surface represent the stored data. The data is read by shinning a beam of plane polarized light through the surface, the plane of the light being rotated when shone through a recorded domain. This is known as the magneto-optic effect, which can be detected.

Many drive manufacturers of optical disc systems are now attempting to make drives that are compatible with the three different media formats, making the purchase of such systems less of a risk.

One further hybrid method makes use of optical tracking techniques to improve the performance of a floppy disk so that it is capable of storing 20 Mbytes of data. Essentially the main restriction on storing large amounts of data on a floppy disk is media instability. In effect the disk wobbles, making it very difficult for the read/write head to accurately track the data on the disk and hence limiting the data density. Therefore, to fully exploit the inherent storage capabilities of a floppy disk requires an improvement in tracking technology.

A particular solution to this problem involves etching optical tracking guides on to a $3\frac{1}{2}''$ disk. As the optical guides are etched on to the disk they cannot be erased. The tracks created are soft sectored, allowing any magnetic recording technique to be used. The optical guides can then be tracked using a cheap infra-red LED, instead of a laser diode: hence the data can be very accurately tracked. The modified disks are only slightly more expensive than normal floppy disks; however, they offer the same performance as a 20-Mbyte hard disk and have the added advantage of being removable.

6.5 Summary

This chapter has attempted to give a broad overview of the types of storage that are available to the microprocessor designer. Semiconductor memories have been dealt with in more detail as hardware designers are more likely to be faced with the problem of producing a suitable memory system for a microprocessor. Magnetic storage drives have been covered in less depth as these devices are extremely specialized to design and will probably be bought "off the shelf" to fit into equipment. However, what each of the various media has to offer and how they store information is very valuable for understanding the limitations in their use and how they may be employed in a real design. Optical storage techniques have been introduced, since they will probably become very popular in the near future. This is an introductory text and is therefore broad in its coverage. Further in-depth information can be obtained from the following books.

6.6 Further reading

1 AMD (1990). *PAL Device Data Book—Bipolar and CMOS*.
2 Clements, A. (1987). *The Principles of Computer Hardware*, Oxford Scientific Publications.
3 Downton, A. C. (1988). *Computers and Microprocessors Components and Systems* (2nd edn.), Van Nostrand Reinhold.
4 Hall, D. V. (1985). *Microprocessors and digital systems*, McGraw Hill.
5 Lewin, D. (1980). *Theory and Design of Digital Computer Systems*, Nelson.
6 Mano, M. M. (1988). *Computer Engineering Hardware Design*, Prentice-Hall International.
7 Nashelsky, L. (1972). *Introduction to Digital Computer Technology* (2nd edn.), Wiley.
8 Stone, H. S. (ed.) (1975). *Introduction to Computer Architecture*, Science Research Associates.

6.7 Problems

6.1 List the advantages and disadvantages of both semiconductor memory and magnetic storage.

6.2 How many bits do the following memories contain?

    (a) $6K \times 1$;   (b) $4K \times 8$;   (c) $64K \times 1$;   (d) 64×4;

    (e) $4M \times 1$;   (f) $4M \times 4$;   (g) 512×1;   (h) $1G \times 1$?

6.3 Fill in the missing information in the following table.

Number of address bits	Number of data bits	Memory size (in bits)
?	?	$4K \times 1$
10	4	?
?	8	$8K \times ?$
32	1	?
?	?	$1G \times 4$

6.4 For each of the following number of locations in address space, give the address range of the memory in decimal and hexadecimal:

    (a) $1K$;   (b) $4K$;   (c) $16K$;   (d) $64K$;   (e) $1M$.

6.5 Describe the major differences between a static and dynamic memory cell with regard to structure and method of addressing the data.

6.6 In Figure 6.3, if point $B = 0$ and $A = 1$, which of the transistors Q_1 and Q_2 would be turned on and what states would the bit lines take up when the memory cell was accessed for a read?

6.7 Tabulate the various advantages and disadvantages of each type of read-only memory.

6.8 Draw a diagram to show the crosspoint transistors required to program the MOS ROM with the data 1011 1000. Repeat the exercise for a bipolar ROM.

6.9 How many $4\,K \times 1$-bit memory chips are required to provide a full complement of memory for an 8-bit processor with 16 address lines? How many chips would be needed if $8\,K \times 8$-bit memories were used?

6.10 How large are the memory blocks for the following systems, if a 4-line-to-16-line decoder is used to produce a block decoding scheme:

(a) 16-bit address; (b) 20-bit address; (c) 32-bit address?

6.11 Calculate the number of devices required to populate the following memory map of an 8-bit processor, given that $1\,K \times 8$-bit ROMs and $4\,K \times 1$-bit RAMs are available.

```
FFFF ┌─────────────┐
     │    ROM      │
F000 ├─────────────┤
     │             │
     │    Free     │
     │             │
0FFF ├─────────────┤
     │    RAM      │
0000 └─────────────┘
```

6.12 Using the memory map from the previous question, design a full decoder using random logic. Also show how the memories will be connected.

6.13 Design the decoder of question 6.11 using partial decoding and draw the resulting memory map.

6.14 Design a partial decoder for the following memory map using random logic and 3-line-to-8-line decoders when necessary. Assume that only $4\,K \times 8$-bit blocks of RAM and ROM are available.

```
FF FFFF ┌─────────────┐
        │   RAM₂      │
FF 1000 ├─────────────┤
FF 0FFF │             │
        │   ROM₂      │
FF 0000 ├─────────────┤
        │             │
        │    Free     │
        │             │
00 FFFF ├─────────────┤
        │   RAM₁      │
00 1000 ├─────────────┤
00 0FFF │             │
        │   ROM₁      │
00 0000 └─────────────┘
```

6.15 Produce a decoder for the memory map in question 6.14 using a PROM. Show the bit table that is required. Assume that only 512×8-bit PROMs are available for the decoders.

6.16 What are the sizes of the logical and physical address spaces for the following designs?

Processor address lines	Memory address lines
16	10
16	16
20	16
32	20

6.17 In Figure 6.26 describe how RAM_{10} could be accessed. What values need to be loaded into the decoding hardware?

6.18 A bank-selection scheme similar to that of Figure 6.27 uses a mapping table that has 16 locations and is 10 bits wide. If a 20-bit logical address space is available, what is the maximum physical address space that can be accessed? Is this physical address space currently practical?

6.19 A processor is designed using the mapping-table arrangement of Figure 6.27. What values must be in each of the 16 locations if the following 20-bit addresses are to be accessed from the 16-bit address:

(a) FF000, F000;   (b) 08000, 1000;

(c) 10111, 2111;   (d) AA6AE, 66AE?

6.20 Given the virtual-memory scheme shown in Figure 6.28, what must the translation table contain if the following logical addresses are to be mapped into the following physical blocks:

(a) 1F 0000, 7th block;   (b) 10 0000, 6th block;

(c) 00 FF10, 3rd block;   (d) 0B 0001, 1st block?

6.21 A virtual-memory system has been designed such that it is only capable of storing 4 pages of data at any one time. Using the FIFO algorithm, determine what pages will be left in the memory at the end of the following page sequence.

A B C D E A F A G H B C F A

If an LRU algorithm is implemented as follows, what pages will be left in the memory at the end of the above sequence? A 2-bit counter that can count from 0 to 3 is associated with each page table entry. The page with the highest count is always chosen to be swapped out. The count for this page entry is then reset to zero; all other page counters are incremented. If the page is already present then the count for that page entry is reset to zero and any counters which were less than this page count are incremented. If the maximum count is reached for two or more pages then the first page in the sequence is selected to be swapped out. What are the values of the counters for each page at the end of the sequence?

6.22 A computer contains a write through cache with an access time of 15 ns and a main memory with an access time of 250 ns. If the hit rate is 85% and 70% of the computer's memory accesses are reads, then what is the average read access time? What is the overall average read and write access time?

6.23 Using the tag-memory cache shown in Figure 6.29 (d), what are the contents of the cache after the following memory accesses have taken place?

Memory address	Data
00 0000	AAAA
01 0010	FE00
3F 10A0	ABCD
01 0000	0101
30 2000	ABCD

6.24 If the following data is to be stored on magnetic media, draw diagrams to show the recording current and playback voltage if NRZ1 and MFM are used:

(a) 100 0001;   (b) 111 1000;

(c) 101 0101;   (d) 010 1010.

6.25 A magnetic drum has 256 tracks with 64 sectors per track and 1024 bits per sector, how many bytes can be stored on the drum and how many bits are needed to address a sector?

6.26 A drum is spinning at $2500\,rev\,min^{-1}$ and it is using a scanning head which can step from track to track in 1 ms. What is the maximum latency time? If there are 128 tracks, what is the average seek time?

6.27 A hard-disk system is designed with 9 disks and 256 tracks per disk. If each track contains 10 sectors each of 512 bits, how much data can be stored on this disk?

6.28 Given the following statistics about a single-sided/single-density soft-sectored floppy-disk drive, calculate the following information:

(a) the maximum access time to the data;

(b) the maximum rate in Kbits per second at which the raw data will be read from the disk;

(c) the maximum formatted storage available on the disk;

(d) the maximum data density in bits per inch.

Rotational speed	$300 \, \text{rev min}^{-1}$	Time for head to settle	20 ms
Number of tracks	40	Time to load data	20 ms
Tracks per inch (tpi)	96	Sectors per track	8
Outer track radius	1.7604″	Bits per sector	256
Seek speed	6 ms track to track		

6.29 A tape drive is capable of storing 1600 bpi using a 9-track head. What length of tape is needed to record 2 Mbytes of data, if each 512-byte block is separated from the next by a $\frac{1}{2}″$ inter-block gap?

6.30 Given the data in question 6.29, how much information (in Mbytes) can be stored on a 2400-ft tape?

7: Input/Output Devices

All microprocessors need an output interface to communicate their actions to the outside world. Usually an input interface is also required, although it is not essential; some simple tasks are independent of external input. Devices such as keyboards, screens, disk memories, printers and so on are known as peripheral devices. These input/output (I/O) devices are used to store data or communicate with humans. Another class of input/output is in instrumentation and control, where the microprocessor interacts with some external system such as a drilling machine or a chemical plant. Microprocessors may also communicate with other microprocessor or computer systems to exchange data.

In this chapter, we will examine the different kinds of input and output devices and how we can connect them to a microprocessor system. We will see how the interfaces are constructed internally, and look at the types of applications where each device is used.

7.1 Connecting peripherals to the microprocessor

7.1.1 Peripheral-interface devices

We cannot normally connect peripheral devices directly to the microprocessor. The connection is usually done by way of an interface chip. An interface has two ports, each with a set of control and data lines. The microprocessor communicates with the interface through one of these ports, sending commands and output data and receiving status reports and input data. The other port connects to the peripheral device. For complex peripherals such as serial communications lines or disk drives, some of the control logic may be incorporated in a special-purpose interface chip. Other peripherals such as lamps and switches require little in this respect and can be controlled by simple general-purpose interface chips.

When the number of devices in a microprocessor system must be kept to a minimum, multifunction input/output devices may be useful. These can contain combinations of RAM and ROM in addition to interface logic. In the extreme case of this, the microprocessing unit (MPU) itself may be integrated into the same device as the memory and I/O circuits.

Many interfaces use an 8-bit data bus, since many I/O operations involve character data which fits conveniently into this size. Microprocessors which operate on words larger than 8 bits often perform I/O using only 8 bits, although some faster peripheral interfaces can use larger data busses. The examples in this chapter are based on an 8-bit I/O data-bus size.

7.1.2 Memory-mapped input/output

An interface device usually contains a set of registers which can be read from or written to by the microprocessor. This is similar to a RAM, except that the registers are also connected to input or output devices, so that they may control and sense external signals. Sometimes, the registers are used to control special interface circuits inside the I/O device such as shift registers, counters and so on. Since the device appears like a memory to the microprocessor, it can be connected to the address and data busses in the same way, as shown in Figure 7.1: this is known as memory-mapped I/O.

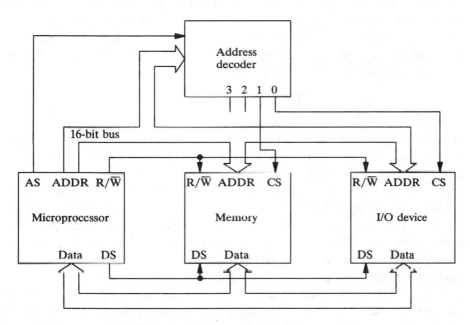

Figure 7.1. A memory-mapped input/output device.

The address decoder enables an I/O device by enabling its CS (chip-select) input when the correct address is on the address bus. Two of the address bits are used to select one of four devices in this case. Data is then read or written when the DS (data-strobe) signal is enabled. If the R/$\overline{\text{W}}$ signal is high, a read operation takes place; if it is low a write operation takes place.

The microprocessor can use the same instructions for input and output as it does for reading and writing memory data. Sometimes a range of addresses is set aside for input and output devices to avoid confusing them with memory devices. An example of such a memory map is shown in Figure 7.2 (a), where the I/O devices are at addresses between 8000 and BFFF.

Since the contents of the registers in an I/O device can be changed by external signals, the data which we see on reading from a particular register

219

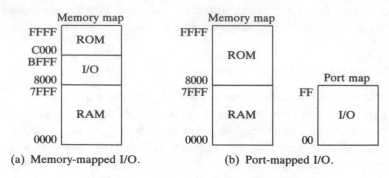

Figure 7.2. Memory and port maps.

will not necessarily be the same as that written earlier. In fact, the register bus may not be bi-directional, since reading from an output device or writing to an input device may not be allowed. This may cause problems with instructions which read data, modify it, then write it back to the same location, such as bit-test/set/reset instructions. In these cases, it may be necessary to keep a memory copy of the data written to a device for later reference.

7.1.3 Port-mapped input/output

Some microprocessors have separate address maps for the memory and I/O devices as shown in Figure 7.2(b). The same address and data busses are generally used (see Figure 7.3), but different control signals are used to enable either the memory devices or the I/O devices. This is sometimes known as I/O-mapped I/O (a rather confusing term) or isolated I/O. We will refer to it here as port-mapped I/O. The MPU sets the M/$\overline{\text{IO}}$ (memory-input/output) signal high to perform an operation on memory and low to perform an I/O operation. Two address decoders are used, one of which is enabled depending on the state of the M/$\overline{\text{IO}}$ signal.

Special instructions must be used to transfer data to and from the I/O devices in this case. This has the advantage that more memory-address space is available, since none of the address locations are occupied by the I/O devices. Also, in microprocessors which have a protection scheme, the special I/O instructions can not be executed by a program which is running in an unprivileged mode. This forces all I/O operations to be done by calling routines which run in a privileged mode and which (hopefully) are designed to prevent misuse of the peripheral devices. For instance, if a program which was not fully debugged could accidentally damage a machine connected to the I/O port, then special checking could be carried out on any I/O operations called by that program.

7.1.4 Inside an I/O device

The simplest possible output device is a single-bit output as shown in Figure 7.4. It can be used to turn a lamp on or off, for instance. It contains a

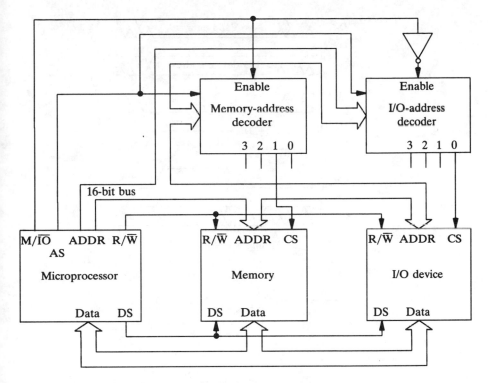

Figure 7.3. A port-mapped input/output device.

D-type latch to store the value of one of the data-bus bits when it is clocked by the microprocessor during a write operation. The data-strobe signal DS is gated by the write signal (R/$\overline{\text{W}}$ inverted) and chip-select signal CS. The address decoder raises CS when the address bus holds the correct address for this output device. The device could be extended to provide as many output bits as the number of bits in the data bus simply by connecting more *D*-type latches.

The simplest possible input device is a single-bit input as shown in Figure 7.5. It can be used to sense the state of an on/off switch, for instance. The read signal R/$\overline{\text{W}}$ and chip-select signal CS are used to enable a three-state buffer when the correct address appears on the address bus during a read operation. This allows the input signal in_0 to drive the data-bus line D_0. The data-strobe signal is not used by the I/O logic, but it indicates the time at which the MPU reads the input data. Like the output circuit above, the number of bits can be extended to the full width of the data bus.

Output data can be written at any time under command of the microprocessor, but input data may only be meaningful at particular times. We can use an extra input line to indicate that valid data is available or that some processing of the input data must be done. This may require the MPU to interrupt its current task and attend to the peripheral device. The interrupt line to the MPU will accomplish this, and it can be driven from a latch in the interface device as shown in Figure 7.6. The latch is set by a pulse on the

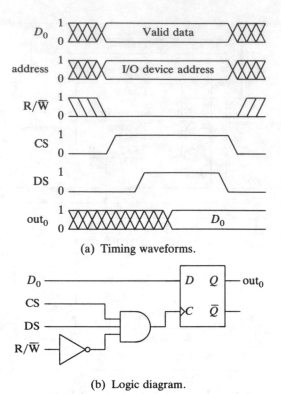

(a) Timing waveforms.

(b) Logic diagram.

Figure 7.4. Single-bit output interface.

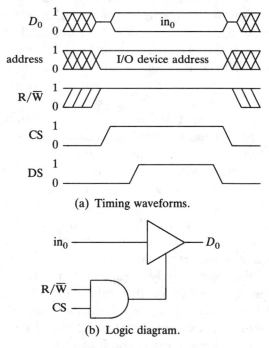

(a) Timing waveforms.

(b) Logic diagram.

Figure 7.5. Single-bit input interface.

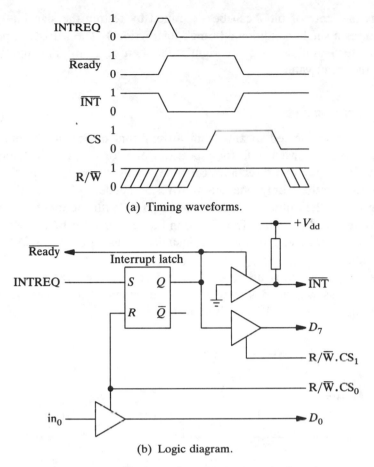

(a) Timing waveforms.

(b) Logic diagram.

Figure 7.6. Input device with interrupt latch.

INTREQ line from a peripheral, indicating that data is available on the in_0 line. This causes the tri-state buffer to turn on, pulling the $\overline{INT}$ line low. The MPU will call a special subroutine when this happens, part of which will read data from the address which enables CS_0. This will simultaneously read the value of in_0 and reset the latch, removing the interrupt request from the $\overline{INT}$ line. The latch output is also fed back to the peripheral as a $\overline{Ready}$ signal to indicate that the MPU is busy processing the current input and is not ready for the next.

You may be wondering why the buffer driving the $\overline{INT}$ line is a tri-state type with a pull-up resistor. The reason is that there may be several devices which can interrupt the MPU. They can all be connected in parallel to the $\overline{INT}$ line since the buffers will form a *wired-NOR* gate and any one which turns on will pull the $\overline{INT}$ line to 0. (Only one pull-up resistor is attached, external to the I/O device.) When the MPU services an interrupt, it can find out which device caused the interrupt by reading from the address which enables CS_1. This will enable the state of the interrupt latch to be sensed by examining bit 7 of the input word. (This assumes that the data is 8 bits wide

223

and thus the state of bit 7 can be examined by testing the sign bit—usually this requires a single conditional-branch instruction.) The MPU can poll each device in turn until it finds the right one, then reset its interrupt latch by reading the input value.

7.1.5 Bi-directional ports

If data is communicated to and from an external device along some wires, two ports could be used, one for input and one for output. In many cases, data is sent in only one direction at a time, so by connecting the output and input ports together, only one set of wires is needed. Of course, we must make sure that the output port does not interfere with the input port by disabling it when reading data. The direction of transfer must be agreed between the devices at each end of the wire, usually by using an extra I/O port for control signals.

A bi-directional port can be made by combining the circuits for input and output ports as described above, with some circuitry to disable the output when an input is required. This is shown in Figure 7.7.

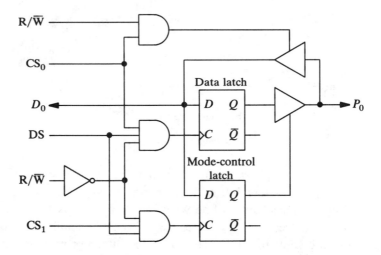

Figure 7.7. Single-bit bi-directional device.

A mode-control latch is included to control the direction of data transfer. This can be controlled by writing data to the address which selects CS_1. A logic 1 written to this latch will enable the output buffer to drive P_0. A read operation will read the value of the data latch if the port is configured as an output or the external signal on P_0 if it is an input.

This type of port may be switched continuously to transfer data in either direction along the same wires. Alternatively, it may be programmed once at power-up to function as a fixed input or output. This apparently wasteful usage is common since the device can perform all the functions of input, output and bi-directional ports. Manufacturers need only produce one type of

224

parallel I/O device, and it can be produced in larger quantities, bringing the cost down.

7.1.6 Addressing multiple registers

As we have seen above, a single peripheral-interface device may contain several registers for data-direction control, interrupt status and possibly others. For instance, there may be two or more ports in a single interface chip. Also, in more complex peripheral controllers, more control and status registers are needed.

Up to now we have assumed that the address decoder will produce separate chip-select signals for each register. This can take up many pins on the device if there are many registers. Instead, only one chip-select signal is used for each device with some extra decoding being done inside the device using some of the address-bus bits, usually the least significant ones. Alternatively, we can use a control latch in the device to select which register will be enabled at a particular address as shown in Figure 7.8. An example of this is the Motorola MC6821 parallel interface device which uses the same address for the data and I/O-direction registers. They are enabled separately by writing to a control register.

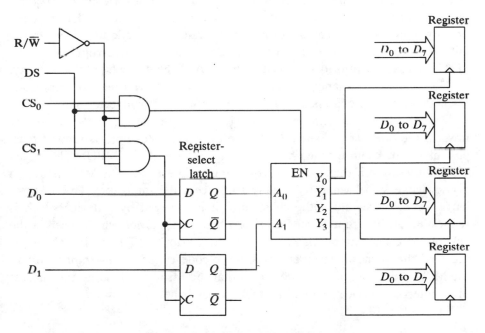

Figure 7.8. Multiple register selection.

Some peripherals use a counter to select a series of registers in sequence. A sequence of words written to the same address will be written into the series of registers, the selection counter being incremented on each write operation. A similar process can be used for reading multiple registers.

225

7.1.7 Status and control registers

We have seen the need for extra latches in the examples above for controlling the direction of bi-directional ports or sensing interrupt status. More complex I/O devices may require many such control and status bits; hence they are normally grouped together into registers which can be read or written as bytes. The MPU selects the appropriate bit by using bit test-and-set instructions. If more than 8 status or control bits are needed, several registers may be used.

7.1.8 Multiple interrupts

When several devices are able to send interrupt signals to the same MPU, some means of distinguishing them must be used. One way is to use a single interrupt input to the MPU, each device being connected to it in a wired-or fashion as in Figure 7.6. A status-register bit is provided in each interrupting device to indicate whether that device has requested an interrupt, so the MPU must read the status register from each device to determine which one requires service. Alternatively, a priority encoder may be used to ensure that only the most important interrupt will be serviced. Each peripheral-device interrupt is fed into the priority encoder which generates a code number for the microprocessor which defines the source of the interrupt. The highest-priority input will override any other inputs, causing its code to be sent to the microprocessor. Instead of polling every device when an interrupt occurs, the MPU can read the priority code and then poll only the device(s) at that priority.

Daisy-chaining is another method of assigning priorities to interrupts. Each I/O device is equipped with some logic which is connected into a chain with the MPU at one end and the devices with decreasing order of priority towards the other end. When a device requests an interrupt, its request must pass along the chain through higher-priority devices to the MPU. These may block the request if they are also requesting an interrupt. The MPU issues an acknowledge signal when it is ready to process the interrupt. The acknowledge signal passes along the chain until it is received by a device which is requesting an interrupt, where it is absorbed. The device which receives the acknowledge first will be that with the highest priority. The device which has received the acknowledge then transmits a code called an *interrupt vector* to the MPU which identifies the device so that the MPU can look up the address of an interrupt-service routine in a table which is held in memory.

7.1.9 Buffer registers

If there is a possibility that new input data may arrive at an interface before the microprocessor has had time to respond to the previous data, some extra registers may be included in an interface to store the data. New data is fed to an input register, then through the chain of registers to the last one, where it is held until it can be read by the MPU. When the MPU reads an item of data, the data in the registers is moved along so that the next item can be read. This is known as a FIFO (first-in-first-out) buffer. A FIFO buffer can improve the data-transfer speed of an interface, despite the fact that the microprocessor can only read the data from the FIFO buffer at the same rate as from the original input. This happens because the microprocessor takes a significant time to switch between an input-service routine and the main program. If this *context switch* takes place for every input item, the transfer rate will be slow. A FIFO buffer allows the interface to group several items of data together so that they can all be dealt with during a single service request.

7.1.10 External interface busses

Sometimes, a peripheral device may be connected indirectly to the microprocessor, via an external instrumentation bus for example. This means that an interface device must be connected between the microprocessor I/O or memory bus and the instrumentation bus. Several devices, such as digital voltmeters, printers, plotters, etc., may be connected to the instrumentation bus, but the microprocessor will see only the bus-interface device in its I/O or memory map. The MPU can access devices individually by writing command and address bytes to the bus-interface device. This scheme has the advantage that the peripheral devices may be located at some distance from the microprocessor without introducing extra cable capacitance to the microprocessor bus. A disadvantage of this scheme is that each peripheral device must contain an instrumentation-bus interface. This may also require a microprocessor inside each peripheral, all of which adds to the cost.

7.2 Interface devices

7.2.1 Parallel interface devices

The parallel interface is the simplest and most versatile interface device. It provides one or more 8-bit input/output ports as described above. Examples of this type of device are the Motorola MC6821 parallel-interface adapter (PIA), the Zilog Z80/PIO and the Intel 8255 PPI.

If each I/O line is to be used in a simple fashion to turn a light on or off, or to sense the state of a switch, then only electrical buffering is required. However, if the device is to be used to send a sequence of bytes to another parallel interface device (e.g. to send characters to a printer) then some extra wires must be used to synchronize the transfer.

One way to do this is by four-cycle handshake signalling. The transmitting device has a transfer-request output REQ which is connected to the receiving device, and the receiving device has a transfer-acknowledge output ACK which is connected to the transmitting device (see Figure 7.9). A transfer is started by the transmitter placing data on the bus and setting REQ to 1. The receiver detects the REQ transition and reads the data from the

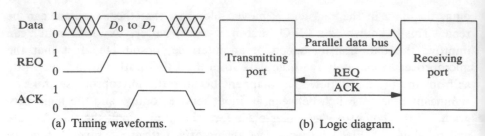

(a) Timing waveforms.   (b) Logic diagram.

Figure 7.9. Handshaking signals.

bus, then sets ACK to 1. The transmitter senses the transition on ACK and removes the data from the bus. The transfer is now complete, but the signal lines must be reset before the next transfer. The transmitter sets REQ to 0, the receiver senses this and sets ACK to 0 and the transfer is finished.

Parallel ports normally include some extra control and status lines for each 8-bit data port. These extra lines are used to perform the REQ and ACK signalling. The handshake sequence may be carried out automatically by the interface chip, or it may be implemented by a program which manipulates the control lines. Bits in the control and status registers are used to do this. If handshaking is not used, these lines can be used as extra I/O bits.

It is sometimes useful to be able to interrupt the microprocessor to read or write a particular port when a peripheral is ready, so an interrupt input is also provided. The interrupt from a particular port can interrupt the microprocessor only if it has been enabled by writing to a mask bit in the control register. When the microprocessor receives an interrupt, it can find out the source of the interrupt by examining a bit in the status register of each port.

7.2.2 Serial-communication interface devices

Serial-communication devices are available for peripherals which require data to be sent along a single pair of wires. This is most often used for connecting printers and visual-display units to a computer or microprocessor, perhaps via a telephone line. It can also be used for communicating between computers as we shall see later, or for recording data onto magnetic tape.

In serial communication, each byte is sent as 8 bits one after another (some transmissions involve only alphabetic characters which can be encoded in as few as 5 bits). It is sometimes necessary to use lines which have a restricted bandwidth, such as telephone lines. For this, we need a modem (*mo*dulator-*dem*odulator) which can convert the logic 0 and 1 signals of the serial data into signals which the line can transmit. In the case of telephone lines, these signals are audible tones of different frequencies. In serial data transmission the receiver must know the time at which each bit starts so that the parallel data may be reassembled. The clock signals in the transmitter and receiver must obviously be at the same frequency for correct operation. In addition, however, they must be in phase so that the receiver samples the signal while it is stable, not when it is changing between bits. This is further complicated by the fact that independent clock signals in the transmitter and receiver can only be matched in frequency to within a certain tolerance; hence the clock signals will inevitably drift slightly in relative phase over a period of time.

Asynchronous serial transmission overcomes this problem by transmitting each word (normally 5, 6, 7 or 8 bits) along with its own clock-phase reference information (see Figure 7.10(a)).

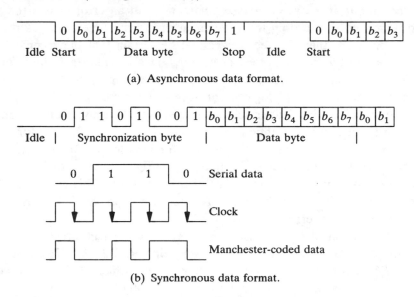

(a) Asynchronous data format.

(b) Synchronous data format.

Figure 7.10. Asynchronous and synchronous serial-data formats.

When the line is idle, the transmitter sends a continuous logic 1 signal. At the start of each word, the transmitter sends a *start bit* which is a logic 0 state lasting one clock period. The transition from 1 to 0 at the beginning of the start bit is used to reset the receiver clock signal and trigger the receiver control circuitry. After the start bit, the data-word bits are sent in sequence at a rate of one per clock period, followed by a *stop bit*. The stop bit is a logic 1 signal which lasts between one and two clock periods. It ensures that even if the last bit transmitted was a logic 0 that there will still be a 1 to 0 transition if the next start bit follows immediately. The length of the stop bit depends on how much time the receiver requires to process each character before it can accept the next.

In practice, the receiver clock signal is generated by a counter which produces one pulse every time it reaches a count of (say) 16. A precise master oscillator signal at 16 times the bit rate is used to clock the counter at the correct speed. The counter is reset by the arrival of a start bit, ensuring that the receiver will sample at the correct time to within 1/16th of a bit period. As long as the master oscillator does not drift by more than a small fraction of a bit period over the length of a word, the data will be correctly received. Serial interfaces often allow the counter to be changed so that it divides the master clock by various ratios, such as 1, 16 or 64. The master clock rate must be changed accordingly to maintain the same bit rate. A division ratio of 64 will synchronize the receiver clock to within 1/64 of a bit period; a division ratio of 1 will only work if the master clock is synchronized to the incoming data.

Errors can occur in the reception of serial data since no communication line is perfect and interference or noise may corrupt the signal. This can be detected by assigning a parity bit to each character at the transmitter and then checking it at the receiver (see Chapter 2). If an error causes the stop bit to be missed (by starting the receiver at the wrong time or by corrupting the stop bit itself) then the receiver cannot be certain that the received data bits were correctly framed into a character. This is called a framing error. Framing errors may be generated by a break in the communication line which causes the data to stay at logic 0 for several character periods. This *line break* situation may be deliberately generated by the transmitter to signal a high-priority request for attention to the receiver. (A common use for this is for a user to stop a program which is in an endless loop.) Reception errors may be caused by the receiver itself if the MPU does not read characters from the receiver buffer register as fast as they are sent. This will cause characters to be overwritten in the buffer register and hence lost. This is known as a receiver overrun error.

The connection of devices using asynchronous serial interfaces is defined in various standards, the most well-known being RS232, RS422, RS423 and RS485. These standards define the voltage levels which should be used to ensure recognition of logic 0 and 1 signals over long lines. They also define the signals to be used to control links over telephone lines using modems. Modems and other devices require various handshaking and control lines in addition to the serial data. For example, a modem may indicate that the telephone is ringing and the MPU may answer it automatically. A printer may indicate that no more characters should be sent if it has run out of paper or cannot keep up with the speed of transmission. The *request-to-send* (RTS) and *clear-to-send* (CTS) signals can be used as handshaking signals to control the flow of serial data. The flow control can be achieved without these extra signals by sending special characters. The software which controls the interface can monitor the input data for XON and XOFF characters. If an XOFF is received, no more characters are sent until an XON is received.

Many interface devices are available for asynchronous communication, for example the Motorola MC6850 ACIA as shown in Figure 7.11. It contains four registers, two at each of two address locations. The read/write signal distinguishes between the two at each address. When a byte is written to the data address, it is stored in the transmit buffer register and sent as serial data on the TXD output. When a byte of serial data is received, it is stored in the receiver buffer register and can be read from the data address by the MPU. The MPU can also write to the control register and set up the correct number of stop bits, the state of modem-control lines and so on. The ACIA can be programmed to interrupt the MPU when serial data has been received, or when it is ready to send more serial data.

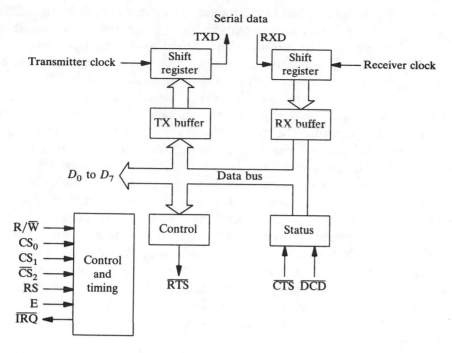

Figure 7.11. Asynchronous communications interface.

Asynchronous data transmission is inefficient, since the start and stop bits transmit no information but use at least 2 bits for every 8 data bits. Synchronous transmission removes the need for these by sending the transmitter-clock signal along with the data. This can still be achieved using a single transmission line if the data is encoded with the clock signal to produce a Manchester-encoded signal as shown in Figure 7.10. Since no start and stop bits are present, the beginning of the first word must be signalled by sending a framing code. This is a sequence of bits which cannot occur in normal data. Once a framing code has been sent, it can be followed by a sequence of many data words ended by a terminating code. The framing and terminating codes are provided as special control characters in the ASCII and EBCDIC character sets. The frame is started with the SYN (synchronization) character and terminated with the ETX (end-of-text) character. Codes such as Bisync require two SYN characters at the start of a message. These are needed because a loss of synchronization may cause the last bits of one character to combine with the first bits of the next to form a false SYN character. (The division between the characters cannot be known once synchronization is lost.)

If a control character occurs as part of the data which is to be sent, the receiver must not interpret it as control information. This can be done by sending control characters as two-character sequences using the DLE (data-link-escape) character. In this mode, control characters are interpreted as data unless they follow a DLE. If a DLE actually occurs in the data, it is sent as DLE DLE. The receiver strips off the control information and passes on the data characters to the program.

If the data is not supplied to the transmitter quickly enough by the MPU, a transmitter-underrun condition occurs. One option is that the transmission may be aborted and a termination sent. Alternatively, the transmitter may send SYN characters to maintain synchronization, since some time would be taken to recover if there were a break in the transmission.

Protocols such as Bisync treat data as a sequence of characters. The SDLC format treats it as a sequence of bits; hence any number of bits per character may be used. The synchronizing pattern 0111 1110 (known as a flag) is used to start and end each transmission. If a transmission follows another immediately, the end flag of the first can be the start flag of the second. The sequence of six ones in this pattern is its distinguishing feature, so data must not contain more than five ones in a row. This is achieved by *bit stuffing*, where a zero is automatically inserted by the transmitter after five ones in a row. The receiver removes these bits so that the data is restored.

Error checking in synchronous transmission depends on the data type. If ASCII characters are being sent, then parity bits can be used in the same way as for asynchronous transmission. Extra error checking is usually included by adding a block-parity character at the end of the data. Each bit of this character is the parity bit for all the bits in the same position in every data character in the block.

An alternative error-detection scheme is the cyclic-redundancy check (CRC). The details of the operation of this code are beyond this text, except to say that it forms a 16-bit check code from the data using a shift register and some exclusive-or gates. This type of code is more likely to detect errors of the type encountered on communication lines than the simple parity system.

An example of a synchronous device is the Motorola MC6854 ADLC as shown in Figure 7.12. This can operate in SDLC or Bisync modes or in other modes determined by the user. Internal circuitry automatically senses flag or SYN codes, generates and checks the error-detection code and performs zero-insertion and deletion in SDLC.

7.2.3 Counter and timer interfaces

Counting of pulses and measuring of time or frequency can be carried out by software using the microprocessor clock cycle as a time reference, but this occupies the CPU for a large amount of time when it could, perhaps, be doing something else. Counter/timer peripheral devices are available to carry out this function independently of the main CPU. Once again, the device contains a set of registers which can be read or written by the microprocessor. The contents of each register can be incremented or decremented by a pulse from an external source or by a clock signal generated internally from the CPU clock by a frequency divider (i.e. a counter). This enables the number of pulses, period or frequency of the external signal to be measured.

An example of such a device is the Motorola MC6840 programmable timer. It contains three counters which can be programmed to operate in

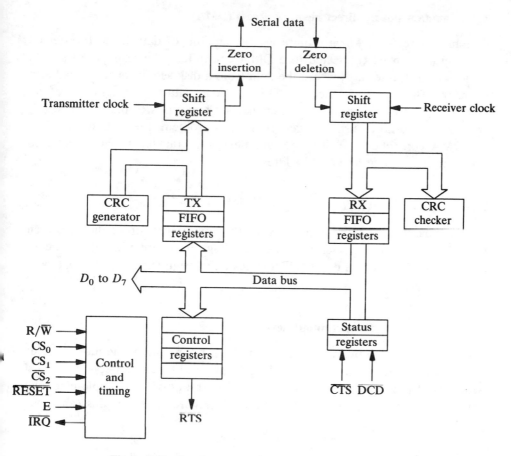

Figure 7.12. Synchronous serial-communications interface.

different modes. Starting values can be written to registers which accompany the counters. When a timer receives a pulse at its gate input, it will load the starting value from its register, then decrement it by one for each pulse received on its clock input. When the counter reaches zero, it can interrupt the microprocessor and generate an output pulse.

Another application of a timer is the provision of a real-time clock, where the timer is programmed to interrupt the CPU every, say, $\frac{1}{100}$th of a second. The CPU interrupt-service routine will then keep track of the seconds, minutes, etc., by incrementing the current time stored in memory by $\frac{1}{100}$th of a second on each interrupt. Since this is a common application, special-purpose interface devices are available for real-time clocks. An example is the Motorola MC146818 which has its own crystal clock and counters for the date and time. The circuit is designed to operate from a battery when the power for the rest of the system is turned off so that the correct time is maintained. The battery is also used to maintain the contents of a small amount of RAM which may be used to store data which must not be lost when the power is turned off.

233

7.2.4 Input/output by direct memory access (DMA)

Peripheral interfaces which require a large amount of data to be transferred very quickly can be connected to a DMA controller. An example of such a device is a disk interface, where the data from a disk sector must be read into memory. The speed at which the CPU can load data from the interface and store it in memory may not be fast enough to keep up with the rate at which data arrives from the disk. To get around this problem, the CPU can instruct the DMA controller and disk-interface device to transfer a whole sector to (or from) memory at a selected address.

7.3 The user interface

A major part of the input/output task is the communication of data between you, the user, and the microprocessor. This group of interface devices contains a wide range of different technologies and design variations to cope with different tasks.

7.3.1 Keyboard and switch input devices

Data can be entered by the user in a variety of ways, one of the most popular being a keyboard. This can be anything from a full typewriter-style keyboard with special function keys down to a single push-button. A simple electrical switch can be used for each key as shown in Figure 7.13, but this can cause problems.

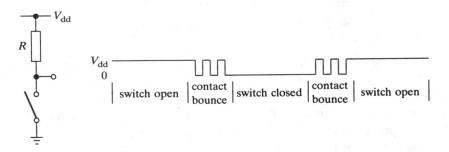

Figure 7.13. Single-pole-switch input device.

When the switch contacts open or close, mechanical vibration can cause the electrical circuit to be opened and closed several times: this is known as *contact bounce*. In order to ensure that each key depression does not send several pulses to the microprocessor, the switch must be *debounced*. This can be done electrically by using a switch with three states as shown in Figure 7.14. Two of the states correspond to the *off* and *on* positions in which one of the circuits is closed. The third is the state in which the moving contact is in the middle, and neither circuit is closed. If we use a flip-flop which sets when $A = 0$ and resets when $B = 0$, it will remain stable during the change-over period when both A and B are pulled to logic 1 by the resistors. A simple latch circuit such as that

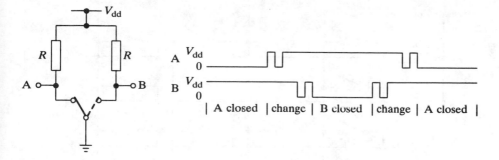

Figure 7.14. Double-pole-switch input device.

shown in Figure 7.15 (a) can be used to produce a clean logic 0 or 1 signal from this type of switch. The version in Figure 7.15 (b) is more economical in components, and uses a resistor to weaken the feedback path in the latch so that the switch can force it to 0 or 1. When the switch is between contacts, the buffer output holds the input stable via the resistor.

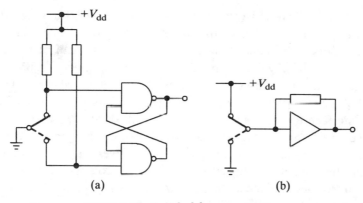

Figure 7.15. Switch-debounce circuits.

An alternative method of debouncing is to ignore further switch transitions for a short period after a transition has been detected, to allow the contact-bounce oscillations time to decay. The timer can be a hardware device or it can be implemented as part of the software. Only one timer is required for a whole keyboard if only one key is depressed at a time.

Keyboards with many keys would require a large number of input ports if they were connected directly to a parallel interface device. Normally, the keys are operated at a rate no greater than normal typing speed, around 10 keystrokes per second. The keys can therefore be split into groups which are scanned in sequence at a rate faster than the maximum keying rate. The keys are arranged electrically in a matrix of rows and columns as shown in Figure 7.16 (the physical arrangement can be different, e.g. a typewriter keyboard). Each column is enabled in turn by turning on one of the transistors X_0 to X_3, thus pulling one column wire to logic 0 while all the others are open circuit. A switch closure in the selected column will cause the corresponding row wire

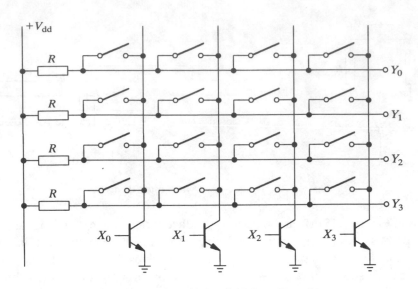

Figure 7.16. Multiplexed keyboard matrix.

(Y_0 to Y_3) to be pulled down to 0, provided that the switch resistance is small compared to the resistor R. All the other row outputs will be pulled to logic 1 by the resistors. The state of each column of switches can be sensed by the MPU by reading the Y values in parallel while enabling one X bit at a time.

Rollover is the term used to describe the situation where a new key is pressed before the previous one(s) has been released. Keyboards with a separate interface for each key can cope with this provided that the software is designed to do so. Multiplexed keyboards can produce incorrect outputs if more than two keys are pressed together, since several row and column wires may be shorted together. The software which controls the keyboard interface must therefore *lock out* the data from the keyboard until some of the keys have been released.

The key switches may be constructed from metal contacts which are operated mechanically by the key top. Alternatively, the contacts may be made from ferromagnetic material and encased in a sealed glass tube. This is known as a reed switch, and it can be operated by the magnetic field from a small permanent magnet attached to the key top. Since the contacts are sealed, they are protected from corrosion by the atmosphere—a common source of switch failure. Low-cost keyboard contacts may be made from rubber which has been made moderately conductive by the inclusion of graphite particles. Conductive rubber contact pads can be pressed down by the key top directly onto a pair of tracks on a printed circuit board. Even the key tops themselves can be moulded from a sheet of rubber which is imprinted with the legends for the key functions. A further variation is the use of a thin plastic membrane backed with metal foil which can be pressed through the holes in an insulating sheet to make contact with a circuit board.

Since electrical contacts are affected by corrosion and wear, contactless keyboard mechanisms have been developed. One example of this is the

Hall-effect key switch which senses the magnetic field from a magnet in the key top by a special semiconductor sensor. Capacitance can also be used to sense a key depression if a metal plate in the key top forms part of a variable capacitor. The capacitance change when the key is depressed can be detected by the interface circuitry since it will change the impedance of the capacitor to an a.c. signal.

7.3.2 Pointing devices

When the user wishes to indicate a particular location on a screen for editing text or graphics, a cursor marker can be stepped across the screen by pressing control keys. This can be rather tedious; so various input devices are available to provide this type of input. The joystick is simply a lever whose tip can be moved in two dimensions by the user. The lever operates switches or other sensors which provide information to the MPU on the position of the lever. A flat plate which can be tilted is sometimes used as an alternative.

Two wheels with position sensors can be used to move the cursor in the X and Y directions. This can be extended by using a tracker ball which moves the wheels indirectly when it is rotated in a panel mounting. If this is inverted, so that the ball touches the table top and the mounting is moved by the user, this is known as a mouse.

A location on the screen can be indicated directly by using a light pen which senses the light from the screen. Since the picture is formed by a scanning spot, the position of the pen can be determined if the time at which the spot illuminates the pen is known.

If the screen is equipped with a sensor which can detect the location of a finger, a position may be entered by simply touching the screen. The sensor may use infra-red or ultra-sonic beams to detect the presence of the finger.

7.3.3 Graphical input

If positional information is to be fed in from a source other than the screen, this can be done by tracing the required shape on a bitpad or digitizer. The position of a pen or puck can be sensed by a magnetic field or ultra-sonic waves. The puck may incorporate a transparent area with crosshairs so that drawings placed on a base board may be traced to digitize their shapes.

7.3.4 Display output devices

Displays can range from simple on/off indicators to high-definition multi-color graphics displays.

DISPLAY TECHNOLOGIES

Early display devices used incandescent lamps or gas-discharge tubes. The incandescent lamp gives off light by heating a thin conductor by an electric

237

current until it glows. This consumes a lot of power, much of which is wasted as heat. This type of display is still used in some situations where high brightness is required. The gas-discharge tube is a sealed glass tube which contains a gas such as neon at low pressure. A voltage applied to electrodes in the tube will produce a glow around the negative electrode. The voltage required is usually of the order of 100 V which is much higher than the supply voltages of around 5 V or 12 V used for most microprocessor logic devices.

Vacuum fluorescent displays (VFDs) contain a phosphor material which gives off light when struck by a beam of electrons. The electrons are emitted from an electrically heated cathode filament and are accelerated towards the phosphor on symbol-shaped anodes by applying a positive voltage to them. The electrons travel through a fine metal grid which has a variable voltage controlled by the logic circuitry. The light from the phosphor can be turned on or off by controlling the voltages on the grid and anode.

The cathode-ray tube (CRT) uses a similar technique to the VFD except that the electrons are formed into a thin beam and then accelerated by a very high voltage (from 1 to 30 kV) before striking a phosphor-coated screen. This produces a bright spot on the screen which can be moved horizontally and vertically by an electric or magnetic field. The brightness of the spot can be controlled by varying the voltage on the cathode. A display can be produced by moving the spot over the screen so fast that the eye sees a continuous image. This can be assisted by using a phosphor with a long persistence, which means that it continues to glow some time after it has been struck by electrons. The spot can trace out a shape directly on the screen, producing a vector display. Alternatively, it can repeatedly scan the screen from top to bottom in a fixed pattern of horizontal lines, producing shapes by making the spot brighten at the correct times. This is known as a raster display.

Light-emitting diodes (LEDs) are special semiconductor diodes which emit light when a current passes through them. They are much more efficient than incandescent bulbs in converting electrical energy to light, but have a limited light output. The light emitted by LEDs is colored, red being the most common.

Liquid-crystal displays (LCDs) do not give off light themselves, but reflect or transmit light from an external source under control of an electrical voltage. A thin layer of liquid-crystal material is trapped between two plates of glass which have transparent conductive electrodes deposited on their inner surfaces in the shape of the required symbols. When a voltage is applied to these electrodes, the tiny crystals in the liquid crystal will rotate to line up with the electric field, causing a rotation in the plane of polarization of the light transmitted through the device. If the device is viewed through a polarizing filter, light which has had its polarization rotated will be stopped by the filter; hence that region of the display will appear darker. Since LCDs do not produce light, they can operate with very low power inputs. Chemical decomposition can occur in the display over a period of time if direct current is used; hence low-frequency a.c. signals are used to drive LCDs in practice.

If a display is to be used to show numerals or text, it must contain several indicators for each character position. In a purely numeric display, each position could use ten indicators corresponding to the numerals $0, 1, 2, \ldots, 9$. Each indicator could be shaped like the appropriate numeral, but it is difficult to superimpose these so that they all appear in the same position. The gas-discharge display uses thin metal electrodes in the shape of each numeral. The electrodes are arranged one behind the other so that those at the back are viewed through the gaps in those at the front. Alternatively, incandescent displays have been used in which the numerals are projected onto a translucent screen from behind.

By stylizing the numerals to some extent, it is possible to create them by displaying a selection from a set of 7 segments as shown in Figure 7.17. The BCD code for each numeral is also shown along with the segments required to display it.

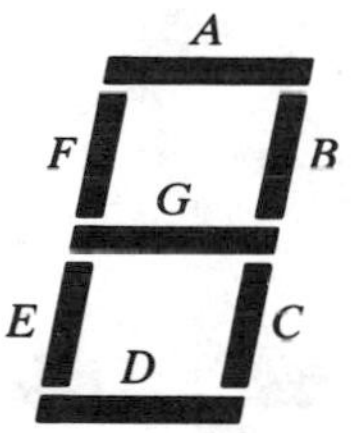

Display	Code	*ABCDEFG*	Display	Code	*ABCDEFG*
0	0000	1 1 1 1 1 1 0	5	0101	1 0 1 1 0 1 1
1	0001	0 1 1 0 0 0 0	6	0110	1 0 1 1 1 1 1
2	0010	1 1 0 1 1 0 1	7	0111	1 1 1 0 0 0 0
3	0011	1 1 1 1 0 0 1	8	1111	1 1 1 1 1 1 1
4	0100	0 1 1 0 0 1 1	9	1001	1 1 1 1 0 1 1

Figure 7.17. Seven-segment display format.

Manufacturers produce a range of decoders which perform this code conversion. The seven-segment arrangement removes the need to superimpose display elements. Technologies for displaying the segments include incandescent filaments, LEDs, LCDs, electromagnetically moved shutters, VFDs and gas-discharge display panels.

The range of characters can be extended to include all the upper-case symbols of the alphabet and some special symbols by using 14 segments as shown in Figure 7.18. The stylized appearance of the characters in segmented displays can be overcome to some degree by using a matrix of dots as shown in Figure 7.19. The minimum matrix size for alphanumeric work is 7 by 5 as shown, although larger numbers of dots will produce characters with a more pleasing appearance. This type of display can be implemented by any of the technologies described above.

DISPLAY MULTIPLEXING

The wiring between a display and its decoder/driver circuits may become cumbersome if many characters are to be displayed. One way of reducing the

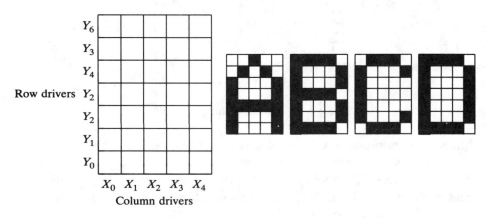

Figure 7.18. Fourteen-segment alphanumeric display.

Row drivers

X_0 X_1 X_2 X_3 X_4
Column drivers

Figure 7.19. 7×5 dot-matrix character-display format.

number of wires is to incorporate a separate decoder/driver in each character position with some address selection logic. This is expensive, however, and the circuitry may increase the cost and may not fit into the space available between the character positions. An alternative is to multiplex the display elements so that only a small number are driven at any time. An example of this is the four-digit, seven-segment LED display shown in Figure 7.20. One digit is enabled at a time by turning on one of the transistors from X_0 to X_3. The correct segments for that digit position are illuminated by switching on some of the transistors A to G. The next digit is then selected and the segment data changed to the correct value for that digit. When the last digit has

240

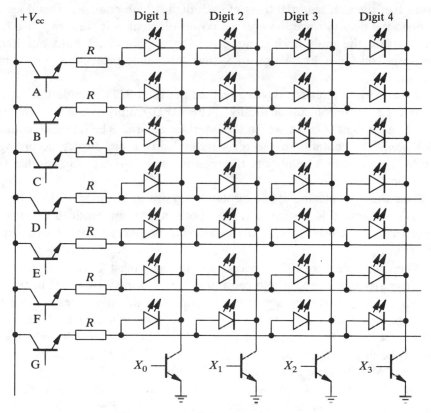

Digit 1   Digit 2   Digit 3   Digit 4

+V_{cc}

A

B

C

D

E

F

G

X_0 X_1 X_2 X_3

Figure 7.20. Multiplexed LED display.

been displayed, the cycle repeats from the first again. Note that current-limiting resistors R have been included to control the LED brightness. These must be connected in series with the segment-driver transistors. If they were connected to the digit drivers instead, the digit brightness would vary depending on the number of segments which are lit in that position.

When a display is multiplexed, the rate at which the drive is switched between groups of elements must be high enough so that the averaging effects of the display and the eye make the display appear continuous. If the display elements are split into a matrix of rows and columns, only one column of which is active at a time, this means one wire with a driver circuit is needed for each row and column. For example, a seven-segment, eight-digit display would require $7 \times 8 = 56$ driver circuits if not multiplexed, but only $7 + 8 = 15$ if multiplexing is used.

The interface required for a non-multiplexed LED seven-segment display is simply a parallel interface device with some appropriate buffers to provide the correct voltage and current levels for the display elements. A multiplexed display can be driven in this way, with separate drive circuits for the digit and segment inputs. Software can be used to switch on only one digit driver with the appropriate segment drivers for that digit. Since only one digit is on at a time, a higher drive current than for non-multiplexed displays is used to

241

ensure that the average brightness of each digit is high enough. This does not overload the display since the average power dissipation is still low. If a fault were to stop the multiplexing, however, the display could burn out if one digit were constantly lit. To guard against this, some protection circuits may be used.

Display-driver interface devices such as the Intel 8279 provide this type of multiplexed operation automatically. The MPU simply writes the required data for each digit to a register in the interface device. The interface continuously scans the contents of the registers and drives the display accordingly. The MPU can write a number to the registers and then carry on another task while the display operates.

Since the multiplexing of the display has much in common with multiplexing a keyboard, this device performs both tasks. In particular, common driver circuits are used to scan the columns of both key switches and display segments.

When multiplexing LCD devices, special care must be taken, since the a.c. drive signal required would cause unselected display elements to become active. These displays use a mode of multiplexing in which alternate positive and negative pulses are fed to the row drivers. This will only work for a limited number of rows, up to about four. Above this, the time for which each part of the display is pulsed becomes too small. Interface devices such as the Intersil ICM 7231A are available to drive multiplexed LCDs.

CRT displays are a special case of multiplexed displays, since they can function with only one display-brightness control bit. The action of scanning the spot in a raster on the screen allows each point on the screen to be selected in sequence and made bright or dark by the data input. Grey scale and color displays can be implemented in a similar way by using a binary number to control the spot brightness of the electron guns.

The scanning of the spot is carried out automatically by the circuits inside the video monitor, but these must be kept synchronized with the display controller so that when the spot arrives at a particular part of the screen, the display controller produces the correct brightness information for that point. The spot is assumed to travel at a constant speed along each line; so the controller simply clocks out the brightness data at a constant rate for each line. The end of a line is signalled by a short synchronizing pulse generated by the controller. At the end of a complete screen frame, a longer synchronizing pulse is generated. These pulses can be sent on the same wire, since they are distinguished by their duration. They can even be combined with the brightness information since they never occur when the spot is visible.

Interface devices such as the Motorola MC6845 (see Figure 7.21) provide all the circuitry required to produce the serial data and synchronization signals for a CRT display. The device can be programmed with parameters such as the number of lines on the screen and the width of the synchronization pulses. A CRT controller cannot hold enough data in its internal registers to store the display data; so this is obtained from external memory devices. The display-memory address is switched between the microprocessor address bus

242

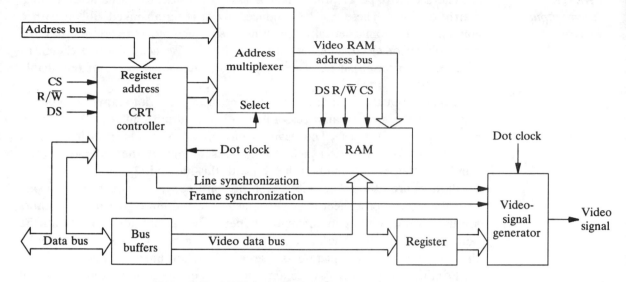

Figure 7.21. CRT-controller interface.

and the CRT-controller scan counters by a multiplexer. While the CRT beam is scanning the display area, the memory contents are read out sequentially to the video processing circuits where they are turned into dot patterns. If a memory bit corresponds directly to a single dot on the screen, this is known as a bit-mapped display (this is used for graphics). If the memory data is used to select a dot pattern for a character from a decoder (such as a ROM) then it is known as a character-mapped display.

When the CRT beam is retracing to the beginning of the next line or frame, the MPU may control the memory address bus to read or write data. If the MPU addresses the memory outside this interval, then interference may be seen on the screen as a *snow* effect.

7.3.5 Hard copy

Printer mechanisms are usually of two basic types. In the first kind, the character shapes are moulded onto the surface of a drum, belt, ball or wheel. These are struck by an electromagnetically driven hammer to press an inked ribbon onto the paper. In the line printer, the characters for a whole line of text are printed in one rotation of the drum, which has a complete set of characters round its circumference at each position along the line. The correct character is printed at each position by pressing the paper from behind at the correct time as the drum rotates. The characters can be produced similarly using a belt driven between two wheels. These devices are capable of high-speed operation.

In ball- or wheel-type printers the ball or wheel moves relative to the paper, striking each character individually. The ball must be rotated around two axes to select the correct character, the wheel around only one. These devices can produce high-quality output, similar to a modern office typewriter.

243

The second type of printer mechanism constructs each character from a matrix of dots. These can be produced by striking an inked ribbon with a column of electromagnetically driven needles which move across the page, or by a column of electrically heated pads which cause heat-sensitive chemicals to darken in special paper. These types of printer are capable of reasonable quality at a low price.

Very-high-quality printing can be done using the dot-matrix method by using a laser beam to scan across a light-sensitive drum similar to that used in photocopying machines. The beam is modulated to produce an image in a similar way to a television picture. Dot resolutions of the order of several hundred dots per inch can be achieved with this method.

Plotters are used to produce graphical hard copy. They can move a pen over the surface of a sheet of paper attached to a drum or flat plate. Motors are used to move the pen and/or the paper in two dimensions, and to lift the pen from the paper. The position of the pen can be controlled accurately by the computer to draw a picture as a series of straight lines.

Printing and plotting devices are normally equipped with parallel or serial interfaces with handshaking control. Data is sent in encoded form (usually ASCII characters) over the interface. Some printer mechanisms are designed to be integrated directly into a microcomputer system without an intermediate interface. These allow the MPU to control directly the paper or print-head movement and printing operations.

7.4 Data communication and storage

The data which is in a microprocessor system must often be stored for future use or communicated to another microprocessor. Specialized interface devices are available for each type of storage device or communication line.

7.4.1 Data storage

Data can be stored on magnetic disks or tapes as explained in Chapter 6. Large tape storage devices often have one read/write head for each data bit; hence a parallel interface device can be used. In smaller tape storage devices and disk storage devices, the data is recorded serially using a single read/write head. A clock signal is encoded with the data, hence the interface device must be a synchronous serial one.

A disk-drive interface such as the uPD765 from NEC includes a serial interface, along with logic to control signals for moving the head from track to track and to sense the positions of the sectors as the disk rotates. This particular device is intended for use with floppy-disk drives which have a lower data-transfer rate than hard-disk drives. Most of the functions such as reading or writing a block of data and formatting a disk can be carried out by the interface while the microprocessor does another task.

7.4.2 Data communication

If we wish to send data from one microprocessor to another, the easiest way is to connect together two parallel ports so that data written to one can be read from the other. Handshaking can be used to control the rate of transfer.

If the microprocessors are a longer distance apart, the cost of cable for parallel connection becomes large, making serial data transmission more economical. Unfortunately, long-distance communication lines cause problems since they introduce errors and transmission delays into the system. For these reasons and others, information is normally sent in blocks known as *packets*. Each packet has an error-detecting code which can be checked by the receiver. If a packet contains an error, it is discarded by the receiver. The receiving station then transmits a request for a repeat of the packet until it receives it correctly.

If a packet is lost altogether (perhaps because its framing code was corrupted) then the receiver might receive the next packet correctly without realizing that data has been lost. One way to prevent this is to require the receiving station to acknowledge each packet by transmitting a message indicating whether a retransmission of the previous packet or the next packet is required. This is fine on short links, but the delays caused by long-distance links (especially satellite links) cause a long delay between packets. An alternative way is to insert a sequence number in each packet which is incremented for each new packet. A missing packet can then be detected by a break in the sequence at the receiver. The acknowledge messages from the receiving station can then include the sequence numbers of the packets to which they refer.

The sequence number uses up space which could be used for data; so it must be limited in range. This means that we will run out of sequence numbers sooner or later in a long transmission. It is possible to use old sequence numbers again, once the packets which contained them have been correctly received. The sequence numbers are therefore used cyclically, with the largest being followed by the smallest again. The transmitter must be careful not to re-use a sequence number until it is certain that the packet which last used it has been acknowledged as correctly received. This prevents the receiver from interpreting a new packet with a re-used sequence number as a retransmission of an old packet (the retransmission request may not have reached the transmitting station due to the link delay).

Since many links carry data in both directions, the acknowledgement and retransmission requests can be inserted into packets which are being transmitted anyway. At the start of a communication, special packets must be used to initialize the sequence counters and identify the type of message which will be sent. Special packets are also used to terminate or interrupt transmissions.

If packets are lost, a situation may arise where both ends of the communication link are waiting for a response from the other end. This deadlock can be broken by starting a timer at the transmitter. If the packet has not been acknowledged when time-out occurs, the packet can be re-transmitted.

In a network, several computers can be connected together so that they share the same communication line. This requires that packets contain address information which specifies their destination, so that other receivers can ignore it. Information about the source of the packet must also be included at the start of the communication so that the called machine can send acknowledgements and data to the correct place.

Sharing of a line can cause problems if several machines wish to communicate at the same time. The packets can be sent by each machine to a central network processor on separate lines so that they can be distributed to the correct destinations. The network processor must then have an interface for each machine which connects to it, and it must be re-programmed every time a machine is added, moved or removed. It may be a significantly expensive device and, moreover, it will stop the whole network from functioning if it breaks down.

A more flexible system can be devised in which each machine transmits packets on the same cable. If a station wishes to transmit, it first listens to its receiver to make sure that the line is not busy. This is called CSMA (carrier-sense, multiple-access) operation. If two packets are transmitted simultaneously, they will be corrupted and errors will be detected at the receiver. The non-arrival of the data will cause a re-transmission after some time has elapsed. A quicker way to do this is for the transmitter to monitor the line for other transmissions which *collide* with its own. This mode of operation is called CSMA-CD (CSMA with collision detection). If a collision occurs, the transmitter aborts immediately, since there is no point in sending the rest of the message. The Ethernet® (IEEE 802.3) standard network operates on this principle. Manchester-encoded data at 10 Mbits per second is sent from each station on the same coaxial cable.

Collisions can be avoided altogether if a token-passing method is used. The token is a message which gives the recipient permission to transmit a packet of data. It must then send the token to the next machine in the chain, which can then transmit and pass on the token. Eventually, the token passes round all the machines and repeats the circuit continuously. If a machine has no data to send, it simply passes on the token. If the token is corrupted and lost, the network will be idle until a new token is generated. If several machines spot the missing token and try to generate new tokens, then the multiple tokens must be detected and the token re-generation must start again. Token-passing is normally used on ring networks where there is a natural choice for the next machine to receive the token, namely, the next machine on the ring.

7.5 Sensors and actuators

When a microprocessor system is required to control or monitor (for example) an industrial process, it must be able to convert between digital data and the process parameters of temperature, pressure etc. Sensors are needed to measure parameters and actuators allow the microprocessor to control them. Some parameters may be analog in nature: for example, the speed of rotation of a shaft can take on any value in a continuous range between some limits. Conversion from analog to digital values will be required if this data is to be processed by a microprocessor. Since a digital value can only increase by steps, there is some loss of precision in the conversion. For example, if a voltage between 0 and 7 V is converted to a binary number between 000 and 111, then 2.2 V will appear as the same number as 2.4 V, namely 010.

Some sensors and actuators can produce digital values directly: for example, the number of units manufactured on a production line must be a positive integer which can be measured by a counter. There is no such thing as "half a unit" for example. This means that the microprocessor does not lose precision in processing the information.

We use a digital interface device where a quantity can be represented digitally and an analog-to-digital converter (ADC) or digital-to-analog converter (DAC) where the value of the signal can vary continuously over a range. The number of bits processed by the converter (i.e. the precision) depends on the application; the more bits used, the higher the precision.

7.5.1 Analog sensors

TEMPERATURE MEASUREMENT

Temperature can be measured by converting it to a voltage or current, using the fact that the resistivity of certain materials is strongly dependent on temperature. A resistor made from such a material will have a resistance which varies according to the temperature. Unfortunately, this is not a very linear variation. If two different metals are joined to make a thermocouple, a voltage will appear between them. This voltage depends linearly on the temperature, although it is quite small (approximately $50 \mu V$ per degree Centigrade). The voltage across a forward-biased semiconductor diode is an exponential function of temperature, but by using some additional circuitry, it is possible to use it to produce a voltage which is linearly proportional to absolute temperature (PTAT). The whole circuit can be integrated onto a single chip, making a PTAT sensor with an output voltage which is accurately proportional to the temperature in kelvins.

STRAIN GAUGES

When a force is applied to a material, it expands or contracts. This can be measured by attaching a thin metal resistor to the material so that as it stretches: its length will increase and its cross-sectional area will decrease; hence its resistance will increase. Conversely, if it shortens, its resistance will decrease. Two or four strain gauges are normally used together in a voltage divider or bridge circuit so that changes in temperature do not affect the output signal.

247

PRESSURE MEASUREMENT

Pressure in liquids and gases can be sensed by applying the pressure to a flexible diaphragm. There will be a force on the diaphragm which is proportional to the pressure difference between the two sides. The strain caused by this force can be measured as described above. The strain gauges can be integrated onto a single chip which also acts as the diaphragm, provided that the liquids or gases in use are not likely to damage it.

LIQUID-LEVEL AND FLOW-RATE MEASUREMENT

A simple way to measure liquid level is to measure the position of a float using a displacement transducer. The level could also be detected by the resistance change between two electrodes when a conductive liquid covers them. If the liquid is non-conducting, a capacitance change may be detectable when it enters the space between the plates of a capacitor. A proximity detector may be used to sense the position of the liquid surface.

Liquid flow may be measured by inserting a constriction in the path of the liquid. The pressure difference across the constricted section will be proportional to the flow rate and can be measured by a pressure transducer. Alternatively, we can insert an impeller in the flow so that it rotates at a rate proportional to the flow rate. The rotation rate can be measured by a magnetic or optical method so that no contact between the transducer and the fluid is required.

ULTRA-SONIC POSITION MEASUREMENT

High-frequency sound waves (known as ultra-sound) can be reflected by objects to provide an indication of their presence. If the delay between transmission and reception is measured, the distance can be calculated if the speed of sound is known. If the beam is directional, the information can be used to collect information on the position of several objects by scanning the beam.

Ultra-sound can be produced by applying an alternating voltage across a piezo-electric ceramic or crystal plate. The electric field will cause the molecules in the plate to vibrate at the frequency of the input voltage. If the plate size is chosen so that it resonates like a bell at this frequency, then a large ultra-sound output can be obtained. Detection of ultra-sound can be accomplished similarly, since a vibration of the plate will cause a voltage to appear across it. This can then be amplified and processed.

LVDT DISPLACEMENT TRANSDUCERS

The linear variable differential transformer (LVDT) is used to sense position in one dimension. It is a transformer which is wound on a tube containing a moving magnetic core. Each end of the tube is wound with a primary coil so that the magnetic fields from the two coils will act in opposite directions. These are driven by an a.c. signal which is then coupled through the moving core to a secondary coil which is in the center of the tube. If the core is in the center, the signals from the opposite ends of the tube will cancel out. If the core moves to one side, the signal from that side will increase and will be proportional to the displacement. If the core moves in the opposite direction, the signal will increase, but with the opposite polarity. The size of the displacement can be detected by rectifying and smoothing the signal from the secondary coil. The direction can be found by comparing the phase of the signal with that of the signal driving the primary.

PROXIMITY DETECTION

Frequently, it is necessary to sense the presence of an object without physical contact. Proximity sensors produce an output signal when an object is near. If the object is conductive or ferromagnetic, the sensor can operate by detecting the change in the properties of an inductor when the object enters its magnetic field. A capacitive sensor can detect objects which affect the electric field around it. If the object blocks or reflects light or infra-red radiation, an optical detector can be used. These include a light source and a photodiode or phototransistor. When light reaches the photosensitive element, it passes a current. When the light level decreases, the current falls. Reflective or opaque markers may be attached to objects to provide a good signal.

ELECTRICAL VOLTAGE AND CURRENT

When a large voltage or current is to be measured, it must be reduced in size so that it is within the range of the ADC. Resistors can be used to reduce voltages by dropping the excess voltage across a large series resistance. If an alternating voltage is being measured, a transformer with a large turns ratio can be used to reduce the voltage.

Current can be measured if we convert it to a voltage by passing it through a resistor. If a small resistor is used, a large current will produce a small voltage. If the current is a.c., a current transformer can be connected in series with the circuit. The current will be reduced according to the turns ratio. (Note that a current transformer has more turns on the secondary than on the primary for this purpose.) It may not be convenient to break a circuit to insert a series resistor or current transformer. Some current transformers have magnetic cores which can be split so that they can be clamped round a wire without breaking it. This can also be used to measure direct current if the magnetic field in the core is sensed by a Hall-effect transducer.

VOLTAGE ISOLATION

Isolation must be used between signal sources which are at a high voltage and microprocessor interfaces. Opto-isolators can provide the isolation, but their outputs are not a linear function of their inputs. This can be overcome by using some extra circuitry or software to cancel the non-linearity, or alternatively the signal can be converted to a frequency or pulse width, since this will be unaffected by the opto-isolator. Once the signal has been converted in this way, a transformer can also be used to provide the isolation.

Since the isolated input-transducer circuit may require electrical power, this can be provided by a transformer-coupled a.c. power source. The a.c. source can be generated from a d.c. power supply by an oscillator if necessary. The a.c. power can be converted back to d.c. by diode rectification and smoothing capacitors. The oscillator will usually operate at a high frequency since this means that the transformer and capacitors can be made smaller.

7.5.2 Digital sensors

POSITION ENCODERS

The position of a shaft can be sensed digitally by attaching a disc which is marked with light and dark bands. These can be sensed by transmitting or reflecting a light beam to a photosensitive detector. If the pattern is simply alternate light and dark bands, the position can be sensed by counting the number of pulses from the detector to determine how far the shaft has rotated. If the direction can be reversed, two detectors are needed. The direction can be sensed from the order in which pulses are received by the two detectors. If absolute positioning is required, a third detector can be arranged to sense a pattern which produces only one pulse per rotation of the shaft.

Alternatively, the bands may be arranged in the form of a code where dark = 0 and light = 1. Each detector senses a pattern which corresponds to one bit of a word which is an encoded shaft angle. A simple binary code is not sufficient here, since the bits may not change simultaneously as the shaft rotates. This would cause the position code to be meaningless in the transition region between codes. In a Gray code only one bit changes between adjacent codewords; hence this type of code is used instead of binary.

The same principles can be applied to the measurement of distance in a straight line by using a long patterned strip instead of a wheel.

SPEED MEASUREMENT

Speed can be measured by recording the rate of change of the output from a position encoder. The simplest type of encoder can be used, in which a fixed number of pulses is produced for each unit of distance. A counter/timer interface can be used to measure the pulse rate as described previously.

A magnetic sensor may be used to measure the speed of a toothed steel gear wheel. A coil placed near the teeth in the presence of a magnetic field will generate a voltage pulse every time one tooth passes it. This type of sensor is very robust and cannot be obscured by dirt and grease in the same way as an optical sensor. One disadvantage is that the output pulses are very small at low speeds.

7.5.3 Analog actuators

ALTERNATING-CURRENT AND DIRECT-CURRENT MOTOR DRIVES

An analog signal may be used to control the power supplied to an electric motor to achieve speed or position control. In the case of a d.c. motor, the current in the windings must be varied; a.c. motors also require the supply frequency to be controlled. Power semiconductor switching devices can control these values under control of the MPU. The output speed and position will not only depend on the power fed to the motor, but also on the mechanical inertia of the load, friction, external forces and so on. Speed and position information can be fed back from sensors to the MPU and used to ensure that the motor is supplied with the correct power to maintain the correct speed and position. Care must be taken that this feedback loop does not cause oscillations, since this is a well-known problem in feedback control systems.

SERVOMECHANISMS

If the feedback of position and speed information controls the motor directly by analog means without going through the MPU, then the actuator is a servomechanism. The MPU needs only to produce a voltage proportional to the required speed or position, and the servomechanism will take care of the rest.

7.5.4 Digital actuators

Some types of actuator are digital in nature. These are more easily interfaced to a microprocessor, since no analog-to-digital conversion is needed.

POWER SEMICONDUCTOR SWITCHES

Power transistors can be used to switch large currents on and off to control various devices. They can operate in a linear mode where the output current is proportional to the input. When the output current is large, the transistor must dissipate a large amount of power as heat when it is used in this way, since it carries the same current as the load and may also have a large voltage across it. If the transistor is used in a switching mode, the transistor will have a large voltage across it when no current flows; hence little power is dissipated. When it is fully switched on, a large current will flow through it but there will be a small voltage across it; hence again little power is dissipated. Transistors operate most efficiently when switching in this way, provided that they are controlling loads which do not require continuous current.

Care must be taken when switching current to devices which are inductive, such as motors or other electromagnetic devices. When the transistor switches off, a large reverse voltage can appear across it as the current in the inductor falls. This may damage the transistor; so a diode is usually placed across the inductor to short-circuit this reverse voltage.

251

RELAYS

Electromagnetic relays are switches which are operated by an electromagnet. A parallel-port output line is not usually powerful enough to control the current for the electromagnet, but a single transistor can provide the necessary amplification. The relay-switch contacts can control relatively high voltages and currents. Relays are, however, likely to suffer problems with contact corrosion and wear.

A solid-state relay has no moving parts. It contains a light-emitting diode which can be driven by a logic signal and a light-sensitive power semiconductor which is activated by the LED. Since the light passes through an insulating separator, the logic signal driving the LED is isolated from any high voltages on the semiconductor switch.

SOLENOIDS

A solenoid is an electromagnet coil which can exert a force on a magnetic core, pulling it into the coil. The position of the core is either in or out: it is not easy to hold it in between. A typical use of a solenoid is to withdraw the bolt in a lock.

STEPPING MOTORS

The stepping motor is an extension of the concept of the solenoid in which several coils are used to cause a toothed magnetic wheel to rotate in steps. The current to each coil is switched in sequence to produce rotation in the desired direction, or to hold the rotor solidly in position. It cannot exert a large force, but it can position the shaft accurately. This is ideal for use in equipment which requires position control, such as a plotter.

7.6 Analog interfaces

When analog sensors and actuators are connected to a microprocessor, conversion must take place between analog and digital signals. This process is not always straightforward, however, and we will now examine some of the problems of interfacing these devices in more detail.

7.6.1 Analog signals

SAMPLING AND ALIASING

When a microprocessor operates on an analog signal, it must sample the value at separate intervals in time. The rate at which the samples are taken limits the accuracy with which operations can be carried out on the signal. If the signal changes too fast, the sampled version may bear little relation to the real signal as shown in Figure 7.22. The analog signal shown is a sine wave. As the number of samples per cycle of the sine wave decreases, the version

252

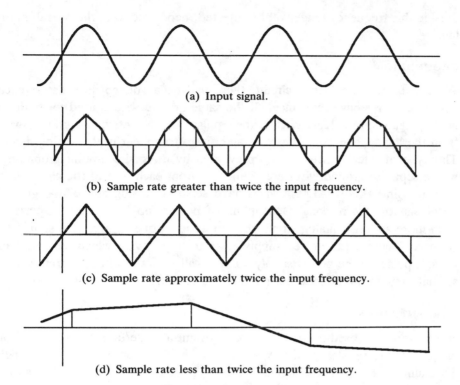

(a) Input signal.

(b) Sample rate greater than twice the input frequency.

(c) Sample rate approximately twice the input frequency.

(d) Sample rate less than twice the input frequency.

Figure 7.22. Aliasing of sampled signals.

reconstructed from the samples becomes less like the original. In fact, the sampling rate must be at least twice the frequency of the maximum frequency of the input signal. This is known as the Nyquist limit.

If an input signal with a higher frequency than half the sampling rate is applied, it will appear as if it is a low-frequency signal, as can be seen from our sine-wave example. This effect is known as aliasing. It can be prevented by using an analog filter to remove frequencies higher than the Nyquist limit before sampling takes place.

NOISE

Signals from the analog world are usually accompanied by noise. This is generated by a number of sources, such as thermal motion of atoms in the transducers and amplifier circuits. Noise is a signal with a randomly varying amplitude which often has an average value of zero if considered over a long time interval. This means that it if it is present in addition to a wanted signal, it can be reduced by taking the average value of the noisy signal over a number of samples. If the average value is non-zero but known, this offset can be subtracted after averaging. This offset may drift; so we must update its value periodically by re-calibration using a known input signal.

Another feature of noise is that its signal power is spread over a range of frequencies. If the frequency range of the wanted signal is restricted, then noise can be reduced by using a filter which only passes this range. Noise outside this frequency range will be rejected, hence reducing the overall noise level.

GROUNDING

When a current flows in a circuit, there will be a voltage drop across each wire since wires have resistance. If the same ground wire is used for both an analog signal and a logic circuit, then pulses of current in the ground wire from the logic circuit may cause voltage pulses to appear in the analog signal. This type of interference may be prevented by using single-point grounding, where separate ground wires are connected from each part of the circuit to a central point. Many ADC devices have separate analog and digital ground terminals for this reason. Decoupling of power supplies is also important, even more so in an analog circuit than in a digital one. Capacitors should be connected across the power supply near to any device which can produce current pulses in the power supply. These will smooth out any current surges so that they do not affect other parts of the circuit.

CURRENT LOOPS

If long wires are used between a transducer and its interface, the voltage drop caused by the resistance of the wires may be large compared to the signal. To maintain accuracy, the signal can be converted to a current which is proportional to the transducer output. The interface can convert this current back to a voltage by passing it through a resistor. Provided that the transducer current source can provide enough power to compensate for the losses in the wires, this will eliminate the voltage-drop problem. (This works because the proportion of voltage lost by resistance effects is greater than the proportion of current lost by leakage, since wire insulation is normally very good).

DIFFERENTIAL SIGNALS

Problems can arise when a signal must be sent between places which do not share a common ground voltage. The ground voltage in separate buildings is normally slightly different, since earth has a relatively high resistance. If a single wire is connected between such places, the signal voltage relative to ground will be different at each end. Connecting a ground wire will not help, since current will flow in it and cause an interference voltage in the signal. This problem can be overcome by using two wires to send the signal. The signal is sent as a voltage difference between the two wires. At the receiving end, a difference amplifier is used to recover the signal. Although the voltage on each wire will be different relative to ground at each end of the wire, the voltage difference will remain constant. Any interference signal which affects both wires will be cancelled by this method, so the wires are normally twisted together to make sure that any external electric or magnetic field affects both wires equally.

254

Damage can occur to interface circuits which are directly connected to equipment which can generate or transmit electrical energy. For instance, the electric utility supply can contain pulses which are generated by other equipment attached to the same distribution system. Long communication cables may pick up energy induced from nearby power cables, or even lightning strikes. Damage can be prevented by using transient-protection circuits between the microprocessor interface circuits and the external equipment. Simple resistor-capacitor filters will absorb very short pulses. Longer pulses can be restricted by using silicon diodes which turn on when their forward voltage exceeds approximately 0.6 V. Zener diodes can be used for larger voltages. Special resistors can be obtained which have a high resistance at low voltages and a low resistance when the voltage exceeds a certain limit. If these are connected across the input to an interface, any high-voltage pulses will be short-circuited to ground.

If the transient problem is a serious one, a transformer or opto-isolator may be required.

OPERATIONAL AMPLIFIERS

Analog signals may require some processing before conversion. Most of these functions can be carried out by a high-gain differential amplifier (known as an operational amplifier) with a few external components. Small signals may be amplified by the circuit shown in Figure 7.23, where the voltage amplification is set by the ratio of two resistances. Similar circuits can be made to carry out the operations of filtering, rectification, integration, differentiation and so on.

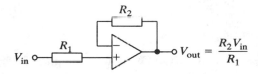

Figure 7.23. Operational amplifier.

ANALOG MULTIPLEXING

A single ADC or DAC can operate on several analog signals if it is connected to an analog multiplexer or demultiplexer as shown in Figure 7.24. The analog signals are switched by field-effect transistors (FET) such that only one is connected at a time.

The output voltage level is held by the charge stored on a capacitor in between output samples. When an output sample arrives, the analog switch turns on, allowing the DAC output to charge the capacitor to a new voltage level.

255

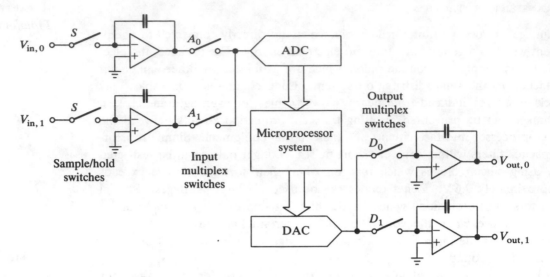

Figure 7.24. Analog multiplexing and sample/hold circuits.

The input channels are sampled one at a time in sequence. If it is important to sample all the inputs simultaneously, their values at a particular moment can be stored on capacitors in a *sample-and-hold* circuit and then conversion can take place for each signal in turn.

7.6.2 Digital-to-analog conversion

Digital signals can be converted into analog signals by a variety of methods depending on the precision, speed and cost requirements. Usually, the output analog signal is a voltage.

WEIGHTED RESISTORS

The digital-to-analog converter (DAC) shown in Figure 7.25 produces a voltage which is proportional to the binary number presented at its inputs.

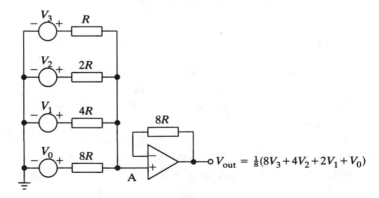

Figure 7.25. Weighted-resistor DAC.

256

We assume that each bit of the word is represented by a voltage which is switched between two precise values to represent logic 0 and 1. These voltages can be converted to currents by connecting a resistor to each bit as shown. Since the bits in the binary number are weighted in the ratio $1, 2, 4, 8, \ldots$, etc., the currents are weighted in the same way by using resistance values in the ratio $1, \frac{1}{2}, \frac{1}{4}, \frac{1}{8}, \ldots$, etc. These currents are summed at point A in the circuit, where they feed into a common return circuit. The current sum here is thus proportional to the binary number at the input. The current can be converted to a voltage by feeding the currents into a low-impedance amplifier input.

VOLTAGE DIVIDER

An alternative DAC is shown in Figure 7.26.

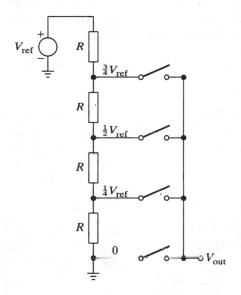

Figure 7.26. Voltage-divider DAC.

Here, a reference voltage is applied to a chain of resistors which causes the same current to flow in each resistor. Since the resistors all have the same resistance, the voltage across each is also the same. A set of electronic switches (i.e. transistors) is used to select one tap to connect to the output. The input binary number is decoded by some logic to determine which switch should be turned on.

R-2R

The DAC shown in Figure 7.27 uses only two values of resistance, namely R and $2R$, in the form of a ladder network. In the weighted-resistor network above, the current for each bit is determined by the size of a resistance. In this network, the current from a particular bit is divided by two at each stage

257

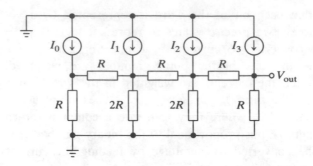

Figure 7.27. *R-2R* DAC.

along the ladder by splitting it into two equal parts, one of which is returned directly to ground, the other passed to the next stage. The bit currents are therefore divided by a power of two which is dependent on the number of stages from the end of the ladder. This provides the correct weighting so that the current sum is proportional to the binary number. This can be fed into an amplifier as before.

CAPACITOR TRIMMING

Some high-precision DACs use capacitors to store the reference voltage for each bit. This means that the value can be trimmed by a control circuit to maintain the precision of the converter during drifts in temperature, supply voltage and so on.

MARK/SPACE RATIO AND PHASE CONTROL

If a single bit is pulsed alternately to logic 1 and 0 at a fixed repetition rate, the average voltage of the output will vary according to the time spent at 1 compared to that at 0, i.e. the mark/space ratio. The pulses can be smoothed out by an integrator circuit as shown in Figure 7.28. The output voltage will be proportional to the mark/space ratio of the pulses, provided that the time constant of the integrator is sufficient to smooth out the individual pulses. It is easy to write a program to generate these pulses using a single parallel-port output bit. Alternatively, a programmable-timer device may be used.

If the pulse-repetition frequency is twice that of an a.c. power source such as the electric utility supply, the power supplied to a load can be varied by using a thyristor or triac device as shown in Figure 7.29. The thyristor only conducts current in one direction; the triac can conduct in either direction, since it essentially contains two thyristors back-to-back. It follows that a thyristor can control only half of the power available from an a.c. supply. Each device is normally non-conducting until a small current pulse is received at its trigger electrode. The device then turns on like a switch and can only be turned off by removal of the supply voltage. This happens naturally at the end of each half-cycle of an a.c. supply.

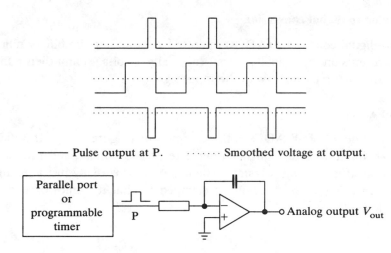

—— Pulse output at P.   ······ Smoothed voltage at output.

Figure 7.28. Pulse-width modulation DAC.

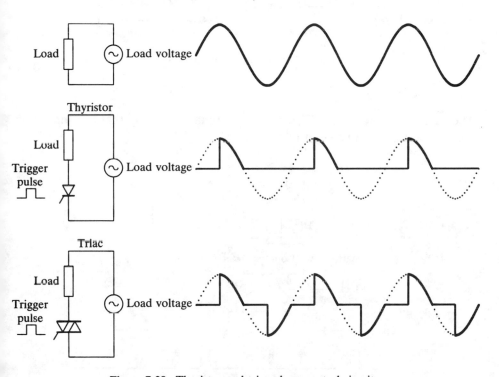

Figure 7.29. Thyristor and triac phase-control circuits.

By triggering the triac at different times during a half-cycle, the width of the current pulse fed to the load can be varied. Normally, we rely on the inductance and/or thermal and mechanical inertia of the load to integrate the pulses. It should be noted that some form of isolation is required between the high-voltage triac terminals and the low-voltage control circuit to prevent damage and electric-shock hazards.

7.6.3 Analog-to-digital conversion

Analog-to-digital converters (ADCs) operate by estimating the binary number which represents an input analog signal (normally a voltage) and then refining this estimate by comparing it with the input signal.

THE COMPARATOR

A basic component of all ADCs is the comparator (Figure 7.30). It indicates whether an input voltage is higher than a reference voltage and produces a logic 1 output if it is, a 0 if it is not. The comparator is really just a high-gain amplifier which has its output voltage clamped so that it cannot exceed normal logic levels.

Figure 7.30. Comparator operation.

RAMP CONVERSION

The ramp-converter ADC (Figure 7.31) estimates initially that the input voltage is the minimum value of its range.

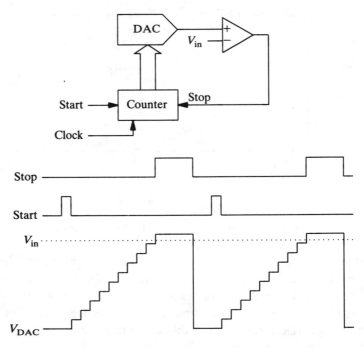

Figure 7.31. Ramp ADC.

The estimate is normally stored in a counter which allows it to be reset and incremented. When the conversion is started, the counter is reset to zero to produce the minimum output voltage from the DAC. The comparator output indicates whether the estimate is greater or less than the input value. If it is less, the counter is incremented by one and the process is repeated. This continues until the comparator indicates that the estimate is greater than the input or the maximum value of the range is reached. The final estimate is then read from the counter.

Alternatively, the ramp voltage can be obtained by an analog ramp generator which is synchronized to the counter. This allows precise conversion without needing an expensive DAC with many bits.

TRACKING CONVERSION

A tracking converter (Figure 7.32) is very similar to a ramp converter except that the initial estimate is the final estimate from the previous conversion and the estimate is incremented or decremented at each step depending on whether the estimate is less or greater than the input.

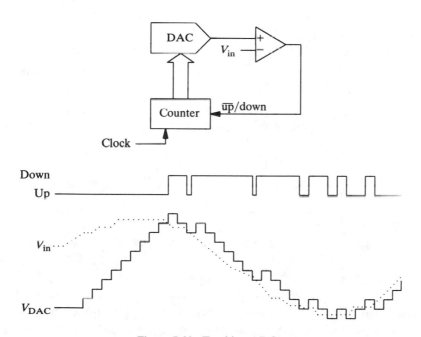

Figure 7.32. Tracking ADC.

The initial behaviour of this converter is similar to the ramp converter, since no previous conversion value is available. After this, provided that the input signal does not change by an amount equivalent to more than one step, the estimate will *track* the input signal continuously. If the input is constant, the estimate will oscillate by one step around the correct value since the counter must always increment or decrement at each step.

SUCCESSIVE APPROXIMATION

The successive-approximation ADC uses an initial estimate of one half of the full conversion range. The comparator then indicates whether the input value is in the upper or lower part of the range. If it is in the lower part, the new estimate is one quarter of the full range (the center of the lower half) and if it is in the upper part, the new estimate is three quarters of the full range (the center of the upper half). This process is then repeated until the estimate is close enough to the input value. This is a fast method, needing only 8 clock cycles for an 8-bit conversion. The logic, though, is more complex than a counter.

FLASH CONVERSION

The flash converter uses estimates of all possible values at the same time. This requires one comparator for each possible value. The input voltage is compared with reference voltages from a tapped voltage divider as shown in Figure 7.33.

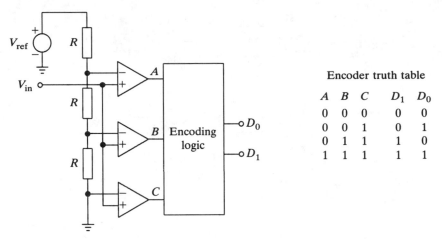

Encoder truth table

A	B	C	D_1	D_0
0	0	0	0	0
0	0	1	0	1
0	1	1	1	0
1	1	1	1	1

Figure 7.33. Flash ADC.

All the comparators which are comparing with a reference voltage below the input value will produce a 0 output, all those above will produce a 1. Some encoder logic is necessary to convert this set of outputs into a binary number.

Flash conversion is very fast as its name suggests, although it uses a lot of comparators which limits the number of bits which can be converted. Television-picture signals need a high sampling rate, and flash converters are ideal for converting them to digital form.

VOLTAGE-TO-FREQUENCY CONVERSION

If an analog signal changes slowly enough, it can be measured by using it to control the frequency of an oscillator. The oscillator period or frequency can then be measured by a counter/timer interface or by a parallel port with a software timing routine. The oscillator which performs the conversion can be placed some distance away from the interface, since its output signal is effectively a digital one and hence not susceptible to interference and noise in the same way as an analog one. It is also easier to process such a signal if voltage isolation is required, since a simple opto-coupler or transformer may be used.

An example of such a circuit is shown in Figure 7.34, where a low-cost timer IC is used to produce an output square wave whose period depends on the resistance of a thermistor. This allows the MPU to measure the temperature by comparing the period with a calibration table. Since no ADC is used, the circuit is small and of low cost.

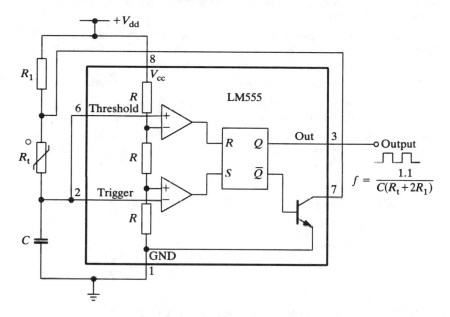

Figure 7.34. Voltage to frequency converter.

7.7 Summary

There are basically three parts in an input/output system: the interface device, the peripheral device and the software driver. In this chapter, we have covered a broad range of each of these, but the list cannot be exhaustive. It is hoped, however, that the general principles of operation of those covered will be applicable to a much wider range of devices.

7.8 Further Reading

1 *Binary Synchronous Communications—General Information* (GA27-3004, File TP-09), IBM Corporation.
2 Craine, J. F. and Martin, G. R. (1985). *Microcomputers in Engineering and Science*, Addison-Wesley, Wokingham, England. An introductory text on interfacing microprocessors, including details of transducers and analog interfaces.
3 McNamara, J. E. (1982). *Technical Aspects of Data Communication*, Digital Equipment Corporation, Massachusetts.

4 Metcalfe, R. M. and Boggs, D. R. (1976). ETHERNET: Distributed packet switching for local computer networks, *Communications of the ACM* **July**, 395–404.

5 *Microcomputer Components*, Motorola Inc., 1979.

6 *Microprocessors and Peripherals*, NEC Electronics, 1986. Some useful reference data on peripheral-interface ICs. Other manufacturers' data books would be useful too, but are too numerous to mention.

7.9 Problems

7.1 Design an interface for the keyboard matrix shown in Figure 7.16. Write an assembly-code subroutine to scan this keyboard and return a value in a register between 0 and 15 corresponding to a key which has been pressed.

7.2 Write an assembly-code subroutine which calls the keyboard routine in problem 7.1 and debounces the output.

7.3 Design an interface for the multiplexed seven-segment LED display shown in Figure 7.20. Write a subroutine to initialize the display and one which may be called from a timer-generated interrupt routine to scan the digits.

7.4 Write a program to measure the width of a pulse on one bit of a parallel input port.

7.5 Design an interface for the dot-matrix display shown in Figure 7.19. Write a program to display the letter "A" on it.

7.6 Given the definition of the registers in a serial interface below, write a pair of subroutines which will send and receive a block of data over this interface. Remember to include error-detection features.

INTERFACE-REGISTER DEFINITIONS

| Address | Register | Contents | | | | | | | |
---------	----------	D_7	D_6	D_5	D_4 D_3 D_2	D_1	D_0
F000	Transmit	←	–	–	data	–	–	→
F001	Receive	←	–	–	data	–	–	→
F002	Control	TIE	RIE	0	0 0 0	0	0
F003	Status	RIN	TIN	0	0 0 0	RRF	TRE

CONTROL/STATUS-BIT DEFINITIONS

TRE. When TRE = 1, the transmitter register is empty: data may be sent by writing to it. This will cause TRE to go to 0 until the data has been sent.

RRF. When RRF = 1, the receiver register is full: data has been received and may be read from it. After the data has been read, RRF will go to 0.

TIE, RIE. To enable interrupts when TRE goes to 1, TIE must be 1. Similarly, to enable interrupts when RRF goes to 1, RIE must be 1.

TIN, RIN. When an interrupt is generated by the serial interface, TIN will be 1 if TRE generated the interrupt and RIN will be 1 if RRF generated it.

7.7 An unshielded cable operating at a high voltage has its temperature measured by a thermistor. Show how this information could be fed to a microprocessor.

7.8 A stepping motor has four windings labelled A, B, C and D. The current in the windings must be switched as follows to make the motor rotate.

Step	Winding			
	A	*B*	*C*	*D*
0	off	on	off	on
1	on	off	off	on
2	on	off	on	off
3	off	on	on	off
4	(repeat from step 0)			

Design an interface for this motor and write a subroutine which will move it through a given number of steps at a given stepping rate.

7.9 Write a program which will display the time of day on the screen of a terminal, given that a periodic interrupt signal is available.

7.10 Explain the difference between port-mapped and memory-mapped I/O.

7.11 Using the serial interface described in problem 7.6, write an interrupt-driven routine which will:

(a) read a character from the serial input when the receive register is full and store it in a queue;

(b) take a character from a queue and send it on the serial output when the transmit register is empty.

Write subroutines which will:

(c) take characters from the receive queue, returning an error code if the queue is empty;

(d) place characters in the transmit queue, returning an error code if the queue is full.

7.12 Draw a circuit for an *R*-2*R* DAC. Derive the output voltage for the circuit in terms of the current-source inputs.

7.13 A microcomputer has a DAC attached to an output port. The DAC output is compared with an external input voltage in a comparator, the output of which is attached to an input port. Write subroutines to implement analog-to-digital conversion in software using the following algorithms:

$$\text{(a) ramp ;} \qquad \text{(b) tracking;} \qquad \text{(c) successive approximation.}$$

7.14 Design the encoder logic circuit required for a flash ADC to produce a 2-bit output.

7.15 Consult data sheets on commercial interface devices and discuss how the internal registers are arranged.

7.16 Design a circuit which accepts 4 interrupt inputs and drives a single interrupt-request input of a microprocessor. The circuit should also latch a number between 0 and 3 which indicates the level of the interrupt. The code should be that of the highest-numbered active input line, all lower inputs being ignored.

Write an interrupt-service routine which processes interrupts from this circuit, choosing an appropriate service subroutine from a table according to the level of the interrupt.

7.17 Data is to be exchanged between two microprocessor systems using an 8-bit bus with a handshaking protocol. Design such an interface and write subroutines to transfer bytes of data on the link.

8: High-level Languages

8.1 Introduction

Programming in assembly language has several disadvantages. Although it can produce code which is fast and takes up little memory, it is difficult to write large programs to perform complex tasks. Long sequences of instructions may be required to perform relatively simple tasks, making the code slow to write and difficult to debug.

Another problem with assembly language is that it is usually specific to a particular microprocessor. If we are writing a program which must run on a variety of different processors, a great deal of effort may be needed to re-code and debug each implementation.

A high-level language is designed to overcome these difficulties by allowing the programmer to describe complex actions in a concise and readable way. A compiler program then translates this into machine code for the required microprocessor. A high-level language provides a higher level of abstraction from the hardware details than assembly language. Programmers do not need to know about the number of registers and so on: the compiler takes care of the detail at this level. As a result, programs written in this way are more portable than assembly-code programs; we can run the same program on several different microprocessors by using different versions of the compiler to translate the program.

Two high-level languages will be studied in this chapter, namely Pascal and C, with the emphasis on Pascal, since C can be rather cryptic to read. In the limited space available, we can do no more than get a rough grasp of the features of these languages, and the reader is encouraged to study one of the numerous textbooks on these languages for further information.

8.1.1 Compilers and interpreters

A language may be compiled directly into machine code, resulting in a fast execution time. The penalty is that the compilation process may take some time. Alternatively, a language may be interpreted. Interpreters read the source code of the language one statement at a time, executing it as they go. We never actually get a machine-code representation of the program: it is executed by calling a sequence of interpreter routines.

Interpreters have the advantage that the user can find errors in the program more easily, since the source code which produced the error can be immediately identified. Unfortunately, the execution of the program can be rather slow compared to a true compiler. If a compiler and an interpreter are available for the same language, we can debug the program using the interpreter and then compile the final working version. A good compiler will allow various levels of optimization, where more time can be spent on compilation to produce more efficient code. Either the speed of execution or the size of the code can be optimized, but normally not both. The level of optimization chosen will depend on how many times the compiled program will be used and whether speed or memory space is critical in an application.

An intermediate between compiling and interpreting is that the source code can be compiled to an interpreted pseudo machine code. The pseudo code is at a lower level than the source code, allowing a compact representation of the program. On the other hand, each pseudo-code instruction represents an operation requiring many machine-code instructions. A pseudo-code interpreter program reads the pseudo code and executes an appropriate machine-code subroutine for each instruction.

Debugging information, such as the source-code line numbers, can be carried through to the compiled code. This information takes up extra memory space, so most compilers allow it to be omitted once the program has been debugged. A compiler which runs on the same processor for which it generates code is called a native compiler. Very small microprocessor systems may be incapable of running programs as large as compilers. Code can be compiled on another system (perhaps a mini or mainframe computer) to run on the microprocessor. This is known as a cross-compiler. Compilers which generate pseudo code can be run on any machine, provided that a pseudo-code interpreter exists for the target microprocessor.

8.1.2 Examples of high-level languages

A large number of languages exist for use on microprocessors, most of which originated on larger computers. FORTRAN (FORmula TRANslator) is a compiled language with a long history dating back to the early days of computing, resulting in the standard version of 1966. It is still widely used, since many large programs (and programmers!) have become dependent upon it. Many of the deficiencies of the original language were overcome by a revision in 1977, and the 1988 revision has resulted in a further improvement. Very efficient compilers have been developed for FORTRAN; hence it is often used for large number-crunching tasks.

BASIC (Beginners All-purpose Symbolic Instruction Code) was originally developed as an interpreted language for teaching the fundamentals of programming to students. It has gained widespread acceptance through the mass production of personal computers featuring the interpreter as a standard feature. Unfortunately, it suffers from the same problems as assembly code when large programs are written; the code becomes difficult to understand. It is also quite slow, although faster compiled versions are available. Many purists now believe that BASIC encourages bad habits in novice programmers.

Pascal was also developed as an educational language, though it is designed to encourage the student to form good habits. It is usually compiled

267

to machine or pseudo code, and is a direct descendent of the earlier language Algol. The ISO (International Standards Organization) standard definition of the language ensures compatibility with a wide range of compilers, but there are some deficiencies in the standard, notably in the area of file I/O and character-string handling. Most commercial compilers include extra features in these areas, though in forms which may differ.

The C language was developed to provide a language which has the power and portability of languages such as Pascal, while retaining the efficiency of assembly code. It contains many features similar to Pascal, and various *shorthand* constructs which reduce the amount of typing needed to enter a program. There is no direct dependence on a particular machine-code instruction set, but many of the operations can be directly mapped onto a single machine-code instruction on most microprocessors.

8.1.3 Structure of a high-level language

The input file for a compiler is a readable text file. The compiler analyzes this by splitting it into symbols or *tokens*. This stage is called *lexical analysis* and it strips off irrelevant characters such as spaces. The tokens used in a language are usually symbols which can be classed as follows:

one-character symbols	e. g. $+$, $-$, $=$, $($, $)$;
two-character symbols	e. g. $:=$, $<=$, $++$;
keywords	e. g. **begin**, **end**;
identifiers	e. g. x, *sum*;
number and string constants	e. g. 2, 3.14, "hello".

A keyword is a symbol of the language, while an identifier is defined by the programmer. Comments are usually recognized at the lexical-analysis stage and discarded.

The compiler then checks to make sure that the tokens are in a correct sequence. This is known as *syntax analysis* or *parsing,* since the language is checked against some grammatical rules and classified into phrases in the same way as a sentence in a language such as English can be parsed into a subject clause, verb, etc.

The last stages are the semantic analysis of the program, where the meaning of each phrase is analyzed, and code generation, in which it is translated into a sequence of machine-code or pseudo-code instructions. This is known as the object code.

Some compilers take their input through a pre-processing program. This is usually used to provide a macro facility, which allows text strings to be defined and substituted in the source code. This allows a program to be *customized* for a particular application, or even a particular compiler version. For example, if the same program were to run on two different microprocessors with different terminal screen sizes, the screen size could be substituted with an appropriate value by the pre-processor. The program would need to be compiled separately for each version, but there would be a saving in program size and speed, since no decisions depending on the screen size would need to be taken while the compiled program was executing. In general, it is better to do operations at compile-time rather than at run-time if at all possible.

Post-processors may also be used, since a compiler may not directly produce machine code. It may produce assembly code, leaving some of the work to be done by an assembler. Pseudo-code can also be turned into machine code if a suitable post-processor exists.

8.1.4 Object-code linkage

When a program is compiled, the object code usually needs to be linked to other routines and data before it can run. For instance, the subroutines for input and output or mathematical functions may be held in a library file since they are used so often. When we write a very large program, it may be useful to split it into sections which are compiled separately and then linked together as object code. This reduces the compilation time and helps to make the task more manageable by splitting it up. A linker is a program which combines object code from different sources to make a single program.

It is no good splitting up a program, of course, unless each section can communicate with others; this is normally done by making certain symbol definitions global, so that the linker may substitute the correct value. For example, if a subroutine is defined in one file and used in another, its name must be available to the linker so that all references to that name may be substituted with the same address.

In C, the routines and variables declared outside a routine are global and may be referenced by separately compiled routines. In Pascal, there is no standard way to do this, but many compilers allow the body of a procedure to be replaced by the word *extern* or *external* to indicate that the procedure is in another file.

It is possible to link code from different sources, provided that the parameters are compatible—most compilers from the same vendor will support this. For example, a Pascal program might make use of a library of C procedures, provided they are compiled separately and then linked.

8.1.5 The operating-system interface

Most high-level languages are designed to operate in an environment controlled by an operating system which provides input/output facilities etc. There may be system-dependent features which are usually made available by procedures. For example, it may be necessary to use such a system call to find the co-ordinates of a pointing device, or to change the baud rate of a serial I/O device.

Sometimes, a compiled program must run on a system which is too small to support a disk operating system. In such cases, a skeleton operating system is used to provide the bare minimum of support. In this environment, some of the features such as file input/output may be highly restricted (e.g. only text I/O to a terminal) or non-existent. All I/O routines must then be provided by the programmer, making use of direct access to memory and I/O ports or assembly-code subroutines.

8.2 The Pascal language

To illustrate the structure of a high-level language, we will examine some of the features of Pascal, starting with an example to show the structure of a program.

8.2.1 A simple program

The Pascal program shown below adds three numbers together and prints the result.

```
program add(input, output);    {a simple Pascal program}
var
    a, b, c, sum: integer;
begin
    a := 3; b := 5; c := 7;
    sum := a + b + c;
    writeln('Sum of ', a:1, ',', b:1, ' and ', c:1, ' is ', sum:2, '.');
end.
```

The **program** keyword is always used at the start. It is followed by the identifier *add* (the name of our program) and then by a list of file identifiers in brackets. A file is used to refer to an input or output device. The default files *input* and *output* are normally connected to the keyboard and screen, respectively. Other files may be specified here if a disk is to be used to store data. The text between braces "{" and "}" is a comment and is ignored by the Pascal compiler.

A set of declarations follows the program heading. In this case we define some storage locations for variable data using the **var** statement. There are four named locations for variables, all of type *integer*. This normally means a 16- or 32-bit memory location, the contents of which will be treated as signed binary numbers. Pascal is a strongly-typed language, which means that it places certain restrictions on what you can do with variables of different types. For instance, you may not use a variable of type *integer* in a place where another type is expected.

The statements which make up the main body of the program are enclosed between the **begin** and **end** keywords. These are used wherever a series of statements must be grouped together to be treated alike. Statements are separated by semicolons within the block, and the block is terminated by a period to indicate the end of the program.

The first three statements are used to assign values to *a*, *b* and *c*: the assignment operator := means "calculate the value on the right and store it in the location named on the left." We can see that this means that the next statement performs an addition of *a*, *b* and *c* and places the result in the variable sum.

The last statement of the block is used to print the answer. It is, in fact, a call to a procedure (a Pascal name for a subroutine) called *writeln*. The parameters are printed on the *output* file, followed by a new-line marker. The quoted strings of characters are printed directly. The variables are printed as decimal numbers, the number of digits wide being specified by the number after the colon. These *field widths* may be omitted, in which case a default width will be used. The output will therefore be

Sum of 3, 5 and 7 is 15.

We can compare this Pascal program with an assembly-language program to do the same task.

```
      BEGIN MOVE.F #3,A          Load constants into memory
            MOVE.F #5,B
            MOVE.F #7,C
      *
            MOVE.F A,R0          Do the addition
            ADD.F  B,R0
            ADD.F  C,R0
            MOVE.F R0,SUM
      *
            LDA.F  STR1,R0       Print the answer
            JSR    PRINT_STRING
            LDA.F  A,R0
            MOVE.F #1,R1
            JSR    PRINT_INT
            LDA.F  STR2,R0
            JSR    PRINT_STRING
            ...    ...           etc.
            ...    ...
      *
      A     DS.F   1             Variable storage
      B     DS.F   1
      C     DS.F   1
      SUM   DS.F   1
      *
      STR1  DC.B   'Sum of ',0   String constants
      STR2  DC.B   ',',0
      STR3  DC.B   ' and ',0
      STR4  DC.B   ' is ',0
      STR5  DC.B   '.',0
```

We assume here that the PRINT_STRING and PRINT_INT subroutines are available to print a null-terminated string or a decimal integer, respectively; the address of the item to be printed is loaded into *R*0 before calling the routine.

PRINT_INT receives the field width in $R1$. Note that the assembly-code version is quite long and requires the user to think in detail about which registers data will be stored in, etc. Once assembled, however, it may be faster and take up less memory than the compiled Pascal version.

8.2.2 Procedures and functions

Procedures and functions are Pascal subroutines. We have already seen an example of a procedure call in using *writeln*, a standard procedure which is built into Pascal. Functions are like procedures, except that they return a single value (in addition to any other actions) so that they may be used in an expression. For example, the statement

$$y := r*sin(theta);$$

uses a call to the standard function *sin* to compute the sine of an angle (in radians). The result is then multiplied by r and assigned to y.

Procedures and functions can be defined by the user. They must be declared before the **begin** of the block in which they are used. Here is an example of a user-defined procedure.

```
program userproc(input, output);
var
    a, b, c, sum: integer;

    procedure add(x, y, z: integer; var s: integer);
    var
        temp: integer;
    begin
        temp := x + y + z;
        s := temp;
    end;

begin
    a := 3; b := 5; c := 7;
    add(a, b, c, sum);
    writeln('Sum of ', a:1, ',', b:1, ' and ', c:1, ' is ', sum:2, '.');
end.
```

The procedure heading includes the name *add* and a list of formal parameters in brackets. Parameters are variables which are used to exchange data between the called procedure and the caller. Each formal parameter will be replaced by an actual parameter when the procedure is called. The first three parameters x, y and z will be assigned values from a, b and c, respectively, when the procedure is called further on. If x were changed within the procedure, however, the value of a would not change; x is a local copy of its value. To allow a value to be returned in a parameter, we use the keyword **var** as shown for s here. Since this procedure returns only one value, it could have been declared as a function, thus:

function *add*(*x*, *y*, *z*: *integer*): *integer*;

· · ·

· · ·

sum := *add*(*a*, *b*, *c*);

The variable *temp* is used in the procedure, somewhat redundantly, to show that procedures and functions may have their own variable declarations. In fact, they may also have procedure declarations. The variables declared within a procedure are not available to the main program: a storage area is allocated to them on entry to the procedure and then released at the end. Main program variables are, however, available within procedures: they are said to be *global*. The situation becomes more complicated when procedures are declared within procedures. There are rules known as *scope rules* to determine the parts of the program in which declarations will be recognized. Here is a skeleton program to illustrate the point.

```
program scope(input, output);
var p: integer;

    procedure a;
    var q: integer;

        procedure b;
        var r: integer;

        begin
            body of b
        end;

    begin
        body of a
    end;

    procedure c;
    var s: integer;

    begin
        body of c
    end;

begin
    body of program
end.
```

Procedures *a* and *c* are declared in the main program, procedure *b* is declared inside procedure *a*. This means that procedures a and *c* can be called from the main program body, from the body of *c* or from the body of *a*. Procedure *b* can only be called from *a* or itself. Note that it is quite permissible for recursion to occur, that is, for a procedure to call itself, provided that the recursion is terminated on some condition.

273

The declaration of variable p is global to the program; it can be accessed from any procedure or the main body of the program. The scope of variable q is more limited: it is local to a and global to b. The variable r is local to b and s is local to c; these are not accessible to other parts of the program.

When a procedure is entered, storage space is set aside for its local variables. If recursion occurs, a version of this local storage is created anew for each level of the recursion.

If the same identifier is declared twice, once globally and once locally, the local declaration takes precedence in the block where it is declared.

8.2.3 Symbolic constants

When a number is used repeatedly throughout a program, it is useful to be able to define it as a constant and give it a name. This means that any changes to the value at a later date has to be made in only one place. Constants may be declared at the start of a program or procedure, for instance as follows.

$$\textbf{const}$$
$$pi = 3.14159;$$
$$maxvalue = 1000;$$

8.2.4 Data types

So far, we have considered only the integer type. Pascal supports several standard types which are described below:

integer	an integer;
real	a floating-point number;
char	a character in the symbol set (usually ASCII);
boolean	a logical true or false value.

There are also several user-defined types, such as the following:

subrange	e.g. 1..31;
enumerated type	e.g. (*red, blue, green*);
array	e.g. **array** [1..10] *of integer*;
set	e.g. *set of char*.

The subrange is merely a restricted range of some other basic type such as *integer*. The enumerated type is a list of symbolic values which may be taken by a variable of that type. An array is a group of variables of the same base type which can be be manipulated together. An **array** variable x of the type in the example above would consist of ten integer variables, namely $x[1], x[2], x[3], \ldots, x[10]$. These can be used to represent matrices. The set type is like the one which we know from set theory in mathematics, containing zero or more members from some universal set. Each member can occur at most once in a set. In a Pascal set, the universal set contains all possible values of the base type given in the set declaration: here, it is the set of all characters. Pascal allows various set operations, including union, complement and intersection.

If a type is to be used frequently, it may be declared at the start of a block and given an identifier which is then used to represent it. Variables can then be declared using that identifier, e.g. as follows.

type
    color = (*red, blue, green*);
    dayofmonth = 1..31;
var
    Screencolor: *color*;
    holiday: **array** [*dayofmonth*] **of** *boolean*;

A number of different types may be grouped together to be treated as a single item. The **record** type declaration can be used to create such a type, e.g. as follows.

type
    date = **record**
        day: 1..31;
        month: 1..12;
        year: *integer*;
    end;
var
    birthdate: *date*;

Assignments can be carried out directly between variables of type *date* here, or the individual components of the record can be accessed by the field names, e.g. as follows.

birthdate.day := 24;
birthdate.month := 6;
birthdate.year := 1958;

Alternatively, the **with** statement may be used to abbreviate this, as follows.

with *birthdate* **do**
    begin
        day := 24;
        month := 6;
        year := 1958;
    end;

Records may be required to store different kinds of data in the same variable. This can be accomplished by the variant record, where the data is interpreted in different ways, depending on the field name by which it is accessed. A tag field may be used to indicate the usage in a particular case, e.g. as follows.

type
  kindofshape = (*circle, rectangle*);
  shape = **record**
        Xorigin, Yorigin: *integer*;
        case *kind*: *kindofshape* **of**
          circle: (*radius*: *integer*);
          rectangle: (*length, width*: *integer*);
    end;

The value of the tag *kind* changes the interpretation of the record data. The *radius* and *length* parts of the record would map onto the same storage space, while *Xorigin* and *Yorigin* would be separate.

8.2.5 Input and output

Pascal uses files for input and output. A file may be stored on disk or tape, or it may represent a device such as a terminal or keyboard. In any case, data is exchanged either as text (e.g. ASCII) or binary data. A file variable is used to refer to a file, and is declared as follows.

var f: file of integer;

This represents a file of binary integer data. Before use, the file must be opened using the *reset* or *rewrite* procedure. For instance,

reset(f);
read(f, i, j, k);

will read three integers *i*, *j* and *k* from file *f*. Writing to a file is done similarly; for instance,

rewrite(f);
write(f, i, j, k);

will write three integers.

    If data is to be in a human-readable form, a special type of file is used:

var t: text;

which is equivalent to:

var t: file of char;

    Some compilers treat text files slightly differently from files of *char*; however, this need not concern us here. Once opened, a text file may be read or written using the *read*, *readln*, *write* or *writeln* procedures. The *write* procedure is like *writeln*, but does not move to a new line afterwards. Similarly, the *readln* procedure reads past the end of a line, while *read* does not. All data is converted to/from formatted text when these procedures are used on text files.

The standard text files *input* and *output* are declared at the start of the program as described previously. If a file used in the program is not declared there, it is temporary and exists only while the program is running. If it is declared, it is assumed that some correspondence between Pascal files and real files or devices is made by an operating system which starts the program. This is somewhat unsatisfactory; hence most compilers allow the use of a procedure such as *assign*, thus:

CHAPTER 8
High-level Languages

$$assign(f, 'fdname');$$

where the string 'fdname' is the name of a file or device known to the operating system.

8.2.6 Pointers

A powerful feature of Pascal is the ability to allocate and de-allocate storage space while the program is running. This is done by using pointers—variables which point to other variables. In effect, the content of a pointer variable can be thought of as the address of the variable to which it points. (In practice, this depends on the compiler: it need not be exactly equivalent.)

A pointer variable or type can be declared for any variable type (except a file), and it is then specific to that type, e.g. as follows.

$$\textbf{var } p: \uparrow integer;$$

The symbol ↑ (sometimes @ or ˆ is used) indicates that *p* is a pointer to an integer. Declaring a pointer allocates storage for an address, but does not actually create a space to store the data; a space must first be allocated by the *new* procedure. This takes a pointer as a parameter, and reserves a new memory location for a variable. On return from the procedure, the pointer points to this variable. For instance,

$$new(p);$$
$$p \uparrow := 123;$$

makes the pointer *p* point to an integer which is then assigned the value 123. To refer to the newly created variable, we can now use the pointer *p* as follows.

$$j := p \uparrow;$$

The variable *j* is assigned the value of the variable pointed to by *p*, namely 123. A special value of **nil** can be assigned to a pointer if it does not point to a variable. When the user has finished with it, the storage space may be returned to the system for re-use by calling the *dispose* procedure. This takes a pointer as a parameter and deletes the variable to which it points, as follows.

$$dispose(p);$$

Pointers can be used to create data structures such as lists or trees.

The following program reads in a sequence of three numbers and prints them in reverse order.

```
program reverse(input, output);
type
    listptr = ↑ listitem;

    listitem = record
                    itemdata: integer;
                    nextitem: listptr;
               end;
var
    i, data: integer;
    item, head: listptr;

begin
    head := nil;

        for i := 1 to 3 do
        begin
            read(data);                    {read a number}
            new(item);                     {create a new list item}
            item ↑ .itemdata := data;      {put the number in it}
            item ↑ .nextitem := head;
            head := item;                  {chain it into the list}
        end;

    item := head;                          {point to the start}

    while item <> nil do                   {while item not nil}
        begin
            write(item ↑ .itemdata);       {print the number}
            item := item ↑ .nextitem;      {move down the list}
        end;

    writeln;                               {start a new line}
end.
```

This program builds a list from dynamically-created variables of type *listitem*. The data is read in and each new variable is added to the head of the list by assigning a pointer to the previous list to its *nextitem* field. The tail of the list is terminated by a record with a **nil** pointer. After all the data has been read in, the pointer *item* is pointed to the head of the list and then each item is printed in turn as the pointer is moved along the tail. The list which results from the input "7 6 5" is therefore as given in Figure 8.1.

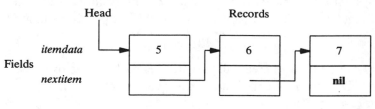

Figure 8.1

8.2.7 Conditional statements

Sometimes, we want to carry out an action only if a condition is met. This can be achieved with the **if** statement as follows:

$$\textbf{if } x < 0 \textbf{ then } x := -x;$$

which takes the absolute value of variable x. The assignment is carried out only if the expression $x < 0$ is true. We may wish to carry out one of two alternative actions depending on the test, in which case we add an **else** clause, for example as follows.

> **if** (*year* **mod** 4) = 0
> **then** *write*('leap year')
> **else** *write*('not a leap year');

The condition in both cases is, in fact, an expression which is of type *boolean*. If the result is true, the **then** clause is executed, otherwise the **else** clause is executed. Any Boolean expression can be used, including operators such as **and, or, not** and so on. We could, in fact, replace the *write* statements above with blocks of statements delimited with **begin** and **end** if we wished.

If there are more than two alternative outcomes then the **case** statement can be used, e.g. as follows.

> **case** *daynum* **of**
>   1: *day* := 'Sunday';
>   2: *day* := 'Monday';
>   3: *day* := 'Tuesday';
>   4: *day* := 'Wednesday';
>   5: *day* := 'Thursday';
>   6: *day* := 'Friday';
>   7: *day* := 'Saturday';
> **end**;

Here, there is a choice of seven possible values for *daynum*, each of which has an assignment statement associated with it. If the value of *daynum* is outside this range, an error occurs, though some compilers allow an **else** or **otherwise** clause at the end of the case statement to specify some default action.

8.2.8 Loops

Many repetitive tasks can be carried out by a program loop. This can be constructed using a label at the start and a **goto** at the end, similar to assembly language. (Labels in Pascal must be integers and are declared at the start of a block.) For example:

> $i := 1;$
> 99: *writeln*(i);
> $i := i+1;$
>   **if** $i <= 3$ **then goto** 99;

will increment and print the value of i three times while i is less than or equal to 3.

The **goto** statement is generally not used by good programmers except where a loop must be terminated suddenly, for example when an error occurs. It tends to make programs difficult to understand, causing transfer of control to a statement elsewhere without apparent reason.

A better set of alternative loop structures is provided in Pascal. When a simple ascending or descending sequence of numbers is required, the **for** statement is used, e.g. the following.

```
program forloop(input, output);
var
    i: integer;

begin
    for i := 1 to 3 do
    begin
        writeln('Testing ', i);
    end;
end.
```

Here the value of i is incremented from 1 to 3, and the block of statements between the **begin** and **end** is executed once for each value of i.

Loops may have terminating conditions which are more complicated than this. The **while** ... **do** and **repeat** ... **until** statements are used in these cases. The loop can be terminated by any Boolean expression in these cases, as in the following example.

```
program whileloop(input, output);
var
    i: integer;

begin
    i := 1;
    while i <= 3 do
        begin
            writeln('Testing ', i);
            i := i+1;
        end;
end.
```

The loop is executed as long as the Boolean expression $i \leq 3$ remains true. If this expression is false initially, the loop is never executed. Sometimes, a loop must be executed at least once before the condition is tested. In this case, we use a **repeat** loop.

```
program repeatloop(input, output);
var
    i: integer;
```

```
begin
    i := 1;
    repeat
        writeln('Testing ', i);
        i := i+1;
    until i > 3;
end.
```

Note that the termination of the loop occurs when the Boolean expression $i > 3$ becomes true; conversely the **while** loop terminates when its expression becomes false.

8.3 The C language

In order to compare the Pascal and C languages, the program examples used for Pascal will now be repeated in C. The addition program below looks similar to the Pascal version.

```
int a, b, c, sum;  /* a simple C program */
main()
{
    a = 3; b = 5; c = 7;
    sum = a+b+c;
    printf("Sum of %1d, %1d and %1d is %2d.\n", a, b, c, sum);
}
```

Note that the **program** header is not present. The integers a, b, c and sum are declared as type int before the start of the program statements. In C, programs are in fact procedures; they are distinguished from other procedures by being called main as shown here. The empty parentheses following this indicate that there are no parameters to the main program. The Pascal **begin** and **end** keywords are replaced by braces { and } in C. The comment delimiters in C are /* and */ as seen here.

The assignment of values to variables is again similar to Pascal, but here the symbol = is used for assignment.

The standard procedure printf is used to write the results to the standard output device. The first parameter is a string which contains the text to be printed, along with some formatting directives. Here, the form %1d indicates that a decimal integer value of one digit width should be substituted. The values are taken from the remaining items in the parameter list. At the end of the string, the code \n indicates that a new line should be started.

8.3.1 Functions

In C, procedures and functions are not treated differently. A function may be called like a procedure, in which case its returned value is discarded. If a procedure is called like a function, its returned value will be undefined. The program below uses a procedure called add.

```
int a, b, c, sum;

       add(x, y, z, s) int x, y, z, *s;
       {
       int temp;
       temp = x+y+z;
       *s = temp;
       }
main ()
{
       a = 3; b = 5; c = 7;
       add(a, b, c, &sum);
       printf("Sum of %1d, %1d and %1d is %2d.\n", a, b, c, sum);
}
```

The types of the parameters x, y, z and s are declared after the procedure heading, but before its body. The declaration *s indicates that s is a pointer: this is equivalent to the use of a **var** parameter in Pascal where a value must be returned from a procedure. When the procedure is called from main, this parameter is given as a pointer to the variable sum by writing &sum.

The scope rules for C are simpler than for Pascal, since procedures cannot be declared inside other procedures. This means that there are only two types of variable scope; those declared outside the procedures which are global and those in each procedure which are local. Variables declared outside main in our first C example in 8.3 are global and accessible by other functions. Since there are no other functions there, the variables could have been made local to main by placing the declaration inside its body.

8.3.2 Constants

Symbolic constants in C are declared using the pre-processor command #define, as follows.

```
#define PI 3.14159
#define MAXSIZE 1000
```

Thereafter, wherever the string PI is encountered in the program, the pre-processor will substitute 3.14159. This command can also be used to define macros, but this is beyond our scope in this chapter.

8.3.3 Data types

As in Pascal, variables in C have types. The basic types available are the following:

int	an integer;
float	a floating-point number;
char	a character code (i.e. a byte).

Unlike Pascal, however, the types are not checked by the compiler; a separate type-checking pre-processor program is required. User defined types are also available, as the following examples show.

```
enumeration    e.g. enum {red, blue, green};
array          e.g. int x [10];
```

Names may be given to user-declared types, as follows.

```
typedef enum {red, blue, green} color ;
color screencolor, bordercolor;
```

The C equivalent of the record is the structure type, such as the following.

```
typedef struct
       {int day;
        int month;
        int year;
        } date;

date birthdate;
```

The components are accessed as a complete record or by field name, as follows.

```
birthdate.day = 24;
```

The variant record in Pascal becomes a union in C, as for instance the following.

```
typedef enum {circle, rectangle} kindofshape;

typedef struct
       {int xorigin, yorigin;
        kindofshape kind;
        union {
               struct {int radius} circdata;
               struct {int length, width} rectdata;
               } shapedata;
        } shape;
```

8.3.4 Input and output

Like Pascal, C uses files for input and output. The FILE type is not a standard part of the language, but is defined in a standard library. The file stdio.h is included at the start of the program by a pre-processor command to set up several definitions for file input/output. A file variable is then declared as a pointer as follows.

```
#include <stdio.h>
FILE *f;
```

Now f represents a file of bytes. The interpretation of the file as text or

283

binary data depends on the user. Before use, the file must be opened using the fopen function, as follows.

```
f = fopen("fdname", "r");
fread(&i, sizeof(int), 1, f);
```

will read an integer i from f. The r parameter to fopen causes the file to be opened for reading. The fread function takes four parameters. The first is a pointer to a storage location in the program and the second is the size of the data item to be transferred. This is conveniently provided by the sizeof function which returns the number of bytes in a variable of type int here. The third parameter is the number of items to be transferred, and the fourth is the file pointer. Writing to a file is done similarly:

```
f = fopen("fdname", "w");
fwrite(&i, sizeof(int), 1, f);
```

will write an integer i to file f.

If data in the file is to be in a human-readable form, the procedures fscanf and fprintf are used. All data is converted to/from formatted text by these procedures.

The standard text files stdin, stdout and stderr are set up automatically at the start of the program for input, output and error messages, respectively. They are normally attached to the keyboard and screen, but can be redirected by the operating system which invokes the program. There are other standard I/O library procedures too numerous to list here.

8.3.5 Pointers

A pointer to a variable type can be declared as follows.

```
int *p;
```

The variable p is now a pointer to an integer. To point to a variable i, we assign a value to the pointer:

```
p = &i;
```

The pointer may now be used to refer to the value of i, as follows.

```
j = *p;
```

There is no standard way to allocate new storage for use with pointers in C. Some compilers provide an allocate procedure which performs this function.

8.3.6 Conditional statements

The "if" statement in C is written as follows.

```
if (x < 0) x = -x;
```

Note that the parentheses around the condition are mandatory. An *else* clause can be included, as follows.

```
if (year%4 == 0) printf("leap year");
else printf("not a leap year");
```

Note that the Boolean expression must be in parentheses and that the == symbol is used to compare for equality, since = is used for assignment.

For more than two alternatives, the switch command is used similarly to the **case** in Pascal, as follows.

```
switch (daynum)
    {
    case 1 : day = "Sunday";       break;
    case 2 : day = "Monday";       break;
    case 3 : day = "Tuesday";      break;
    case 4 : day = "Wednesday";    break;
    case 5 : day = "Thursday";     break;
    case 6 : day = "Friday";       break;
    case 7 : day = "Saturday";     break;
    default: day = "Invalid";      break;
    }
```

The break statement is needed to prevent execution of the statements following the one which is selected by the switch, since the switch statement allows execution of all the remaining statements in the list following the one selected.

8.3.7 Loops

Labels and goto statements can be used to create loops in C, as follows.

```
        i = 1;
label:  printf ("%d", i);
        i++;
        if (i <= 3) then goto label;
```

The i++ statement is equivalent to i = i+1 here.

The **for** statement exists in C too, but it is more flexible than in Pascal. Here is our loop example again.

```
main()
{
    int i;
    for (i = 1; i <= 3; i++)
    {
     printf("Testing %d\n", i);
    }
}
```

The parentheses after for include three statements: the first is executed before the loop starts and is used here to initialize i to 1. The second expression is evaluated and, if it is false, the loop terminates. If it is true, the statements in the body of the loop are executed and then the third statement in the parentheses (i++ here) is executed before repeating the loop test. Since any expressions can be used in the for statement, many variations are possible.

The **while** loop also exists in C, and is used similarly to the Pascal version, as the following example shows.

```
main()
{
        int i;
        i = 1;
        while (i <= 3)
            {
            printf("Testing %d\n", i);
            i++;
            }
}
```

The **repeat** loop in Pascal is replaced in C by a do ... while loop, as follows.

```
main()
{
        int i;
        i = 1;
        do
            {
            printf("Testing %d\n", i);
            i++;
            }
        while (i <= 3);
}
```

The condition at the end must be true for the loop to continue, unlike the Pascal **repeat** statement.

8.4 Interfacing high-level language to assembly code

It is easier to do most tasks with a high-level language rather than assembly code, but sometimes a combination of the two is better. Assembly code provides easy access to peripheral-device registers for writing input/output routines. It is also faster and may be used to speed up time-critical parts of a program.

High-level languages may include features which allow manipulation of memory and input/output data, but these are not standard and will vary from one compiler to another. For example, functions and procedures may be provided to read and write memory locations directly, as follows.

286

```
i := peek(a);  {get data i from memory address a}
poke(a, i);    {put data i to memory address a}
```

Similar procedures will be required for input/output devices if their addresses are mapped separately.

In some cases, machine code may be inserted "in-line" with high-level language (HLL) statements by some suitable format. The user must be careful to observe any restrictions on the use of registers and memory used by the compiled HLL code.

The most universal way to combine a HLL and assembly-code program is to call external procedures from the HLL program. The external assembly code and HLL routines are assembled/compiled separately, then linked together. The HLL program then calls assembly-code subroutines to carry out low-level tasks. This normally requires that data be communicated between the HLL and assembly-code routines. An easy way to do this is to use registers to hold the data, since these can be easily accessed from assembly code. In the HLL, special variable names may be reserved to refer to the registers. Memory locations may be used to pass data, provided their addresses are known to both assembler and HLL. One way to do this is to reserve memory locations within the assembly-code routine, then use routines such as peek and poke, as mentioned previously, to access these locations from the HLL. Alternatively, a HLL variable address may be passed to the assembly-code routine by a HLL function which computes the address of a variable.

All of these methods of exchanging data are somewhat unsatisfactory, since the HLL programmer must know the details of assembly-code programming; it would be better if the assembly-code routines, once written, could be used by programmers who have no such knowledge. This can be accomplished by using the HLL conventions for passing parameters in the assembly-code routines. Before studying this, it is necessary to know something of the way in which data is stored in a language such as Pascal.

There are three types of variable storage, namely, static storage, dynamic storage and heap storage. Static variables, which are global to the whole program, remain at fixed memory locations throughout the program. Dynamic variables are local variables which are created on entry to a procedure and then destroyed on exit. Both of these may be stored on the stack, or the static variables may be given a fixed area of memory. The heap is a pool of memory which is allocated by the new procedure. Management of the heap is a complicated task, but in its simplest form it may simply be a second stack. Requests to *dispose* of storage may not have any effect in simple storage-management systems, but more complex ones will allow re-use of such storage by subsequent calls to *new*.

When a procedure is called, the parameters and local variables are stored on the stack along with various registers, including the program-counter return address. The state of the stack after a procedure call is as given in Figure 8.2.

287

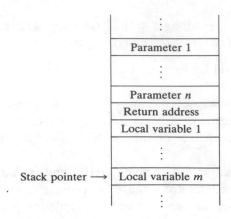

Figure 8.2 The state of the stack after a procedure call.

When the procedure is called, the parameters 1 to n are evaluated and pushed onto the stack. (Their addresses are pushed if they are **var** parameters.) The procedure is then called as a subroutine, pushing a return address onto the stack. At the start of the procedure, an offset is added to the stack pointer to reserve space for local variables 1 to m. These variables can be referenced by indexed addressing using the stack pointer. However, if the stack is used within the subroutine, it may be difficult to keep track of the stack pointer's value. We can make things easier by using a register as a *base pointer* to point to a fixed place on the stack. (In a small microprocessor, the base pointer may be held in memory.) The base pointer is loaded from the stack pointer on entry to the procedure. Since one procedure may call another, the base pointer from the previous procedure must be preserved. This can be done by pushing the old value onto the stack.

The information on the stack relating to a procedure call is known as a stack frame, and the offsets from the base pointer to each variable or parameter will be determined by the compiler. If a procedure needs to access a global variable, however, a problem arises. The base pointer in each stack frame can be used to find the stack frame of the procedure which called the current one, and this can be used to trace the nested procedure calls on the stack. Unfortunately, this does not identify the frame in which a global variable resides, since procedures can call each other (or themselves) to any depth of nesting, providing the scope rules are obeyed. To get round this, we need to keep a pointer to the stack frame for the procedure in which the current procedure is declared. This frame then contains a pointer to the frame for the next level, and so on until the variables global to the whole program are reached. The pointer used for this is called the static link, and must be preserved on the stack when a new stack frame is created.

Because the use of global variables is somewhat complicated, it is not usual for them to be referenced within an assembly-language routine. The parameters are used to pass data to and from the routine and must be in the format used by the HLL. Local variables may be stored as in the HLL code, by adding an offset to the stack and using the base pointer to reference them.

To remove them at the end, all that is required is to load the stack pointer from the base pointer. Manipulation of the stack in this way is done automatically by assembly-code instructions in some microprocessors (e.g. LINK and UNLK in the Motorola MC68000).

To avoid the need to reference local storage using indexed addressing, fixed memory locations may be used. The disadvantage here is that the routine will no longer be re-entrant, which means that it may not be used recursively or by routines which may interrupt one another. If the assembly-code routine is to be used re-entrantly, then it must store its local variables on the stack.

To illustrate the writing of an assembly-code routine for use with a HLL, we will define a hypothetical Pascal compiler for the NeMiSys microprocessor. The parameters will be passed as full words on the stack, the first parameter having the highest address (i.e. pushed first on the stack). Value parameters will be passed as actual data and **var** parameters as the addresses of the data. We will use $R5$ as the frame pointer.

Since a compiler does not actually exist for this machine, we will need to simulate its action. Let us assume that the procedure *proc* is declared in Pascal, thus:

procedure *proc*(*x*: *integer*; **var** *y*: *integer*); *external*;

and is then called as follows.

$$proc(23, a);$$

Assuming that our compiler generates assembly code with labels which correspond to the identifiers in the original program, it would translate this to the following code.

```
MOVE.F  #23,-(SP)   push 1st parameter
LDA.F   A,-(SP)     push address of 2nd parameter
JSR     PROC        do the (external) subroutine
ADD.F   #8,SP       discard the parameters
```

Since eight bytes are pushed on the stack as parameters, these are discarded by adjusting the stack pointer at the end. (Our machine uses a 32-bit address; so four bytes are needed to store the address of each parameter.) We can now write an assembly code routine which links to this Pascal program.

```
PROC MOVE.F R5,-(SP)        preserve old frame pointer
     MOVE.F SP,R5           create new frame pointer
     SUB    #4,SP           reserve local storage on stack
*
     MOVE.F -12(R5),4(R5)   copy 1st parameter to local store
     MOVE.F -8(R5),R4       get address of 2nd parameter
     MOVE.F 4(R5),(R4)      copy local store to value of 2nd parameter
*
     MOVE.F R5,SP           remove stack frame
     MOVE.F (SP)+,R5        restore old frame pointer
     RTS
```

289

The first group of statements sets up the stack frame, the middle group carries out some operations on the parameters and then the last group removes the stack frame. When the middle group of statements is executing, the stack will appear as in Figure 8.3.

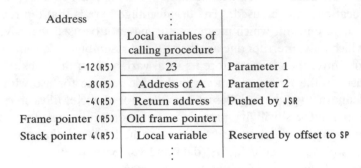

Address		
	Local variables of calling procedure	
-12(R5)	23	Parameter 1
-8(R5)	Address of A	Parameter 2
-4(R5)	Return address	Pushed by JSR
Frame pointer (R5)	Old frame pointer	
Stack pointer 4(R5)	Local variable	Reserved by offset to SP

Figure 8.3 The state of the stack after the second group of instructions.

8.5 Summary

We have seen how a high-level language such as Pascal or C provides a shorter and more comprehensible way of expressing complex operations than assembly code. Of course, there may be cases where only assembly code is able to do a particular task, such as handling I/O device interrupts. In most cases, however, the limits of memory size and execution speed which forced programmers to use assembly code in the past have been pushed to the point where it is necessary to justify the decision *not* to use a compiler.

In this chapter, we have covered the main features of two languages in a very short space, and this can be no more than a brief introduction. You are recommended to study the texts listed at the end of this chapter for further information.

8.6 Further reading

1 *American National Standard FORTRAN (ANS X3.9-1966)* (1966). American National Standards Institute, New York.
2 Amsbury, W. (1980). *Structured Basic and Beyond*, Computer Science Press, Rockville, Maryland. This book teaches the BASIC language in a way which encourages structured programming, as far as the language will allow.
3 *BSR X3.9 FORTRAN77* (1977). American National Standards Institute, New York.
4 *Computer Programming Language—Pascal* (1982). International Organization for Standardization (ISO) Standard ISO 7185.
5 Edelhart, M. and Nath, S. (1989) *The Brady Book of Turbo Pascal*, Brady Books (Prentice-Hall), New Jersey. An introduction to a version of the language Pascal for the IBM PC user.
6 Goldstein, L. J. (1989) *Hands-on Turbo C*. Brady Books (Prentice-Hall), New Jersey. An introduction to C with special emphasis on a version which is popular on the IBM PC.
7 Grogono, P. (1984). *Programming in Pascal* (2nd edn.). Addison-Wesley, Reading, Massachusetts. A useful beginner's book on learning to write Pascal programs.

8 Harbison, S. P. and Steele, G. L. (1987). *C—A Reference Manual*. Prentice Hall, New Jersey. A good book for learning to program in C, more readable than the book by Kernighan and Ritchie.

9 Kernighan, B. and Ritchie, D. (1988). *The C Programming Language* (2nd edn.), Prentice Hall.

10 McCracken, D. D., (1984). *Computing for Engineers and Scientists with FORTRAN77*. Wiley, New York. FORTRAN for beginners, with examples of applications in science and engineering.

8.7 Problems

8.1 Find out about the FORTRAN language and re-write the example programs from this chapter in FORTRAN. (The differences between the languages will mean that not all the examples can be converted. Specifically, pointers and recursion are not available in FORTRAN77.)

8.2 Repeat the exercise in problem 8.1 for the BASIC language.

8.3 Explain the meaning and usage of the following HLL constructs:

      (a) local variables;   (b) if statements;   (c) for loops;

      (d) while loops;   (e) functions.

8.4 A sequence of integers in which each number is the sum of the two previous numbers is called a Fibonacci sequence, for instance:

$$1, \quad 1, \quad 2, \quad 3, \quad 5, \quad 8, \quad 13, \quad 21, \quad 34, \quad \text{etc.}$$

Write a program in Pascal which will read in the two starting numbers of such a sequence and then print the sequence for as long as the numbers are less than 1000.

8.5 Repeat the exercise in problem 8.4 using the C language.

8.6 Repeat the exercise in problem 8.4 using the NeMiSys assembly language.

8.7 Discuss the difference between:

      (a) cross and native compilers;   (b) compiled and interpreted languages;

            (c) pseudo code and machine code.

8.8 Show the contents of the stack frame produced by calling the procedure *add* in the example shown in this chapter (8.2.2).

8.9 The following Pascal program calls a machine-code subroutine draw to draw a line between two points (x_0, y_0) and (x_1, y_1) on a plotter. Write the machine-code subroutine and describe the effects on the stack when it executes. You may assume that there is a library of machine-code subroutines to move the pen as shown below.

```
program graphic(input, output);

procedure draw(x0, y0, x1, y1: integer); external;

begin
      draw(0, 0, 100, 0);
      draw(0, 100, 100, 100);
end.
```

MACHINE-CODE SUBROUTINE LIBRARY

PENUP	lifts the pen from the paper;
PENDOWN	makes the pen ready to draw;
MOVEPEN	moves the pen or draws a line, depending on whether it is up or down. The new x coordinate is in $R0$ and the new y coordinate in $R1$.

9: Practical Microprocessor Architectures

9.1 Introduction

The 4-bit 4004 processor designed by Intel in the early 1970s is regarded by many as the forerunner of the modern microprocessor. The 4004 was not designed explicitly as the heart of a microcomputer: it lacked the necessary performance. In reality the 4004 was designed by Intel as a cost-effective solution to a customer's problem. The customer, a Japanese calculator manufacturer, wanted to produce a range of calculators that could perform different functions simply by varying the chips that were used. Intel was therefore contracted to produce the required chips. However, a brilliant engineer at Intel, Ted Hoff, realized the immense problems that would be encountered if a different set of chips had to be produced for each type of calculator. To overcome these problems Hoff suggested a very novel solution, at that time, which involved the design of a general-purpose logic chip that would act like the central processing unit (CPU) of a mainframe computer. The *processor* chip could then be programmed to perform a whole range of operations by simply producing a new ROM containing the required program instructions. Like all good ideas the concept was very simple: instead of requiring well over ten chips the new solution needed only 4 chips and was immensely flexible; only the ROM chip needed to be changed to give a completely different set of functions. The 4004, the processor chip, was produced for a customer; therefore Intel had no rights to this new invention. However, as competition in the calculator market increased, Intel renegotiated its contract. For a reduction in price they were allowed to market the new device as the MCS-4 microcomputer. When engineers realized the incredible potential of this new processor chip sales soared, even though it was only capable of handling 4 bits of data at a time. In April 1972 Intel introduced the 8008, the first 8-bit microprocessor; this was followed two years later by the 8080. The 8080 was the first recognizable 8-bit microprocessor with the functionality that is now expected from a modern processor. Intel was quickly joined by other semiconductor manufacturers in producing microprocessors: Motorola produced the MC6800 and Zilog, founded by ex-Intel employees, the Z80: both were 8-bit processors. Thus the microelectronic revolution had started.

The early processors lacked the high-level software and hardware support that is now taken for granted. Generally they were programmed at the machine-code level with minimal support. Also the hardware itself was unreliable since the design engineers using this new technology had very little experience upon which to draw. However, the power of these 8-bit microprocessors was continually refined with the introduction of new devices derived directly from the early machines: for example, the MC6800 gave rise to the MC6802 and MC6809.

The first microprocessor designs tended to be based on the capabilities of the available technologies. However, studies of high-level languages (HLL) such as FORTRAN, Pascal and C highlighted certain features in a microprocessor, that would make the compilation and execution of such programs faster. Ultimately, this led to the division of the microprocessor market into two distinct camps, those who advocate complex-instruction-set computers (CISCs) and those who support reduced-instruction-set computers (RISCs). The CISC processors support high-level constructs such as stack-frame handling and powerful memory-manipulation operations, allowing procedure calls, arrays and case statements, all HLL constructs, to be implemented very easily at the hardware level, normally by only one processor instruction. Unfortunately, these features have some drawbacks in that the instructions can take several clock cycles to execute. However, the programs produced can be quite short due to the almost direct relationship between the microprocessor instruction set and the HLL. In order to handle the complex instructions CISC processors need a complicated control section which takes a comparatively large area of the silicon die to implement (up to 60% in some cases), making the overall design very large. Also the design of a large control section can be very time consuming and prone to error. With such a large number of instructions to choose from, the production of an efficient HLL compiler can be very difficult, since more than one instruction may be able to perform the required operation: in other words the compiler can be spoilt for choice. Consequently the compiler may not always generate the best machine-language code. On the other hand, the RISC philosophy advocates the use of a very simple instruction set to implement only the most basic types of operations. These instructions are normally capable of executing in one clock cycle. Thus a RISC with the same clock speed as a CISC would, in general, execute more instructions in a given time. The drawback is that the size of the programs produced are normally longer than their CISC counterparts as several instruction may be needed to perform a HLL operation. Fortunately as fewer instructions need to be implemented then the controller can be simplified, greatly decreasing both the silicon area required for the processor and possible design errors. Also as there are fewer instructions the compiler can optimize the code far more effectively, possibly using several instructions to emulate one of the more complex instructions in the CISC processors.

The modern CISC processors can handle 16 or even 32 bits of data at a time. These include the Intel 8086 and derivatives such as the 80186, 80286, 80386 and 80486. Motorola has also produced the MC68000 family of processors which include the MC68010, MC68020, MC68030 and MC68040. In general as the technology has improved the number of functional elements that can be included on the chip has increased. This has meant that the chip may contain not only the processor, but a memory-management unit (MMU) and/or a floating-point unit (FPU) capable of dealing with floating-point numbers and trigonometric functions in hardware. The market for RISC processors is only now beginning to expand with nearly all CISC-processor manufacturers taking up this challenge: Motorola has produced the MC88100 family and Acorn computers the ARM processor. It has been estimated that if the current market growth continues then RISC processors could dominate the computer scene in the near future.

One further type of processing element is the microcontroller, which not only contains the processor but also small sections of RAM and ROM and some peripheral lines that can drive external TTL/CMOS lines. Therefore the microcontroller can perform all the major functions of a complete computer that may normally occupy a whole printed circuit board. The microcontroller is quite useful for compact designs that can take advantage of a minimal system to control a particular application. Typical microcontrollers (or single-chip microprocessors) include, for example, the Intel MCS-48 series and the Motorola MC68705.

All of the traditional processors are based on Von Neumann architectures: this essentially describes the way in which the processor fetches the data to be used. In a Von Neumann processor there is only one bus for the transfer of data and instructions, which causes a *bottleneck* at the higher data rates found in modern processors slowing down the whole computer. This has led to much research into different types of processor architecture in order to speed up program execution without necessarily increasing the clock frequency at which the processor operates. A particular avenue of research is parallel architectures where a program is broken into subtasks that can be executed by separate but communicating processors, thereby executing the program in parallel rather than sequentially as in a normal processor. One of the very first tangible results of this research is the Transputer designed by INMOS. Internally the Transputer is like a very fast RISC processor; however, in addition it also posses four high-speed serial data channels that can communicate with neighboring Transputers. Thus many Transputers can be arranged into an array to share the workload. This produces an overall increase in processing speed for each additional Transputer. The size of the array can grow indefinitely without too many restrictions, thereby continuously increasing the processing speed. The Transputer can also contain a FPU that is capable of executing at the same speed as the main processor. The Transputer also uses a special HLL called OCCAM, which has been specifically tailored to this unique architecture. The Transputer will not be discussed in this text; however, it completes the discussion on modern processors and illustrates the progress of designs to date.

The previous discussion has shown the diversity of processor designs and outlined the progress made to date; therefore this chapter will take an in-depth look at just a few microprocessors that are currently available. In each case the internal architecture of the processor and the composition of its instruction set will be discussed. Obviously a processor can only be discussed briefly in a single chapter; however, it should give a flavor of the processors to be found in the real world. The processors discussed will be the 8-bit Z80 and MC6809, 16/32-bit MC68000, the ARM RISC processor and the MCS-48 microcomputer series.

9.2 8-bit processors

The 8-bit processors are now rapidly becoming dated for use in modern personal workstations. However, due to the mature fabrication lines giving high yields and the number of manufacturers now sourcing these components they are exceptionally cheap, giving rise to their use in controllers etc. The two microprocessors that will be discussed in this section are the Motorola MC6809 and the Zilog Z80: these are third-generation 8-bit microprocessors and were designed with the knowledge gained from their predecessors in mind. The problem with designing a new generation of processors is that compatibility must be maintained with the previous devices since their is usually a large investment, in terms of hardware and software, in the old designs, i.e. a large user base. This has led microprocessor manufacturers to maintain *upward compatibility* for new processors. This usually means that the new processor can be used with the older peripheral devices and in some cases run the object code from the older processors; however, source-code compatibility is more likely to be maintained.

When discussing a microprocessor there are two different views that can be taken of its structure. The first aspect of the processor involves its electrical connections, that is, the input and output signals used and the timing relationships between these signals, information vital when designing the hardware for a new computer. The second view of a microprocessor involves the relationship between internal registers and the supported instruction set; this is obviously important for programming the computer. Although the two aspects of a microprocessor have been presented as being quite distinct, in reality this is not the case. For example, the way in which the hardware has been designed may affect the use of certain instructions; similarly certain instructions may require special hardware support.

9.2.1 The Motorola MC6809

The architecture of the MC6809 is directly descended from the original MC6800 processor. Indeed many of the instructions that are used by the MC6809 are compatible with the MC6800 even though the total number of instructions has dropped from 72, in the case of the MC6800, to only 59. Although the number of instructions has been reduced, the number of instructions with different addressing modes has risen from 197 to 1464. This gives the MC6809 a very flexible instruction set, not only for the support of low-level programming, but also for use with high-level languages. However, before looking at the internal architectural aspects of the microprocessor the external hardware features will be examined.

The pin assignments for the MC6809 are shown in Figure 9.1. As can be seen from this diagram the signals break down into the now familiar three-bus structure. There is the 16-bit address bus, 8-bit data bus and the control bus. The control bus in this case is fairly complex; however, it does allow the processor to carry out some advanced operations. The address and data buses are standard and both are capable of entering the high-impedance state when an external device wishes to take over the bus, for example with direct memory-access (DMA) operations.

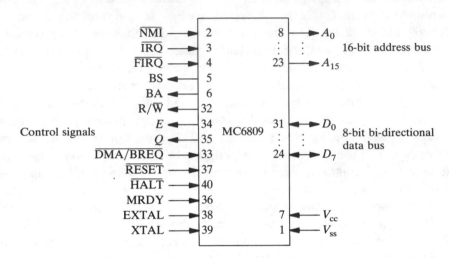

Figure 9.1. Signal-pin assignment for the MC6809.

In order to illustrate the power of the MC6809 it is necessary to discuss the various control signals. In a synchronous computer system, such as the MC6809, it is essential to have a master clock at the center of the system to provide the timing for bus accesses etc. On the MC6809 the timing is achieved by the signals marked EXTAL and XTAL which are used to connect a crystal to the on-chip oscillator: the crystal frequency must be 4 times that of the required bus frequency. The microprocessor is supplied to run at a maximum frequency which should not be exceeded. The two clock signals E and Q are then used to supply the timing signals to external devices. The leading edge of the Q clock signal indicates that the address on the address bus is valid, whereas the falling edge of the E clock signifies that the data should be valid and latched. As the two clocks are in quadrature their edges do not coincide and can be used to correctly access memory devices, all of which require a finite amount of time from initial address application before data can be received or transmitted (see Chapter 6). The two clocks are thus used to assist in data input and output from the processor, the chosen arrangement giving the maximum amount of time between the edges in a single clock cycle. In order to deal with slow memory devices the MRDY input is provided: if this signal is low then the E clock is stretched up to a maximum of 10 μs. Thus, by taking the MRDY signal low the amount of time a memory has to respond to being accessed can be significantly increased. The R/$\overline{\text{W}}$ signal (read/write) is used to indicate the type of operation to be performed on the external device.

In order to start the processor running in a valid state at power up or to place the processor easily into a known state after a program crash, the $\overline{\text{RESET}}$ line is provided. If this pin is taken to logic 0 for at least 8 CPU clock cycles then a reset condition will occur. When a reset occurs the CPU will fetch the 16-bit address located in memory at FFFE_{16} and FFFF_{16}, i.e. at the very top of memory, and place it in the internal program counter. The program located at that location will then be executed. Thus the processor can be directed to start executing a program anywhere in the memory by simply changing the address stored in the reset-vector location. Figure 9.2 shows all eight vector locations at the top of memory which are used to provide storage for addresses to deal with other hardware signals and program instructions.

Address	Vector
FFFF / FFFE	$\overline{\text{Reset}}$
FFFD / FFFC	$\overline{\text{NMI}}$
FFFB / FFFA	SWI1
FFF9 / FFF8	$\overline{\text{IRQ}}$
FFF7 / FFF6	$\overline{\text{FIRQ}}$
FFF5 / FFF4	SWI2
FFF3 / FFF2	SWI3
FFF1 / FFF0	Reserved

The most-significant byte of the vector goes
into the lowest memory-address location

Figure 9.2. Memory map to show the address-vector table of the MC6809.

As well as being able to start the processor executing a program it is also possible to stop the processor by taking the $\overline{\text{HALT}}$ line low. When this takes place the processor will suspend program execution after finishing the current instruction, the processor can remain in this state indefinitely without loss of data: the state is indicated by the two lines BA and BS. The BA and BS signal combination is used by the processor as a coded indication of its internal state. The state of BA = 1 and BS = 1 indicates a halted or bus-grant condition. If BS = 0 and BA = 0 then the processor is running normally. These signals are also used to acknowledge interrupts (see later) (BA = 0 and BS = 1) and to indicate when the processor has executed the SYNC instruction (BA = 1 and BS = 0) and is waiting for external interrupts.

The MC6809 can support three hardware interrupts $\overline{\text{IRQ}}$ (interrupt request), $\overline{\text{FIRQ}}$ (fast interrupt request) and $\overline{\text{NMI}}$ (non-maskable interrupt) (see Chapter 4). The interrupt is generated by taking one of these pins to a low level in the case of $\overline{\text{IRQ}}$ and $\overline{\text{FIRQ}}$ and a negative edge in the case of $\overline{\text{NMI}}$. The non-maskable interrupt (NMI) cannot be inhibited by software (see later); it also has a higher priority than $\overline{\text{IRQ}}$ and $\overline{\text{FIRQ}}$ or any of the software interrupts (SWI). When a $\overline{\text{NMI}}$ is recognized by the processor the entire machine state is saved on the hardware stack; because of this the $\overline{\text{NMI}}$ will not be recognized until the hardware-stack pointer has been loaded after a reset operation. When the $\overline{\text{NMI}}$ is recognized the address stored at vector locations FFFC_{16}–FFFD_{16} (Figure 9.2) is fetched and program execution continued from that point. A fast-interrupt routine is begun when the $\overline{\text{FIRQ}}$ line is taken low, as long as it has not been inhibited. The $\overline{\text{FIRQ}}$ has priority over the normal interrupt since it only stacks the condition-code register and program counter, whereas the response to $\overline{\text{NMI}}$ and $\overline{\text{IRQ}}$ involves stacking the entire machine state, thus slowing down the interrupt-response time. However, $\overline{\text{IRQ}}$ and $\overline{\text{FIRQ}}$ both fetch their service-routine address from the vector table at the locations shown in Figure 9.2; also their interrupt-service routines must clear the source of the interrupt.

The $\overline{\text{DMA/BREQ}}$ line can be used to suspend normal program execution and allow another device to acquire the processor buses, for example, to perform a DMA operation (see Chapters 6 and 7). The device requesting a DMA can pull this line low; then once BS = BA = 1, indicating the data and address buses are in their high-impedance state, the DMA controller will have 15 bus cycles to perform its task before the processor will retrieve the bus for a self-refresh operation.

INTERNAL ARCHITECTURE

The internal architecture of the MC6809 is illustrated in Figure 9.3. This view is more important to the software programmer and is often referred to as the *programmer's model*. There are two general-purpose 8-bit registers (A and B) which are used in arithmetic and data-manipulation operations. Very few instructions treat these two registers differently. Unlike its predecessors the MC6809 can treat the A and B registers as a single 16-bit register called the D register: the A register holds the most significant byte and B the least significant byte. There are several instructions which can handle this double register as a single entity: these include loading (LDD), storing (STD), adding (ADDD), subtracting (SUBD) and comparing (CMPD) data, either immediate or in memory.

The MC6809 contains an 8-bit direct-page (DP) register which is used in some addressing modes to form the most significant byte of the 16-bit memory address. Thus a *page* of 256 bytes can occur in any of 256 address locations. This can not only help save program space by reducing the amount of address data that needs to be specified, but also it can perform a role in protecting data, since each program section can assign a different value to the

298

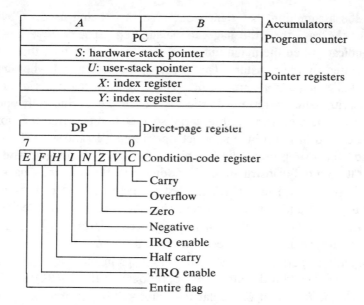

Figure 9.3. Internal architecture of the MC6809.

page register and thus change the base address of its own page. In order to maintain compatibility with the MC6800 the direct-page (DP) register is cleared when a reset occurs. This means the system will use page 0, as the MC6800 does, after reset.

The next four 16-bit registers X, Y, U and S are all termed index or pointer registers. The X and Y registers are general-purpose index registers which are used by certain indexed addressing modes. The X register maintains compatibility with the MC6800 internal architecture; also instructions using this register execute more quickly than if the Y register were used. The MC6809 posses a hardware-stack pointer (S) that is used by the processor when storing the return address from a subroutine call or to stack the machine state after an interrupt. The U register is called the user-stack pointer and can be used like a normal stack to push and pop (pull) data, and possibly for passing data to subroutines, as the hardware makes no use of this register, simplifying data passing etc. As well as the normal stack operations it is also possible to use the U and S registers in a similar way to the X and Y index registers. Hence it is possible to use the MC6809 to support high-level languages which make heavy use of the stack. The program counter is used, as would be expected, to point to the next instruction to be executed. It can also be used as an index register to make the production of position-independent code easier.

The condition-code (CC) register holds the state of the processor at any given time. Each bit of this register represents some aspect of the processing operations. The first four bits, N, Z, V and C represent outputs from the arithmetic and logic unit (ALU) that are directly used in conditional-branch instructions. The N (negative) flag indicates the sign of any result from the ALU: a 1 indicates a negative value and 0 a positive value. This flag is set or reset on the assumption that signed binary data is being used. Obviously if this is not the case then the flag can be ignored. The V (overflow) flag is used to indicate when the magnitude of a result is greater than that which can be represented by the register. The Z (zero) flag is set to 1 when a result is zero, a 0 indicating a non-zero result. The C (carry) flag holds the carry out from any arithmetic operations or the bit moved out during shift operations. The subtract operations invert the carry so that it represents the borrow bit and can be used in multi-byte subtract operations. In the MC6809 the carry bit is also used to represent the 7th bit of the 16-bit result generated by the 8×8-bit MUL (multiply) instruction. In addition to these normal flags there is also the H (half-carry) flag which is used to record the carry between bits 3 and 4 of any 8-bit addition operations. This flag is used by the decimal-adjust accumulator A (DAA) (note that only the A accumulator can be used by this instruction) instruction to correct the result of any binary-coded decimal (BCD) addition operations. The remaining flags in the condition-code register are all involved with the interrupt operations. If the I ($\overline{\text{IRQ}}$) or F ($\overline{\text{FIRQ}}$) bit is set to one then that particular interrupt will be ignored until the flag is cleared. The F bit can be set by an instruction which has the condition-code register as a destination or by $\overline{\text{NMI}}$, $\overline{\text{FIRQ}}$, SWI and on reset. Similarly the I bit can be set by an instruction or $\overline{\text{NMI}}$, $\overline{\text{FIRQ}}$, $\overline{\text{IRQ}}$, SWI and reset occurring. The E (entire) flag when set to 1 indicates that the entire state was placed on the stack, i.e. an $\overline{\text{IRQ}}$ interrupt, or 0 indicates only the PC and CC were stacked, i.e. an $\overline{\text{FIRQ}}$ interrupt. When the RTI (return-from-interrupt) instruction is executed the E bit of the *stacked* CC is used to determine the registers that need to be restored from the stack. Thus the E flag in the processor represents the processor's previous actions. The flags in general are not changed until an instruction is executed which specifically modifies the bit. Some instructions, such as ANDCC (AND with condition-code register) and ORCC (OR with condition-code register), directly modify these flags since the destination of the result is the condition-code register.

ADDRESSING MODES

The MC6809 has a very comprehensive set of addressing modes: there are seven main modes, several of which can use different registers to increase the overall range. Before describing these addressing modes it is worth establishing some terminology. The address that the processor will finally obtain the data from is called the *effective address*. However, the calculation of the effective address may be very straightforward or it may be produced by the addition of several unrelated addresses held in registers or memory.

The first few addressing modes are the simplest and are most often found in other equivalent 8-bit microprocessors. Inherent addressing, as its name implies, simply means that the effective address is contained within the instruction opcode (operation code) itself. Examples of these instructions are ABX (add accumulator B to index register X), DAA, MUL, SYNC (synchronize), which tell the processor the registers to be used without requiring further information bytes. In these cases the effective addresses are specific registers. Instructions such as RTI and RTS (return from subroutine) all use the hardware-stack pointer, yet another register. The instructions CLRA (clear accumulator *A*), ASRB (arithmetic shift right *B*), etc. all involve the accumulators: no further information is required.

Immediate addressing involves extra data other than the opcode. In this case the data to be used by the instruction follows immediately after the opcode itself. It is possible to use both 8-bit and 16-bit immediate data for loading the *A* and *B* registers or the index registers. Examples of these instructions are LDA #$20 (load accumulator *A* with the hexadecimal value 20) and LDX #$F000 (load the *X* index register with $F000_{16}$). The $ is used to indicate hexadecimal values and # is used to indicate immediate data in Motorola assembler mnemonics. The instruction would consist of the opcode followed by one or two bytes of immediate data.

In order to be able to access data stored at any memory location within the memory map of the processor, two addressing modes are available. These are called extended and direct addressing. In these two modes the effective address of the operand is stored as either the next 16-bits (extended) or 8-bits (direct) following the opcode. In the case of direct addressing the upper 8-bits of the address are supplied by the direct-page register; hence only 256 locations are accessible at any one time using direct addressing. An example of the extended addressing mode would be LDD $2000; in this case the two bytes at addresses $2000 and $2001 are loaded into the double accumulator *D*. With direct addressing if the DP register was equal to $AA when the instruction LDB $10 was executed then the byte located at address $AA10 would be loaded into the *B* accumulator. The extended addressing mode can also be used for an indirect address access. In this case the effective address of the operand is loaded from the contents of the address following the operand. For example if LDA [$2000] were executed and the address location $2000 and $2001 contained $AA and $00, respectively, then when this instruction is executed the byte at address $AA00 would be loaded into the *A* accumulator, since the effective address is formed by loading the two bytes at address $2000 and $2001 to form the 16-bit address, in this case $AA00. The assembler syntax [address] is used to indicate indirect addressing to the assembler program. If the brackets were omitted then the instruction produced would indicate, incorrectly, that extended addressing was to be used, without indirection. The formats of the other addressing modes are such that they can be easily determined by the assembler.

The MC6809 has a very comprehensive range of indexed addressing modes. With these modes of addressing the effective-address calculation involves one of the index registers, *X*, *Y*, *U* and *S* and occasionally the PC. There are five basic forms of indexed addressing termed zero offset, constant offset, accumulator offset, auto increment/decrement and indexed indirect. The indexed-addressing-mode variant and register is specified by a byte (called the post byte) following the opcode byte. In the zero-offset mode the effective address of the data is simply the contents of the specified index register; for example LDD 0,X would load the double accumulator D with the contents of the address pointed to by X. An extension to this mode involves the addition of a non-zero positive or negative offset. The offset is represented by a two's complement number, either a 5- (-16 to $+15$), 8- (-128 to $+127$) or 16- (-32768 to $+32767$) bit offset can be used. The 5-bit offset is the most efficient, in terms of bytes needed to represent the instruction, since it can be included in the byte following the instruction (post byte). The 8-bit and 16-bit offsets require an extra byte or 2 bytes, respectively, on the size of the overall instruction. The effective address of the data is now formed by adding the offset to the specified index register. The fixed offsets can be replaced by an accumulator A, B or D; therefore the offset can be altered during program execution as the value to be added is contained in an accumulator. In this case the register to be used is specified by the post byte: again two's complement numbers are used.

The auto-increment/decrement indexed addressing modes allow the pointer to be incremented after it has been used or decremented before use. Therefore the effective address can be stepped either up or down through, for example, tabular data. To cope with 8-bit and 16-bit data the index registers can be incremented/decremented by one or two. These modes are useful for producing user stacks that can be accessed in a similar way to the U and S stacks. An example of this mode is given by LDA ,X+; in this case the contents of the byte at the address pointed to by the X index register will be put into accumulator A and then X will be incremented by 1. The instruction LDD ,--Y will have a similar effect except that the Y index register will be decremented by two before the effective address is used to load the addressed 16-bit value in D. With the exception of the 5-bit offset mode and auto-increment/decrement by 1, all the indexed addressing modes can use indirection as discussed in extended indirect addressing. For example LDD [,Y++] would load D with the contents of the address pointed to by the address in Y and the Y index register would then be incremented by 2.

As well as using the normal index registers the PC can also be used to form an index address. In this case 8- or 16-bit signed offsets can be used; also indirection can be used. This allows position-independent code to be produced, since all data accesses will be relative to the PC, i.e. a fixed number of locations from the current PC location. For example LEAX 20,PCR (PCR represents the PC to the assembler) would load the index register with the address formed by adding 20 onto the current PC, i.e. $X = PC + 20$. The LEA (load-effective-address) instructions are very useful for loading absolute address values into index registers from essentially relative positions. These absolute addresses can then be used by other parts of the program; however, the initial address derivation is performed totally independently of the position of the code within memory.

The relative addressing mode is also used by the MC6809 branch instructions. Like most processors the MC6809 can perform conditional branches based upon the state of certain condition-code bits. This allows the program flow to be redirected in response to the outcome of a particular test, arithmetic operation, logical operation or data load. The branch instructions are extremely useful since they are position independent by virtue of the fact that the absolute branch address is calculated by adding a predetermined signed offset to the program-counter value when the instruction is executed. In most 8-bit processors, including the MC6809, this offset is only 8 bits, giving a range of -128 to $+127$. However, the MC6809 can also perform a long branch instruction which uses a 16-bit signed offset, giving access to the whole memory map. The long branches are distinguished by an L before the normal branch instruction.

The final addressing mode, called register addressing, is used to allow the programmer to specify the same operation on more than one register at a time for certain instructions. This is done by using a post byte following the opcode to encode the information to select the required registers. For example, PULU X,Y,D indicates that registers X, Y and D should be popped (pulled) from the user stack. This technique is also used by the MC68000 (discussed in Section 9.3) for some of its instructions.

The MC6809 is capable of performing some very powerful types of operations with advanced addressing modes; indeed it has a large number of features only found in 16/32-bit processors. Unfortunately it seems to have been eclipsed by the larger MC68000 processors; nevertheless it remains one of the most popular processor for dedicated applications.

9.2.2 The Zilog Z80

The Z80 has just as long a pedigree as the MC6809, although an older design, having descended from the 8080; indeed many of the functions performed by the Z80 are simply there to maintain compatibility, as far as possible, with the earlier processors. As with the MC6809 the Z80 has a whole family of peripheral chips that interface directly to the processor. These peripheral devices perform operations such as parallel input/output (PIO), serial input/output (SIO) and direct memory access (DMA).

PIN ASSIGNMENTS

The pin assignments for the Z80 are shown in Figure 9.4; again the normal three buses are present. The address bus is 16 bits wide giving access to 64 Kbytes of memory space, with the usual 8-bit data bus. However, as discussed later, the Z80 can support 256 bytes of extra memory for use with input/output devices. This is a fairly unique feature and means that the limited main-memory address space need not be fragmented by the addition of interface devices, as would occur in the case of the MC6809 processor.

The control signals are just as comprehensive as those discussed in connection with the MC6809; however, the philosophy behind the Z80 design is slightly different, as will become clearer when discussing these signals. Unlike the MC6809, the Z80 requires an externally generated clock to act as a master timing clock, there being no on-chip oscillator available. Hence

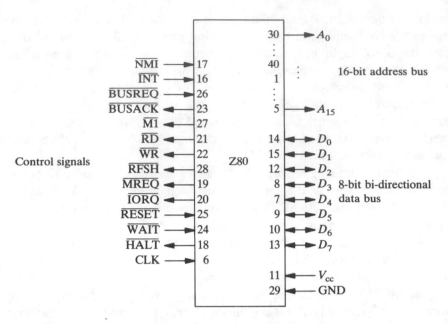

Figure 9.4. Signal-pin assignment for the Z80.

extra chips are required to generate the clock pulses before feeding into the CLK pin. The Z80 is also produced in a range of operating frequencies for different applications. The external output synchronization is given by the $\overline{\text{M1}}$ (machine-cycle-one) line, together with the $\overline{\text{MREQ}}$ (memory-request) line, to indicate the beginning of a fetch/execute cycle. The $\overline{\text{MREQ}}$ line indicates that the address now on the address bus is valid and can be used for main-memory accesses. Either a read or write operation can be selected by taking the $\overline{\text{RD}}$ (read) or $\overline{\text{WR}}$ (write) line low at the appropriate time. In the case of the $\overline{\text{RD}}$ line it is taken low at the same time as the $\overline{\text{MREQ}}$ line. Valid data needs to be available before both of these lines return to logic 1 at the end of the read cycle. However during a write cycle the $\overline{\text{WR}}$ line does not go low until after the processor data bus is stable: it will then remain low until the end of the cycle. From this very brief description it can be seen that the $\overline{\text{WR}}$ and $\overline{\text{RD}}$ lines together perform a similar function to the single $\text{R}/\overline{\text{W}}$ line in the MC6809. However, with the MC6809 the $\text{R}/\overline{\text{W}}$ signal provided no timing information to the external device, only whether a read or write operation was being performed; the E and Q signals provide the timing. As can be seen, the Z80 uses $\overline{\text{MREQ}}$ in conjunction with the $\overline{\text{RD}}$ and $\overline{\text{WR}}$ to provide the timing. Like the MC6809 the Z80 can lengthen the access time to slow memories by taking the $\overline{\text{WAIT}}$ pin low. The CPU will continue to insert wait states, without limit, until the signal returns high.

The Z80 has one further line associated with memory access: this is called the $\overline{\text{IORQ}}$ (I/O-request) line. When this line is taken low, by executing special instructions, it indicates that the lower 8 bits on the address bus hold a valid I/O address that can be decoded to a particular I/O device. This line is used in conjunction with the $\overline{\text{RD}}$ and $\overline{\text{WR}}$ lines to indicate whether a read or write should take place to the I/O devices. This scheme is illustrated in Figure 9.5 which shows how the I/O devices are independent from the main memory. In the MC6809 processor this separate block of I/O devices could not exist outside of the main 64-Kbyte memory block.

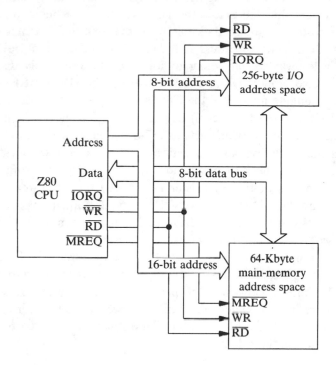

Figure 9.5. Possible memory organization for the Z80.

The $\overline{\text{RESET}}$ pin places the Z80 in a known state to begin execution of the program. When $\overline{\text{RESET}}$ is taken low several internal registers are cleared, including the program counter, and the interrupts are disabled and set to mode 0. As the program counter is set to zero when it leaves the reset state, the processor will begin executing the program from address 0000, i.e. the bottom of memory. The Z80 also supports two interrupt pins: $\overline{\text{NMI}}$ (non-maskable interrupt) and $\overline{\text{INT}}$ (interrupt request). The $\overline{\text{NMI}}$ is like that of the MC6809 in that it cannot be disabled; however, unlike the MC6809 this $\overline{\text{NMI}}$ causes program execution to transfer to a fixed address, in this case 0066_{16}. Thus the $\overline{\text{NMI}}$-service routine must be in this fixed location and cannot be changed; compare this with the MC6809, where the address vector can be altered to point to any location in memory as the start of the $\overline{\text{NMI}}$-service routine. A low level on the $\overline{\text{INT}}$ line will only be recognized if the interrupt is enabled. The effect this signal has on the behavior of the processor is dependent on how the interrupt-mode bits have been set up by the software: this will be discussed later in connection with the internal architecture.

As with the MC6809 it is possible to halt the processor; however, in this case the HALT instruction is used rather than hardware means. When the

305

processor is halted the $\overline{\text{HALT}}$ line goes low to indicate this state. The processor will now execute NOP instructions until a $\overline{\text{NMI}}$ or an enabled $\overline{\text{INT}}$ occurs after which normal processor activities will resume.

In order to allow other devices to gain ownership of the bus and access to memory, etc., the Z80 provides the $\overline{\text{BUSREQ}}$ (bus-request) line. When this line is taken low the Z80 places the address bus, data bus and some of the control lines into a high-impedance state so that external circuitry can control these lines. This state of affairs is indicated to the external devices by the $\overline{\text{BUSACK}}$ acknowledge line going to logic 0. As previously discussed, the MC6809 can also support DMA operations using the $\overline{\text{DMA}}/\text{BREQ}$ line.

One quite unique feature of the Z80 is the support it provides for refreshing dynamic-RAM (see Chapter 6) devices. This is achieved by the provision of an internal register R, which acts as a 7-bit counter, and an external $\overline{\text{RFSH}}$ (refresh) signal. During an instruction-fetch cycle a few clock cycles are required to decode the new instruction and begin its execution; in this idle period the contents of the R register are placed on the lower 7 bits of the address bus and the $\overline{\text{RFSH}}$ line and $\overline{\text{MREQ}}$ line are taken low to indicate that the address bits can be used to refresh the dynamic memories. Therefore no extra hardware is required to refresh the DRAMs and as the processor is using its idle time to perform the refresh it is unnecessary to disable the processor, as would be required in other processor systems, in order to gain access to the buses. However, the penalty for this simple refresh scheme is that the Z80 cannot be halted or placed in a wait condition for any length of time, otherwise the refresh operation would not take place correctly and the DRAMs could lose their data.

INTERNAL ARCHITECTURE

The internal architecture of the Z80 is shown in Figure 9.6. As can be seen from this figure the register set contains similar types to those found in the MC6809. In the Z80 there is only one accumulator register, A, which is used as the primary source and destination for all of the 8-bit arithmetic and logical operations. The other registers B, C, D, E, H and L are all secondary registers for data storage, access to any of these registers being equally easy. However, the use that is made of each of these registers can be quite different. The six registers can be paired up, as shown in Figure 9.6, to produce three 16-bit register pairs. The HL pair is used as the main address pointer for accessing data in memory, since if this pair is used many of the instructions can execute in fewer cycles. There are only a limited number of instructions which deal with BC and DE as register pairs. Therefore these registers tend to be used as storage for operations involving the accumulator.

The Z80 also supplies an alternate set of registers labelled A', F', H', L', B', C', D' and E' which can be swapped with the main register set by executing two single-byte instructions; EX AF,AF' will exchange the accumulator and flag registers, whereas EXX will exchange the three registers pairs. After the exchange all subsequent register accesses will be performed on the active set

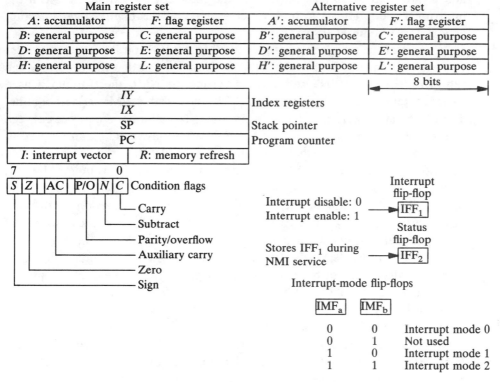

Figure 9.6. Internal architecture of the Z80.

of registers; thus only one set of instructions is required to access all the registers. Usually the alternate register set is retained for fast responses to interrupt requests; it is simply a matter of executing two fast instructions to swap registers, unlike the MC6809 where the interrupt causes the stacking of the machine state, a possibly time-consuming process. At the end of the interrupt-service routine the registers can be swapped back to their original state, thus preserving their contents.

The remaining registers IX and IY provide indexing capabilities similar to those found in the MC6809 X and Y registers. The Z80 only has one stack pointer (SP) to deal with all stack operations. The program counter (PC) is standard and performs the same function as the MC6809 program counter. The R register is used as a counter register for refreshing dynamic memories. Only the seven lower bits are incremented after each opcode fetch, the top bit is set or reset by any instruction which accesses that register. The interrupt-vector (I) register is used to store the top 8-bits of the page of memory containing the interrupt-service routines; this register is used by mode-2 interrupts. The Z80 has three interrupt modes designed to provide compatibility between the 8080 processor, non-Z80 processors and specially designed Z80 peripheral devices. The interrupt mode is determined by three instructions, IM0, IM1 and IM2, which set the bits in the interrupt-mode flip-flops to indicate mode 0, 1 and 2, respectively. Two further instructions DI

307

and EI are used to disable and enable the interrupt by setting the bit in the interrupt flip-flop (IFF_1) to 0 or 1, respectively. Thus control over interrupt action is very comprehensive.

As part of the CPU's response to an NMI the IFF_1 bit is set to 0, in order to disable the maskable interrupt ($\overline{INT}$). However, so that the processor can return to its previous $\overline{INT}$ state after servicing the NMI, the IFF_2 bit is used to store temporarily the "old" IFF_1 bit during the NMI-service routine. Therefore, after a return (RETN) from the NMI routine the processor can restore the IFF_1 bit to its previous value. Certain instructions can be used to interrogate indirectly the state of the IFF_2 bit (see later).

The response by the Z80 to the various interrupt modes is quite different, although the initial stages of the interrupt-service routine are the same. In mode 0 the interrupting device is expected to place an instruction on the data bus. This is normally the restart instruction which will begin a call to one of eight page-zero locations (0000, 0008_{16}, 0010_{16}, 0018_{16}, 0020_{16}, 0028_{16}, 0030_{16} and 0038_{16}). In mode 1 only the 0038_{16} restart location is used and the interrupting device does not have to supply any data. The mode-2 interrupts are specifically used for Z80 peripheral devices. In this case the interrupting device must supply a low-order byte so that, when combined with the I register, it forms a pointer to the address of an interrupt-service routine. Since the location should be a 2-byte address, the value of the lowest address bit must be zero; hence only 128 vectors may be stored in this interrupt table. Once the address in the table has been formed, the interrupt-vector address is obtained and execution of the service routine can then begin from this address. Hence the routine can be placed anywhere within main memory. This also means that different interrupting peripherals can use separate service routines for their particular requirements, rather than having to use a general-purpose routine, as in MC6809, which must determine which device caused the interrupt and then act accordingly. This may slow down the response to an interrupt.

The sets of alternate registers both have their own flag register which contains six flags. Certain flags are quite familiar and should now be accepted as being part of the condition-code or flag register. The C (carry) flag is used to hold the carry out of the most significant bit produced by any arithmetic operations; similarly the shift operations use the carry bit to store the shifted bit. The Z (zero) flag is set to 1 when an operation, arithmetic or logical, generates a zero result; a non-zero result generates a 0. The AC (auxiliary carry) flag is used to store the carry out of bit 3 into bit 4 (compare with the half-carry flag in the MC6809) for use when performing BCD arithmetic. The N (subtract) flag is also used in conjunction with BCD operations when performing the decimal-adjust operation, being set to 1 for subtract instructions and 0 for all add instructions. The S (sign) flag stores the value of the most significant bit of the result after the execution of any arithmetic or logical instructions (compare with the N flag in the MC6809). The P/O (parity/overflow) flag bit serves a dual purpose depending on the last instruction that was executed. With arithmetic instructions this bit is used to indicate an overflow condition as in the MC6809. However, during logical operations or shift operations the P/O bit is used to store the parity of the result, with a 1 indicating even parity and 0 indicating odd parity. If LD A,I or LD A,R (load *A* with *I* or *R*, respectively) instructions are executed then a copy of the IFF_2 flag is made in the P/O flag bit. Each of the flag bits will retain its current value until an instruction specifically alters that flag bit.

ADDRESSING MODES

The addressing modes available on the Z80 are quite extensive and utilize some instructions that can perform block moves of the memory given only the starting address of the block to move, the address to move the block to and the size of the block in bytes. Also certain instructions are capable of manipulating single bits within any one of the 8-bit registers. The simplest form of addressing, after the register modes, is immediate addressing, where the actual data to be used is contained within the instruction. The Z80 can support the immediate loading of 8-bit and 16-bit data. An instruction of this form would be LD BC,00EDH (the H indicating a hexadecimal value, as $ does for the MC6809). This instruction will load the *BC* register pair with the value $00ED_{16}$, *B* taking the most significant byte. The Z80 stores this data in the instruction as least significant byte followed by the most significant byte; therefore if this instruction were viewed in memory the data would appear as ED00H following the instruction. This method of data storage is intended to speed up data transfer along the 8-bit data bus into the CPU registers. The Z80 also supports the direct addressing mode for the transfer of 8-bit or 16-bit data from a 16-bit address location. This is similar to the extended addressing mode of the MC6809. An instruction of this form would be LD HL,(1FFH), load the *HL* register pair from address 1FFH (the () being used to indicate this mode to the assembler). Again the address is stored in reverse format, low byte followed by a high byte. In this case *L* would be loaded with the byte at address 1FFH and *H* with the byte at address 200H. If the *HL* register pair replaces the absolute address in the above brackets then this is termed implied addressing: for example, LD C,(HL) would load *C* with the contents of the address pointed to by *HL*. Similarly LD (HL),B would store the value of register *B* at the address pointed by *HL*. Any one of the seven 8-bit registers can be used to store or load data. Some instructions use *BC* or *DE* as data pointers. Any instruction that uses the (*HL*) register can also use one of the two index registers *IX* and *IY*, the only difference being that an offset can be added to the index register before use, e.g. ADD A,(IY+20H). This implied addressing mode can also be used with the SP (stack-pointer) register, with instructions such as PUSH and POP. Unfortunately only single register pairs can be pushed on the stack (e.g. *AF*, *BC*, *DE*, *HL*, *IX* and *IY*), unlike the MC6809 where several register pushes can be specified in one instruction, with the processor taking care of the order in which the push and pop takes place, thus eliminating the problem of confusing the order. However, it is possible to exchange the two top bytes of stack with the *HL*, *IX* or *IY* registers using for example EX (SP),HL. This is useful for stack-based high-level languages.

As with the majority of microprocessors the Z80 also supports relative addressing. However only an 8-bit relative signed displacement can be used with the Z80: this means that displacements of -126 and $+129$ are possible. The Z80 uses relative jump instructions, e.g. JR (jump relative), instead of the branch instructions used by the MC6809. The conditional jump instruction, like the conditional branch, performs a test on one of the flag-register bits C or Z. The other bit tests are only available with absolute jump instructions. One feature of the Z80 instruction set is the ability to conditionally return from a subroutine call. Most processors only allow a simple return from subroutine e.g. RTS on the MC6809; the Z80 uses RET cc, where cc is a test on one of the condition-code bits. If the condition is satisfied then the return is taken, otherwise sequential execution continues. Of course, a non-conditional return is also available. One further useful instruction is the DJNZ (decrement and jump if not zero) which will decrement the B register each time it is executed until it is zero. If B is not zero after decrementing then the relative branch is taken. This instruction is very useful for performing decrement-and-test loops with relative ease. The counter, register B, is set to the count required and then the loop is entered.

The modified base-page addressing has already been seen in connection with interrupt mode 0. In this addressing mode one byte of the RST (restart) instruction is used to directly call one of the eight locations listed earlier. This is quite useful for executing heavily used routines with the least amount of code; it is also useful, as has been seen, for dealing with interrupts.

One addressing feature that the Z80 possess, that is not present on any comparable 8-bit processor, is the ability to perform block transfers and searches using combinations of the internal registers. The block-transfer operations use the *HL* and *DE* register pairs to point to the areas from which and to which the data will be transferred, respectively. The *BC* register pair is used as a counter to determine the number of bytes that must be transferred. If the LDI or LDIR instructions are used then the *HL* and *DE* registers are incremented and *BC* decremented after the byte-transfer operation has taken place. If the LDI operation is used then the programmer must implement the loop; however, the LDIR instruction will repeat the transfer until the *BC* register is zero. Alternatively, using the LDD and LDDR the *HL* and *DE* register pairs are decremented after each byte transfer. Therefore using the LDI(R) and LDD(R) instructions the memory can be scanned up or down, respectively. There are also block-comparison operations in the form of CPI, CPIR, CPD and CPDR, which instead of performing a transfer simply compare the contents of A with the contents of the address pointed to by *HL*. The outcome of the comparison is recorded as normal in the case of the CPI and CPD instructions and can be acted upon accordingly. The repetitive CPIR and CPDR instructions will terminate the loop if the comparison is successful; thus when the loop terminates in this condition *HL* will be pointing to the byte that is equal to the contents of the A accumulator. If the comparison is not successful then the loop will terminate when *BC* has been decremented to zero.

As has already been seen the Z80 can utilize a separate I/O set of memory locations. Transfers to and from these I/O addresses are effected by the use of the IN and OUT instructions, respectively. Block-transfer operations similar to the above can also be used to transfer blocks of memory to an output port and read data in from an input port to a block of memory. However, with these transfer operations only the *B* register is used to store the count of the number of bytes to be used in the transfer.

9.3 The 16/32-bit processors

The 16-bit, and now the 32-bit, processors have brought about a major revolution in the personal-workstation arena. The immense power that is available from this range of processors far outstrips that available from any of the 8-bit processors. As experience with these processors increases, the likelihood is that they will dominate even the controller market; however, at their current high prices this is unlikely to occur for some time: the 8-bit processor will still be a very popular choice for simple applications. Almost all of the major semiconductor manufacturers now have 16/32-bit processors; Motorola's MC68000 series, Zilog's Z8000, Texas Instrument's 9900, National Semiconductor's NS32000 series and Intel's 8086 series. It is therefore very difficult to choose a suitable device to present in this type of text book. Most of the processors just mentioned have complete books devoted to a description of just one aspect of the processor, whether hardware or software; therefore to do justice to these processors is impossible. However, the next few pages should give some flavor of the computing power available from these devices. The MC68000 processor has been chosen as the example of a typical 16/32 bit processor; this is largely due to the author's experience of the device rather than any other considerations.

9.3.1 The Motorola MC68000 processor

The MC68000 was only possible because of the improvements in silicon technology that allowed a microprocessor of this complexity to be designed and fabricated. Although the MC68000 is termed a 16-bit processor, this only refers to the amount of data that can be placed on the data-bus pins. The internal architecture of the MC68000 is made up of 32-bit registers; therefore in order to fill a 32-bit register two reads from the 16-bit data bus are required, rather like the two-byte reads that are necessary to fill the 16-bit registers in an 8-bit processor. The MC68000 is quite flexible in the size of data it can handle: either 8-bit (byte), 16-bit (word) or 32-bit (long-word) data. The use of byte data sizes allows compatibility to be maintained between the 8-bit processors and their peripherals and also allows character data to be easily handled.

The main memory of the MC68000 is organized differently because of its ability to read/write 16 bits of data simultaneously. In order to perform this operation the memory chips used are arranged as odd and even blocks of memory, with one block connected to the 8 least significant data bits and the other block connected to the 8 most significant bits. Therefore if word data is to be read or written then both memory chips can be accessed simultaneously to read the full 16 bits of data. However, each device is assumed to be two separate addresses: an even address and an odd address. Therefore if a byte read/write is to be performed then either the odd or even address can be used. All opcodes, which are 16 bytes wide, must always lie on an even word boundary so that they can be fetched correctly; similarly all word and long-word data must be aligned with the even addresses. Figure 9.7 illustrates the memory organization for the MC68000.

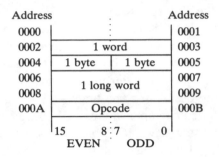

Figure 9.7. MC68000 memory organization.

PIN ASSIGNMENTS

The pin assignments for the MC68000 are shown in Figure 9.8. As can be seen, the buses can be split into the three types; however, the control bus can be further subdivided and shows how the system complexity has increased. The MC68000 can use two types of control bus, depending upon which type of peripheral it needs to control. In order to maintain compatibility with the older MC6800 peripherals a synchronous bus is provided which has the necessary E clock. All the older 8-bit processors use synchronous buses which usually means that the external devices must respond within a certain number of clock cycles unless some pin, such as MRDY, is available to stretch the access-cycle time. The MC68000 also possesses an asynchronous bus which allows the peripheral devices to acknowledge the receipt or availability of data itself; this means that the access can take the minimum amount of time required to complete the transactions. For example if RAM accesses were quicker than EPROM accesses then the MC68000 would be able to read and write data to the RAM in fewer of its own clock cycles than reading EPROM data. Whereas in the case of the synchronous bus the minimum access time is fixed by the processor, if an external device can be accessed more quickly then the processor cannot take this into account. Care must be taken when using the asynchronous bus, since if an address is accessed for which no device exists then the processor could wait indefinitely for the return acknowledgement, unless precautions were taken to prevent this within the

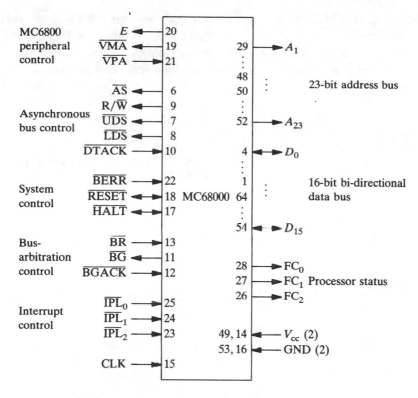

Figure 9.8. Signal-pin assignments for the MC68000.

hardware. As the various pins are discussed it should become clearer as to how the MC68000 can deal with bus accesses.

The MC68000 is such a large chip that two V_{cc} and ground lines are needed to avoid excessive voltage drops along conductors within the chip itself. The CLK signal must be supplied by an external oscillator as no on-chip oscillator is provided. The on-chip clocks and timing are then generated from this incoming clock to provide the necessary control. The clock should never be stopped as this will cause the internal data to be lost; for the same reasons the maximum and minimum pulse widths should not be violated. Currently the MC68000 is available in five frequencies: 4 MHz, 6 MHz, 8 MHz, 10 MHz and 12.5 MHz, with higher clock frequencies to be introduced. When the processor makes a memory access this is called a bus cycle and takes a minimum of 4 clock cycles. In the course of performing an instruction the processor may execute more than one bus cycle. The processor is also available in CMOS, which reduces the power consumption and increases the range of possible applications.

The system control signals allow the processor to be placed in a known state and to deal with hardware-error states. The $\overline{\text{RESET}}$ signal is actually bidirectional: if this line is driven low, along with the $\overline{\text{HALT}}$ line, on power up then the reset sequence will be entered. During this sequence the supervisor-stack pointer (SSP) $A7$ will be loaded with the contents of address

313

$000000 and the PC will be loaded with the long word at address $000004. These addresses form part of the exception-vector table, as shown in Table 9.1, where important addresses that are used during exception processing are stored; $\overline{\text{RESET}}$ is simply one type of the general exception-processing scheme.

Table 9.1. Exception-vector table for the MC68000.

Address (hex)	Vector number	Exception assignment
0000	0	RESET: initial supervisor stack (SSP)
0004	1	RESET: initial program counter (PC)
0008	2	Bus error
000C	3	Address error
0010	4	Illegal instruction
0014	5	Zero divide
0018	6	CHK instruction
001C	7	TRAPV instruction
0020	8	Privilege violation
0024	9	Trace mode
0028	10	Line 1010 emulator
002C	11	Line 1111 emulator
0030	12	Unassigned—reserved by Motorola
0034	13	Unassigned—reserved by Motorola
0038	14	Unassigned—reserved by Motorola
003C	15	Uninitialized interrupt vector
0040 ⋮ 005C	16 ⋮ 23	Unassigned—reserved by Motorola
0060	24	Spurious interrupt
0064	25	Level 1 interrupt autovector
0068	26	Level 2 interrupt autovector
006C	27	Level 3 interrupt autovector
0070	28	Level 4 interrupt autovector
0074	29	Level 5 interrupt autovector
0078	30	Level 6 interrupt autovector
007C	31	Level 7 interrupt autovector
0080	32	Trap #0 vector
0084	33	Trap #1 vector
⋮ 00BC	⋮ 47	⋮ Trap #15 vector
00C0 ⋮ 00FC	48 ⋮ 63	Unassigned—reserved by Motorola
0100 ⋮ 03FC	64 ⋮ 255	User-interrupt vectors

As well as being able to reset the processor by hardware means the processor can also execute the RESET instruction which takes the $\overline{\text{RESET}}$ pin low for 124 clock cycles and will therefore place any external devices in their reset state; however, the internal state of the processor will be unaffected. The $\overline{\text{BERR}}$ (bus-error) line is used to tell the processor when a memory access, i.e. bus cycle, has caused an error, e.g. accessing non-existent memory: this will stop the processor being in a wait state indefinitely. If this line is taken low by the external monitoring hardware then the processor can be forced to retry the bus cycle, provided the $\overline{\text{HALT}}$ line has also been taken low at the same time. This would give the program a chance to recover from a temporary hardware fault; however, if it where permanent then $\overline{\text{BERR}}$ would be reasserted during the re-run and some other action would have to be initiated. The bus error is another example of the exception processing on the MC68000. If the $\overline{\text{HALT}}$ line is taken low without any other signals then the processor can be halted at the end of the current bus cycle. All the three-state lines, e.g. address bus, data bus, etc., go to their high-impedance state. Also if the processor stops itself for any reason, such as a double bus error, then the $\overline{\text{HALT}}$ line will be taken low by the processor to indicate a fault condition to any external monitoring hardware.

The address bus has 23 lines (A_1 to A_{23}) allowing access to 16 Mbytes of memory; there is no external A_0 address line. This arises due to the way in which the memory is handled. Unusually for a processor of this type the data bus is entirely separate from the address bus. The data bus is bi-directional and, like the address bus, can be placed in a high-impedance state so that other devices can gain control of the buses. On some 16-bit processors, in order to reduce the number of package pins the address and data bus are multiplexed, which can increase the complexity of the external hardware in order to deal with the demultiplexing operation.

As previously mentioned, the asynchronous bus allows the MC68000 to be interfaced to memory chips that have differing access times. Of the 5 lines making up this bus, four are outputs and one is an input. The $R/\overline{W}$ (read/write) line performs the same operation as on the MC6809 and indicates whether the processor wishes to read or write data. The $\overline{\text{AS}}$ (address stobe) line is taken low to indicate when the address bus contains a valid address. The two strobe lines $\overline{\text{UDS}}$ (upper data strobe) and $\overline{\text{LDS}}$ (lower data strobe) are used to indicate which data lines contain valid data. If $\overline{\text{UDS}}$ is low then D_8 to D_{15} are valid and similarly $\overline{\text{LDS}}$ low indicates D_0 to D_7 are valid; obviously if they are both low then the complete data bus is valid. These strobe signals are used in conjunction with $R/\overline{W}$ to read or write word and byte information from/to the memory chips. The byte information can either be written to an even or odd byte address, depending on the state of the two strobe lines. In order to determine when the memory access should be terminated, the $\overline{\text{DTACK}}$ (data-transfer-acknowledge) input is provided. When the external device has completed the transfer it takes $\overline{\text{DTACK}}$ low, whereupon the MC68000 will either latch in the data on the data bus if it is a read cycle or begin to terminate the cycle for a write operation. Once the bus cycle has terminated in both cases the accessed device should remove $\overline{\text{DTACK}}$. If the $\overline{\text{DTACK}}$ signal is not asserted then the processor simply waits until either it is asserted or some external monitor asserts the $\overline{\text{BERR}}$ signal. The external monitor could be a simple timer that starts counting at the beginning of a cycle and either stops counting and is reset by a correctly terminated cycle, or after a certain maximum time asserts $\overline{\text{BERR}}$ to indicate an error condition. This maximum time should be chosen to be longer than the slowest device to be accessed. The timer is usually called a *watchdog* timer.

In order to handle the older MC6800 peripheral devices, which cannot deal with the asynchronous bus control, three special control lines are provided. The E (enable) signal is provided to synchronize the data transfers to the peripheral devices (see the MC6809 discussion earlier). If the MC68000 accesses the address of a MC6800 peripheral then the $\overline{VPA}$ (valid-peripheral-address) signal should be taken low so that the MC68000 knows that data transfers must coincide with the enable clock. After the $\overline{VPA}$ signal goes low the processor takes the $\overline{VMA}$ (valid-memory-address) signal low to indicate to the MC6800 peripheral that there is a valid memory address on the address bus; this also indicates when the transfers will be correctly synchronized to the E clock signal.

The 8-bit processors, discussed in Section 9.2, only had a limited number of interrupts available; however, on the MC68000 seven prioritized interrupt levels are possible, with level 1 interrupts having the lowest priority and level 7 interrupts the highest: level 7 interrupts are like non-maskable interrupts on other processors in that they cannot be disabled. Level 0 interrupts are used to indicate no interrupts. The interrupt request is encoded on the $\overline{IPL_0}$, $\overline{IPL_1}$ and $\overline{IPL_2}$ interrupt lines. The response to a particular interrupt is determined by the state of the interrupt mask in the system byte of the status register (Figure 9.9). If the interrupt is accepted then the status lines FC_0, FC_1 and FC_2 all go to logic 1 (Table 9.2) to indicate that the processor is acknowledging the interrupt; address lines A_1, A_2 and A_3 then indicate the interrupt level being acknowledged as a 3-bit code.

Table 9.2. Function-code table.

FC_2	FC_1	FC_0	Function
0	0	0	Unassigned
0	0	1	User data
0	1	0	User program
0	1	1	Unassigned
1	0	0	Unassigned
1	0	1	Supervisor data
1	1	0	Supervisor program
1	1	1	Interrupt acknowledge

Like the Z80 processor the interrupting device can then provide an 8-bit vector on the data bus to indicate the address of its interrupt-service routine in the exception-vector table (Table 9.1): this should be one of the user interrupt vectors. Alternatively, if the device is not capable of providing this vector then one of the autovector interrupts can be used by asserting the $\overline{VPA}$ signal when an interrupt-acknowledge bus cycle is taking place, instead of the normal $\overline{DTACK}$ that would be expected when the interrupting device had been acknowledged and placed its vector on the data bus. Again this provides support for the older peripheral devices without vector registers.

The processor-status lines (FC_0, FC_1, FC_2) are not only used to acknowledge an interrupt but they are also used to indicate the state of the processor during any particular bus cycle, as shown in Table 9.2. These states are valid when $\overline{AS}$ is low and can be used to stop a user program accessing sensitive data if required; this could be true in a multitasking environment. For example, supervisor data space could be differentiated from user data space by using the status lines in the memory-decode equation. Therefore, supervisor and user memory could co-exist at the same physical addresses, although in separate address spaces. Consequently when in supervisor mode (see later) only supervisor memory would be accessible; similarly in user mode only user memory would be accessible. Alternatively, user-mode memory accesses could be prevented from accessing supervisor memory, whereas supervisor mode could access all memory. This would mean that there were no separate address spaces, as above; however, it is a more normal situation. The MC68020 and MC68030 make even more use of the status lines than does the MC68000, mainly for co-processor interfacing and memory-management functions. This type of differentiation between address spaces can also been seen in the discussion of the MCS-48 single-chip microprocessor (Section 9.5.1).

Like all the processors discussed so far the MC68000 is capable of relinquishing control of its buses so that another bus master can take over to perform, for example, DMA operations. In order to be able to gain mastership of the bus there must be some signals which are supplied for bus-arbitration control. The arbitration simply describes the sequence of events that must occur before a new master can take over the bus. The MC68000 provides three pins that can be used to provide this control. The processor itself does not contain any circuits to determine who should control the bus if more than one potential bus master requested the bus simultaneously: this arbitration circuit must be built separately. The start of the cycle to take over the bus begins by the requesting device asserting $\overline{BR}$ (bus request). The processor responds by asserting $\overline{BG}$ (bus grant) to indicate that it will release the bus at the end of the current cycle. Once the $\overline{BG}$ has been asserted the requesting device can release $\overline{BR}$ after $\overline{BGACK}$ (bus-grant acknowledge) is asserted by the requesting device to indicate to the processor that it is now in control, assuming that no other device has taken $\overline{BGACK}$ low and that $\overline{AS}$ and $\overline{DTACK}$ are both high (signifying that the MC68000 has finished with the bus). The processor then removes $\overline{BG}$ and the new bus master maintains $\overline{BGACK}$ asserted until it has completed its task.

INTERNAL ARCHITECTURE

The internal architecture of the MC68000 is shown in Figure 9.9. There are eight 32-bit data registers ($D0$ to $D7$) which can deal with byte, word and long-word data. There are also seven 32-bit address registers ($A0$ to $A6$) which can deal with word and long-word data. All the data registers can undergo any arithmetic and logical operations. There are no special

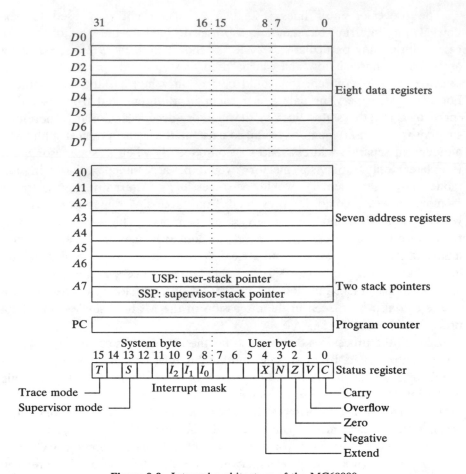

Figure 9.9. Internal architecture of the MC68000.

accumulators, etc.; therefore data operations are very easy to implement. The instruction set is less complex since all data registers are of equal status, no data register having a dedicated function. Operations involving the data registers affect the condition codes in the status byte. On the other hand, operations on the address registers do not affect these bits; hence the address registers can be used in addressing data without affecting the state of the flag bits. This again makes data access far easier since the programmer does not have to be concerned with the status bits that could be affected. Also none of the address registers, $A0$ to $A6$, performs a dedicated function. In fact, both address and data registers can also be used as index registers into memory. The only special-purpose register is address register $A7$ which acts as the user-stack pointer (USP) and supervisor-stack pointer (SSP), depending on the processor state. The program counter is 32-bits wide, even though only 24-bits are used during memory accesses.

As might be expected, the status register is much more complicated and contains 16-bits, as opposed to the 8-bit status register found in the MC6809 and the Z80. It can be considered as two separate parts, the user byte and

318

the system byte. The user byte contains the usual range of flags, whereas the system byte contains information relating to the current state of the processor. The N, Z, V and C flags all behave as described for the MC6809 processor. However, they only refer to the size of operation taking place, either byte, word or long word; for example, a carry out from bit 7 for byte or bit 15 for word or bit 31 for long word would set or reset the carry, that is, these flags represent the result on the chosen *operand* size. The X flag represents the true carry out from certain instructions and is used mainly for multiprecision arithmetic operations. The X bit was included since the C bit often gets used as a general-purpose test flag, when for example a subroutine operation completes successfully; hence a dual role of extension bit for multiprecision arithmetic and test flag were thought to warrant the extra X bit to simplify certain operations.

The lower three bits of the system byte contain the interrupt-mask flags. These flag bits can be set so that interrupts of only a higher priority are recognized. For example, if the bits are set to 101 (level 5) then only level 6 and 7 interrupts will be accepted for processing. However level 7 is equivalent to a non-maskable interrupt and cannot be inhibited. When an acceptable interrupt occurs the mask bits are set to the new level; therefore only higher-priority interrupts can cause the active service routine to be interrupted. The T (trace) bit allows for easier debugging, since if this bit is set to 1 then trace mode will be enabled and after the execution of each instruction an exception-vector address (Table 9.1) will be accessed to determine the address of a debugging service routine. Therefore it is quite an easy matter to single-step through the program code and display the contents of the registers and other important information stored in memory using a user-supplied service routine. A zero trace bit allows normal program execution.

The S (supervisor) bit indicates which one of the two possible states the processor is now in. The processor can exist in two states: either supervisor state ($S = 1$) or user state ($S = 0$). Supervisor state is set after a reset or during exception processing, i.e. interrupts, etc. The user state can be entered by clearing the S bit; however, the S bit cannot be set once in user mode. In supervisor mode certain privileged instructions can be executed, such as STOP, or any instruction which affects the complete status register, such as a MOVE to the status register. The processor state is indicated on the FC_0 to FC_2 lines; hence external hardware can also detect these two states and restrict access to hardware or memory that is only available to the supervisor. This is extremely useful for protecting access to programs and sensitive hardware in a multitasking environment. If a user does try to set the supervisor bit then a privileged exception will occur causing the supervisor state to be entered and the supervisor program to take over. The action of the supervisor program could then be to terminate this rogue user program, thereby barring entry to the privileged supervisor state. The supervisor state and the exception table allow user programs to access sensitive routines in a controlled way, in particular through the use of the TRAP instruction. The TRAP instructions, TRAP #0 to TRAP #15 (vectors 32 to 47), use the exception-vector

319

table to direct access to different supervisor routines, usually from a user program.

As can be seen from the exception-vector table (Table 9.1) the MC68000 can also deal with several different hardware and software problems (e.g. divide by 0) very easily. The exception table also provides vector locations that allow instructions which do not exist to be emulated. This is achieved by placing either 1010_2 or 1111_2 as the first bits of what would be an instruction opcode. When these *instructions* are encountered during instruction execution the vectors in the appropriate locations of the table are used to determine the address of the actual service routine to be accessed. Similarly, not all the 4-bit patterns at the start of an instruction are utilized; hence it is possible to detect when an attempt is being made to execute an illegal instruction and appropriate action can be taken through the use of the exception-vector table (vector 4). This means that the MC68000 can be stopped from executing random data, unlike the 8-bit processors which will continually execute whatever is read as an instruction.

ADDRESSING MODES

Although only 14 addressing modes are available, when combined with the 56 instructions which can utilize almost all these addressing modes, the total number of available instruction types becomes extremely large: well over 1000. Many of the available addressing modes are obviously derived from the MC68000's stable mates. The first addressing mode is implied addressing, where the register to be used is not explicitly mentioned in the instruction; for example, JMP (jump) in which the PC will be affected without specifying it explicitly. Therefore the effective address is entirely specified by the opcode itself. This is similar to the inherent addressing found in the MC6809.

As may be expected, there is a particular addressing mode called register-direct addressing which only uses either data registers or address registers. In these cases data is only transferred between registers, no external memory references being involved. An example of these instructions is MOVE.W D0,D1: this will move the word in $D0$ into $D1$ (i.e. copy the information). Similarly MOVE.L A3,D0 will move (copy) the complete 32 bits of $A3$ into $D0$. The extensions .B (byte), .W (word) and .L (longword) are used to specify the size of the operands involved in the instruction; most instructions can take these extensions. Some instructions can only deal with registers: these include SWAP, EXG and EXT. In fact the SWAP instruction can only swap the upper and lower words of a data register. The EXG instruction can exchange the contents of both address and data registers.

The MC68000 has an immediate addressing mode similar to that of the MC6809. In this mode the data to be moved into the register or memory follows immediately after the instruction; for example, MOVE.W #$1000,D0 would move 1000_{16} into the lower word of register $D0$. There is also a quick form of this instruction which employs some unassigned bits within the opcode itself for storing the immediate data. In the MOVEQ (move-quick) instruction there are eight available bits allowing values from −128 to +127 to be dealt with in only one word, all 32 bits of the destination being affected. The ADDQ (add-quick) and SUBQ (subtract-quick) instructions only have 3 bits available which are used to represent the values 1 to 8: this is ideal for loop structures etc. Address-register immediate addressing must be used with care since any word data (remember only word and longword data can be used with address registers) will be sign extended to the full 32 bits, that is, the top bit of the word (bit 15) will be copied into the top 16 bits of the long word. Thus if the top bit is 0 then all higher bits will be cleared and a 1 bit would cause all the top bits to be set to 1. The feature can be avoided by using MOVE.L to move the data into the register. This sign-extension feature will surface again when discussing short absolute addressing.

In the case of short absolute addressing the low-order half of the effective address follows the opcode in memory. The high-order 16 bits, necessary to complete the address, are formed by sign extending the low-order half of the address. Thus 32 bits of address can be generated in the range $000000 to $007FFF and $FF8000 through to $FFFFFF. Therefore the address values $0000 to $7FFF refer to the corresponding memory locations, but $8000 to $FFFF refer to the highest memory addresses because of the sign extensions. This addressing mode is like the page 0 or direct addressing modes found in the 8-bit processors; obviously addressing these locations is faster since only one word of data needs to be fetched before execution can be completed. An example of this addressing mode would be MOVE.W $1234,D1, which moves the contents of the word at address $1234 into D1. There is also a long address mode so that the complete address space can be accessed. In this case a full 32-bit address (two words of data) follows the opcode: thus no sign extension needs to take place. Usually the programmer need not be concerned with these modes as the assembler will choose the most efficient method, either short or long absolute addressing.

The MC68000 has a wide range of indirect addressing modes, all using the address registers. The term indirect in this context is not true indirection, as was used in the MC6809 processor, since the address of the operand is held in a register and not in the memory. Indirection is specified by the use of () around the address register in the assembly language. For example MOVE.W (A1),D1 would move the contents of the word addressed by A1 into data register D1. There are several different extensions to this one main addressing mode. These extensions allow the MC68000 to deal with such things as tables and arrays in a very straightforward manner.

As with both of the 8-bit processors discussed there are register-indirect modes with postincrement and predecrement available. In the postincrement mode the address register will be incremented by one, two or four for byte, word and long-word operations, respectively, after the data has been fetched from memory. Similarly in predecrement mode the address register will be decremented as above, before it is used to fetch the data from memory. Examples of these modes are MOVE.W (A0)+,D0 for postincrement and MOVE.L -(A6),D2 for predecrement. In the first example the word at the address pointed to by A0

321

will be put in $D0$ and then $A0$ will be incremented by 2, for word data. Similarly, in the second example $A6$ will first be decremented by 4 (long-word data) before it is used to move the contents of the memory address into $D2$.

The indirect mode of addressing can also be used with an indexed displacement. This displacement can either be a fixed 16-bit signed quantity or a fixed displacement plus an index register which can either be an address- or data-register word or long word. In the first mode the signed displacement is added to the address register to form the effective address to be accessed. As the data is 16 bits, a range of $+32767$ to -32768 is possible. This mode is quite useful for accessing the same part of data in a table that may contain several entries. An example of this mode would be MOVE.W D1,10(A1); in this case the effective address equals A1+10 and will be loaded from $D1$. To confuse matters, the displacement is given in bytes even if word or long-word data is being accessed. Using address-register indirect with displacement, the effective address is formed by the addition of three values: the address register, the contents of the index register (data or address register) and a displacement which is 8 bits signed ($+127$ to -128). An example of this addressing mode would be MOVE.L 4(A1,D1.W),D4; the effective address for loading $D4$ would be formed from A1+D1.W+4. This mode can be quite effective for accessing two-dimensional arrays, etc. All of these indirect modes can be used with the program counter replacing the address register in indirection, i.e. ADD $6(PC),D3 and ADD $4E(PC,A3),D3 would have the same effects as above except that the current program counter would be used as the address register. This mode is very useful for producing position-independent code.

The MC68000 can use program-counter relative addressing, allowing position-independent code to be easily written. The branch instruction is the main user of this addressing mode. Some useful branch instructions employ a counter, as well as a test on a bit in the user byte of the status register, to determine whether or not the loop should continue or terminate; hence, using these branch instruction and the postincrement and predecrement modes, it is possible to perform block moves similar to those available on the Z80 processor, although more than one instruction is needed.

Certain of the addressing modes can be used in conjunction with the MOVEM (move-multiple) instruction to store/load multiple registers to/from memory. This instruction takes a list of registers to be stored or loaded from memory and then uses one of the addressing modes to access the memory. For example, MOVEM.L D0-D2/A5,$2000 would move the full 32-bit contents of $D0$, $D1$, $D2$ and $A5$ to memory locations $2000, $2004, $2008 and $200C, respectively. The order in which the registers are stored/loaded from memory is fixed by the processor hardware and not their order in the assembler instruction. The register list occupies 16 bits after the opcode, a 1 in a particular bit position indicating that a register is to be used in the instruction. When used with postincrement and predecrement addressing modes, stack arragements can be maintained by the programmer. Any of the registers can be specified in the instruction, either individually separated by "/", or as a range separated by "-".

The MC68000 itself has spawned many more powerful processors, from the MC68010, MC68020 to the MC68030, which not only increase the functionability of the processor but also increase the speed of the devices; processors capable of running at 33 MHz are becoming available. Currently a 50-MHz MC68040 has been introduced by Motorola, as have more advanced processors from the other manufacturers.

9.4 RISC processors

The previous processors were examples of the so called complex-instruction-set computers (CISC): these processors have an extremely rich instruction set. However, it has been argued by some researchers that this rich instruction set can actually be detrimental to the performance of the overall computer system. They have presented alternative processors which do not have such a diverse instruction set but, nevertheless, can execute the instructions it does use at a very high rate, normally in only a single clock cycle, compared to the several clock cycles required by the CISC processors. These new processors are called reduced-instruction-set computers (RISC) and have become very popular with semiconductor manufacturers over recent years; indeed, most computer manufacturers are involved in producing their own RISC-type machines. The RISC processor to be discussed is the Acorn RISC machine (ARM) announced by Acorn Computers Ltd, in August 1985, making this the newest processor to be discussed in this chapter and one of the first commercial RISC-type systems. This part is also known as the VL86C010 and available from VLSI technology.

9.4.1 The Acorn RISC machine (ARM)

The ARM processor is in many ways like the MC68000 in that is capable of supporting high-level languages and also has a supervisor and user mode to deal with the requirements of a multitasking operating system. However, beyond that the similarities end. The ARM is a true RISC processor: most of its instructions can execute in one cycle, the number of available addressing modes is very small and there are only a limited number of simple instructions.

The ARM is a true 32-bit machine, the bi-directional data bus being 32 bits wide; hence data transfers to/from memory can take place in one operation. Most operations only deal with the full 32 bits (word); however, single-byte data is also handled in order to cope with character information. The address bus is only 26 bits wide, allowing 64 Mbytes of memory to be accessed. This memory is organized as 4-byte words that are aligned on even address boundaries; therefore instruction addresses (an instruction also occupies the full 32 bits) must be multiples of four. In order to simplify the way in which the program counter operates, it has been reduced to 24 bits with the two least-significant address bits being generated by hardware and forced to zero for word accesses. The program counter therefore points to the word data to be accessed and can be incremented by one to move onto the next word. Access to the bytes within the word can be specified by the least-significant address bits from the hardware and another control line on the processor which indicates whether a byte or word transfer is to take place. The ARM is designed to work in conjunction with three other specially designed chips: the memory controller (MEMC), the video controller (VIDC) and the I/O controller (IOC), all of which interface to the ARM to increase its functionality.

Although the ARM can theoretically access 64 Mbytes of memory currently this amount of memory is impractical; hence virtual-memory management (see Chapter 6) schemes can be used in order to improve performance. The use of a virtual-memory scheme with the ARM is greatly simplified by using the MEMC chip. The MEMC chip partitions the memory into three sections: the lower 32 Mbytes called logical RAM, then a section of 16 Mbytes called physical RAM, specially reserved for programs running in supervisor mode. Finally the top 16 Mbytes are allocated to ROM and I/O devices. The ARM can only use memory-mapped I/O like all the other processors discussed, with the exception of the Z80. The MEMC can be programmed to perform an address translation of the incoming logical address to an actual physical address within the real memory chips. Thus the division of memory into groups does not represent the real state of affairs as the pages belonging to each group could be stored anywhere in real memory. The MEMC can handle 128 pages: thus the size of each page depends on the size of the physical RAM, e.g. a 1-Mbyte memory would give 8-Kbyte pages and a 2-Mbyte memory would result in 16-Kbyte pages. If the MEMC translates an address which does not exist in physical memory then it can signal to the CPU that a page fault has occurred, whereupon action can be taken to read in the necessary page and perhaps swap out another page, the choice being dependent on the selection algorithm chosen, e.g. last-recently-used (LRU). Therefore, by using the ARM in conjunction with the MEMC, a demand-paged virtual-memory computer can be easily built. Of course CISC processors, such as the MC680X0, can also use memory-management units to produce demand-paged virtual-memory systems.

The main philosophy behind the RISC approach is to keep things "simple and fast". To make the ARM operations fast, interaction with the CPU for arithmetic and logical operations can only be performed with registers and not directly with memory locations. The ARM has 27 registers, making this easy to achieve, which are all 32 bits wide; however, only 16 registers are visible at any one time, the other 11 registers becoming available in different modes. The organization of the registers is shown in Figure 9.10. Only two of the registers are special purpose and one of those only partially; all other registers are general purpose and can take part in any instructions. The $R14$ register can be used in any operation; however, during one instruction, "branch with link", it is used to store the link, that is, a copy of the program counter from $R15$. Register $R15$ is dedicated to holding the 24-bit program counter and the status information as shown in Figure 9.10. The 24 bits of

	User	FIQ	IRQ	Supervisor (SVC)	
	R0	R0	R0	R0	
	R1	R1	R1	R1	
	R2	R2	R2	R2	
	R3	R3	R3	R3	
	R4	R4	R4	R4	
	R5	R5	R5	R5	
	R6	R6	R6	R6	General-purpose
	R7	R7	R7	R7	registers
	R8	R8_FIQ	R8	R8	
	R9	R9_FIQ	R9	R9	
	R10	R10_FIQ	R10	R10	
	R11	R11_FIQ	R11	R11	
	R12	R12_FIQ	R12	R12	
	R13	R13_FIQ	R13_IRQ	R13_SVC	
Link	R14	R14_FIQ	R14_IRQ	R14_SVC	Partially general purpose (link)
PC/flags	R15	R15	R15	R15	Dedicated (PC)

Detailed organization of R15

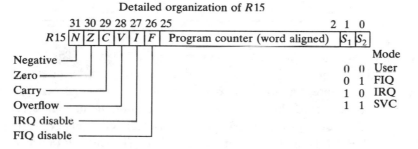

Figure 9.10. Internal architecture of the ARM processor.

the program counter occupy bits 2 to 25 of $R15$, the remaining bits, 0, 1 and 26 to 31, forming an 8-bit status register. The four status bits N, Z, C and V are set according to the results of the operations. Unlike many processors, the setting of the result-status bits by the register-to-register data operations can be controlled by the programmer simply by appending an S onto the end of the instruction mnemonic; this causes the assembler to set a bit in the instruction to indicate whether or not the result bits should be affected. The flag bit is recognized as being true if it is equal to 1. Another feature of the instruction allows for conditional execution, that is, an instruction may only be executed if a given combination of condition flags are set, otherwise the instruction is ignored. Alternatively, the instruction can be executed uncondi- tionally. Again, this feature is determined by appending the encoding condi- tion state onto the mnemonic, which is in turn assembled into the opcode for the instruction. In all, there are 16 different conditions that can be catered for within the instruction.

The remaining flags constitute the system status, the most important part being the mode bits S_1 and S_0 which determine the current executing state of the ARM (Figure 9.10). The supervisor mode is entered after a reset or certain error conditions have been detected: this is very similar to the supervisor mode of the MC68000. The IRQ (interrupt-request) and FIQ (fast-interrupt-request) modes are entered after external interrupt signals are detected. The I and F bits are used to enable/disable acceptance of an $\overline{\text{IRQ}}$ and $\overline{\text{FIQ}}$ request, respectively. When a $\overline{\text{RESET}}$, $\overline{\text{IRQ}}$ or $\overline{\text{FIQ}}$ occurs, the processor executes the instruction located in a vector table in low memory, as shown in Table 9.3, by loading the PC with the appropriate vector address.

Table 9.3. Exception interrupt vector table.

Vector	Mode	I	F	Comments
000 0000	SVC	1	1	RESET location
000 0004	SVC	1	x	Undefined instruction
000 0008	SVC	1	x	Software interrupt (SWI)
000 000C	SVC	1	x	Abort (pre-fetch)
000 0010	SVC	1	x	Abort (data)
000 0014	SVC	1	x	Address exception
000 0018	IRQ	1	x	IRQ location
000 001C	FIQ	1	1	FIQ location

All these conditions disable $\overline{\text{IRQ}}$ by setting the I bit to 1. The $\overline{\text{FIQ}}$ bit is only set by $\overline{\text{RESET}}$ or a $\overline{\text{FIQ}}$ request; therefore an $\overline{\text{IRQ}}$ can be interrupted by an $\overline{\text{FIQ}}$. When entering any one of these modes the registers that can be accessed are changed slightly as shown by the columns in Figure 9.10. Each mode uses its own private copies of various registers, supervisor mode uses $R13\_SVC$ and $R14\_SVC$, IRQ mode uses $R13\_IRQ$ and $R14\_IRQ$ and FIQ mode uses $R8\_FIQ$ to $R14\_FIQ$. This allows the various modes to run without the need to save registers, unless they need to use more than their alloted private registers. Obviously this saves time as normally registers have to be placed on a stack to save their contents for restoration when returning to the user program. The return address is saved in the private $R14$ register for each mode; therefore return from a particular mode simply involves storing $R14$ into $R15$, the program counter. None of the private registers can be accessed from user mode and vice-versa. The current mode of the ARM is indicated by the SPVMD signal line, rather like the FC_0 to FC_2 lines on the MC68000, so that external hardware, particularly the MEMC, can prevent user programs from accessing system data. The MEMC achieves this protection by associating with each page one of four *protection levels*. In user mode, levels 2 and 3 (the highest) are inaccessible, level 1 is read only and level 0 (the lowest) can be read or written. These restrictions do not apply to supervisor mode. The central 16 Mbytes of physical RAM is only accessible to supervisor mode as are certain I/O locations in the top 16 Mbytes, e.g. the MEMC registers themselves. All ROM and non-protected I/O locations are accessible by all modes. If an attempt is made to access these protected locations in user mode then an exception is generated and supervisor mode is entered at one of the abort-vector locations. In fact these abort lines could also be used to deal with virtual-memory management: the program would then have to determine why the abort occurred.

All the instructions are 32-bits long and must be on word boundaries when stored in memory. The instruction itself is split into several bit fields which specify such things as the condition to be tested for, opcode, the sources of the data and the destination of the data. Since a great deal of information is explicitly specified in the instruction itself, the control logic on the ARM can be greatly simplified, allowing a smaller design and higher execution speeds. Usually the contents of each field are produced by the assembler; hence the programmer is generally not concerned with this information. There are five groups of instructions.

1 *Data-manipulation instructions.* These 16 instructions are concerned with performing operations on the internal registers or immediate data and never with external memory.

2 *Load/store single registers.* There are only two instructions in this section available to transfer words or bytes of a register to/from memory.

3 *Load/store multiple registers.* This group of instructions allows several registers to be moved between the ARM and memory simultaneously. These instructions can also be used to perform very complex stack operations, that even some non-RISC processors cannot perform. The MC68000 possesses the MOVEM instruction which is very similar.

4 *Branch instructions.* These instructions provide a convenient means of accessing any of the 64-Mbyte address space. The alternative is to directly alter the contents of the PC by using the group-one instructions.

5 *Software interrupt* (SWI). This provides a means for user programs to access functions that only the operating system should provide; this is similar to the TRAP instructions available on the MC68000 processor. Using SWI causes the processor to access the SWI address at the bottom of memory (Table 9.3). Once the SWI routine has been entered the program must determine the operating-system call to perform; this is achieved by encoding a value in the lower 24 bits of the SWI opcode, which can be read by software decoding of the instruction.

All of the instructions conform to the following standard assembler format.

OP{cond}{s} ⟨dst⟩, ⟨ls⟩, ⟨rs⟩

The OP is the actual assembler mnemonic to be used, for example ADD, AND or SUB. The parts in {} brackets are optional and do not need to be specified as the assembler will use default values. The "cond" is used to specify one of 16 conditions to be tested before executing the instruction. The conditions are specified by two-letter groups, the default being AL (always). If present, the *S* bit causes the instruction to alter the result flags. Most, but not all, instructions then require 3 operands. The first is the destination ⟨dst⟩ of the result produced by the instruction. The ⟨ls⟩ is always a register, like the destination, and indicates which register is to be used in forming the result. The ⟨rs⟩

operand can either be a register, immediate operand or a shifted or rotated register. For example consider the following group-one operations.

1 `ADDEQ R2,R1,R0` will only be executed if the Z flag is true because of the `EQ` conditional specifier. If the instruction is executed then $R2 = R1 + R0$. The result status will not be altered as the S bit has not been set.

2 `ADDS R3,R1,#1` will always be executed and will set the result-status bits. The result of this instruction is $R3 = R1 + 1$. This is an example of immediate (#) data. The immediate data is actually encoded within 12 available bits of the 32-bit instruction. However, rather than simply use the 12 bits to represent either 0 to 4095 or -2048 to $+2047$ signed data, the immediate data is represented as two separate fields of 4 and 8 bits. The 8 bits are used to represent the 256 possible values, the other 4 bits, called the position bits, are used to indicate where the 8-bit value lies within the 32-bit word to be used. The position is given by rotating the 8 bits to the right through the 32-bit word; the number of bits to rotate is given by $2 \times$(position bits). If a value cannot be represented in this way then the assembler will generate an error message; this would occur with any values needing more than eight bits to be represented.

3 `ADD R0,R1,R2,LSL#1`. This instruction will always execute without altering the result-status bits. It will produce $R0 = R1 + 2 \times R2$, since this is the shifted form of the ⟨rs⟩ operand. The type of shift is specified by the fourth operand in the instruction, `LSL` (logical shift left) in this case, and the number of bits to shift, one in this case. There are several different shift instructions available: logical shifts, arithmetic shifts and rotates, both left and right forms being possible. Although in this case an immediate value has been given for the number of shifts, a register could have been used. Also the data is only shifted for the operation, the register, in this case $R2$, being left unaffected.

The two group-two instructions load (`LDR`) and store (`STR`) only have two addressing modes called pre-indexed and post-indexed. With both `LDR` and `STR` words or bytes can be specified, words being the default; a `B` needs to be appended to the instruction for byte transfers. In the pre-indexed addressing mode the address used to access the data is formed by adding a base register to an offset, either immediate or register, before accessing the memory location. An option exists to write the calculated address back into the base register; this is called *write back*. The immediate data used in the offset is 12 bits with one magnitude bit. If a shift or rotated register is used then the number of bits must be immediate, not a register.

In the case of post-indexed addressing the offset is added to the base register after the instruction has executed; obviously in this case the base register is always updated with the calculated address. The `LDR` and `STR` instructions never affect the status bits even if $R15$ is the destination of an `LDR` instruction or a base register. There is a special form of pre-indexed addressing which can be used with the `LDR` instruction and involves the use of the program counter as the base register instead of $R0$ to $R14$. In this form of pre-indexed addressing the assembler needs to determine the offset of the address to be accessed from the value of the PC, the maximum allowed offset is ±4095. A problem also arises since the instructions on the ARM are pipelined, that is, as one instruction is being executed, another has been fetched and is being decoded, while yet another is in the process of being fetched. Consequently the PC is actually pointing to an address two words from the currently executing instruction: fortunately the assembler will take care of this problem. The branch instructions, which use PC-relative addressing to form the branch address, also suffer from this calculation problem as well as needing to deal with the pre-fetched instructions depending upon the outcome of the conditional-branch instruction.

To perform a call to a subroutine a traditional processor might use a BSR (branch-subroutine) or a JSR (jump-subroutine) instruction and save the return address from the subroutine on the stack. However the ARM has no stack, as such, and must save its return address in $R14$ (as do most of the exception and interrupt routines), suitably altered from the contents of $R15$ (which is not actually pointing to the next instruction), after the BL (branch-with-link) instruction, as it is called. To return from the BL instruction is simply a matter of moving $R14$ back into $R15$, using the MOV instruction; the status bits can also be restored by appending S to the instruction. If a call from the subroutine is to be made then action must be taken to preserve $R14$ before the call, otherwise the return address will be lost. If only single levels of subroutine calling are used then this mechanism can provide a very fast call-and-return turn-around time.

The ARM provides several exception mechanisms for dealing with undefined instructions so that unsupported features, that could be implemented in a new ARM processor as a co-processor chip, can be emulated. The system can also deal with addresses which exceed the current 26-bit limit, thus catching problems that may cause wrap around of the system address into other data. One final point concerns the position of the FIQ vector. This vector has been deliberately placed at the end of the exception table so that actual service-routine program code can be placed there, thus avoiding the branch delay associated with the other locations. Also extra private registers are allocated to this mode, obviating the need, hopefully, for saving some registers. All this is intended to make the FIQ have a very fast interrupt-service time for dealing with events that might require such a turnaround.

9.5 Single-chip computers

Normally a microprocessor-based computer system requires several components, e.g. microprocessor, RAM, ROM, I/O devices and any ancillary logic. This is quite a large array of components that are needed to produce even the simplest of systems; the printed circuit board on which the chips are mounted could be equally complex. Occasionally a designer may want to use a microprocessor in a dedicated control application, for example to control a car engine or perhaps a washing machine. Using this approach, if the design of the item to be controlled changes slightly, a simple alteration in the microprocessor program would most probably accommodate the new features. Of course interface circuitry could be needed to convert digital signals to analog voltages and vice versa. Implementation of such a system could be quite costly if it had to be produced in large quantities and had to be rugged and reliable to survive in a hostile environment, negating any advantages gained from the use of a microprocessor in the first place. Manufacturers of microprocessor chips have long realized that a single chip that contained all the main components of a computer could provide a very simple, cheap and reliable design. These single-chip microprocessors usually have a limited amount of RAM and ROM, they also provide digital input and output lines and in some cases several multiplexed analog lines which enter an analog-to-digital (A/D) converter. Therefore, provided the control application does not require an excessively long program, a single chip containing the processing elements can be used. Some single-chip processors only contain ROM which must be programmed during manufacture; however, versions also exist which contain EPROM and can be programmed and erased by the user. Also to make the devices easier to program, in general, the core microprocessor is based on well-known processors already produced by the manufacturer. For example, Motorola produces the MC68705 single-chip processor which is based on the well-known MC6805 processor. Some devices are now being fabricated in CMOS, allowing battery-powered equipment to be easily produced with very few components. The chip to be discussed is the Intel MCS-48 series: these devices are very simple internally but nevertheless illustrate the philosophy behind the single-chip microprocessor architecture.

9.5.1 The MCS-48 series single-chip microcomputers

The various members of this family of chips are all self-contained 8-bit devices with RAM, ROM, 27 I/O lines and the necessary timing/control logic. All the series are pin compatible, the major differences being the size of on-board RAM and ROM (when available); some devices in the series do not contain any internal ROM. The size of internal RAM is quite small, ranging from 64×8 bits for the 8048 to 256×8 bits for the 8050. The ROM, when present, ranges in size from $1K \times 8$ bits up to $4K \times 8$ bits. Therefore by today's standard the internal storage area is quite small; however, these devices are only intended for use in very dedicated applications. To this end the instructions are mainly single bytes, allowing very efficient use of the scarce ROM; a few instructions are two bytes in length. Also the instruction set contains a number of instructions for handling bits and performing arithmetic functions. If the program does exceed the internal ROM storage then up to 4 Kbytes of external ROM and 256 bytes of data memory can be added externally.

The pin assignments for the 48 series of devices is shown in Figure 9.11. The main difference between this and the other processors discussed so far is the absence of any dedicated address and data bus. Instead there are two (port 1 and port 2) 8-bit programmable bidirectional latches that can be used to transfer digital data to control external circuitry. The third 8-bit bus is

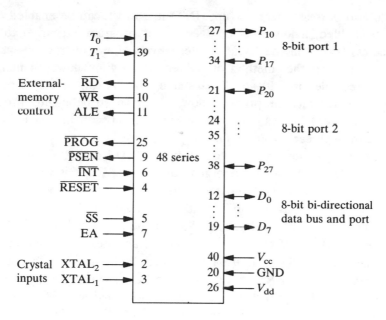

Figure 9.11. Signal-pin assignments for the 48-series chips.

used when external memory is fitted for reading and writing data; alternatively, it can be used as a third port similar to port 1 and port 2. However, this bus is also associated with several control signals which allow it to communicate with external memory very like a normal processor. The $\overline{RD}$ and $\overline{WR}$ lines are used to indicate when this bus port is being used for a read or write operation, respectively, by going low. The $\overline{PSEN}$ (program-strobe enable) is also used to indicate when an external program fetch is taking place. If an external program access is to be made then the 12-bit address from the program counter is split into a low-order 8 bits, which are applied to the bus (D_0 to D_7) and an upper 4 bits, which are applied to P_{20} to P_{23} (port 2), the falling edge of the ALE (address-latch-enable) signal then indicating to the external memory when this address is valid. The PSEN signal then goes low to access the data from memory and on the next rising edge of this signal the data is latched into the processor, completing the external program fetch. Reads and writes to external data memory are achieved in a similar fashion with either the $\overline{RD}$ or $\overline{WR}$ signal replacing the $\overline{PSEN}$ signal for timing purposes. The $\overline{PROG}$ output goes low in order to access external I/O expander circuits. These are used to increase the range of ports available to the main processor; the P_{20} to P_{23} signals are also used in conjunction with this signal. The timing of all these signals is performed by an on-chip oscillator which only needs a crystal connected across $XTAL_1$ and $XTAL_2$ to function. If EA (external access) is high then all memory accesses will be made externally.

The $\overline{RESET}$ signal is used to set the processor to a known state and causes the program to start executing from address 0 in the ROM. The processor also possesses an interrupt ($\overline{\text{INT}}$) line which can be enabled or disabled. If enabled, a low level on $\overline{\text{INT}}$ causes program execution to begin from address 0003_{16} in the ROM. Alternatively, jump instructions can test this line to give another method of redirecting program flow. Similarly the signals T_0 (testable input 0) and T_1 (testable input 1) can be tested by jump instructions to redirect the program flow. To assist in debugging programs, the $\overline{\text{SS}}$ (single-step) line can be used in conjunction with the ALE input to single-step through each instruction.

The internal memory map of the on-board RAM and ROM is shown in Figure 9.12. As can be seen, the organization of the ROM is very simple for the three devices. The lower few locations of memory are used for the various vectors to deal with $\overline{\text{RESET}}$, $\overline{\text{INT}}$ and an interrupt from the internal counter. In the case of the 8048 and 8049 if the 12-bit PC goes beyond the size of the internal ROM then external memory accesses will be automatically made. The internal RAM is used to contain both program data and two banks of registers $R0$ to $R7$; either bank can be selected by using the SEL RB (select-register-bank) instruction. The two lower registers $R0$ and $R1$, of each register bank are used to indirectly address the RAM. Part of the RAM between addresses 8 and 23 can be used as a stack area for storing the return addresses of subroutine calls; however, the stack is limited to only eight levels since an address occupies two bytes. If a stack is not required then this area can be used as normal RAM.

Unlike the processors described so far, the single-chip microprocessors do not, in general, possess a contiguous address spacee. For example, within the MCS-48 processor the RAM and ROM do not occur in the same address space, as both RAM and ROM have an address location 0. Indeed, the MCS-48 processor uses special instructions to access the RAM and ROM portions of its memory array. This situation is very similar to, but not the same as, the address spaces available on the MC680X0 processors (see earlier) where the status lines (FC_0, FC_1 and FC_2) can be used to form completely separate address spaces, which may independently access RAM and/or ROM. However, the MC680X0 processor would still use its normal instructions to access any of the address spaces, when in either supervisor or user mode.

There are very few addressing modes available. Any operations involving arithmetic/logical functions must use an accumulator register (A) unless they involve a simple increment or decrement operation. Hence most address modes involve A as either the source or destination of an instruction. The A accumulator can either read immediate data or read/write any of the registers from the selected register bank and the user memory by use of the $R0$ and $R1$ indirect registers. Of course there is a wide range of instructions that can access the data ports, for either reading or writing data. If the bus port is used then this port can be accessed such that the data is strobed in with the $\overline{\text{RD}}$ and $\overline{\text{WR}}$ lines. Generally the instruction set has been designed to give optimum performance for control applications with most instructions executing in one cycle and only occupying one byte of memory.

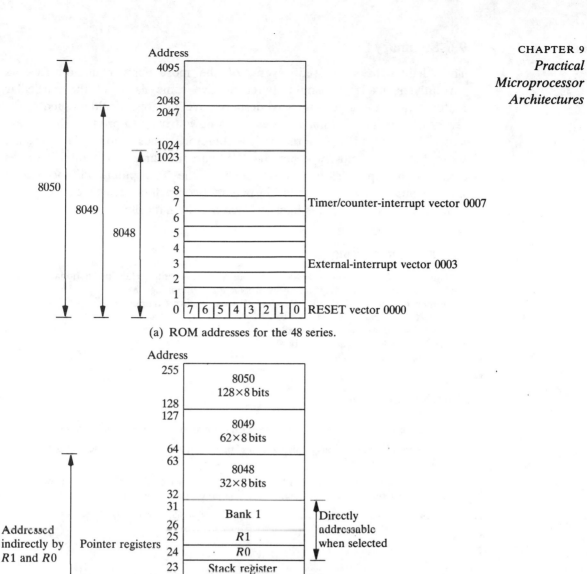

(a) ROM addresses for the 48 series.

(b) RAM addresses for the 48 series.

Figure 9.12. Internal architecture of the 48-series devices.

Although only the 48 series of devices has been briefly discussed, there are many such devices, each offering different functions that could be useful in a control application, for example A/D converters, serial communication links and large memory storage. As device technology improves and the level of integration increases, the functionality of these single-chip microcomputers will almost certainly increase.

333

9.6 Summary

This chapter has discussed several of the most popular microprocessors currently available: it should therefore give some flavor of the particular features that are used and why designs have progressed along such lines. Every manufacturer of microprocessors is now striving to produce faster devices with enhanced characteristics; it is therefore almost impossible to predict what the future holds, whether the RISC architectures will dominate or, as some predict, parallel processors, such as the Transputer, will become a major computing element found in personal computers. All that is certain is that the processing speed is likely to increase dramatically.

9.7 Further reading

1 Cahill, S. J. (1987). *The single chip microcomputer*. Prentice-Hall International.
2 Cockerell, P. (1987). *ARM assembly language programming*. MTC.
3 *Technical notes on the VL86C010 ARM chip set*. Acorn Computers.
4 *The MC6809 data sheet*. Motorola.
5 *The MC68000 data sheet*. Motorola.
6 *The MCS-48 series data sheet*. Intel.
7 *The Z80 data sheet*. Zilog.

9.8 Problems

9.1 List the advantages and disadvantages of the two major types of microprocessor.

9.2 Discuss the implications of using a new processor that is object-code compatible with an existing processor.

9.3 Given the following state of a MC6809 computer, what will be the effect of the following instructions? Each instruction is executed on the initial state.

$$A = \$01, \quad B = \$02, \quad X = \$2000, \quad DP = \$00.$$

	⋮
0010	$03
0011	$02
0012	$07
	⋮
2000	$00
2001	$10
2002	$12
	⋮

(a) LDA #$A0;   (b) ADDA $10 (add operation on the A accumulator);

(c) LDB ,X;   (d) LDB 2,X;

(e) LDD [0,X++];   (f) ADDD $10 (add operation on the double accumulator).

9.4 Given the following state of a Z80-based computer, what will be the effect of the following instructions? Each instruction is executed on the initial state.

$$A = \$01, \quad B = \$05, \quad C = \$00, \quad D = \$10,$$
$$E = \$00, \quad H = \$20, \quad L = \$00.$$

2000	$10
2001	$11
2002	$44

(a) LD A,2CH;   (b) ADD A,B

(c) ADD A,(HL);   (d) LD (2000H),HL;

(e) XOR A (bit-by-bit exclusive-OR operation);   (f) LDI.

9.5 Given the following state of a MC68000-based workstation, what will be the effect of the following instructions? Each instruction is executed on the initial state.

$$D0 = \$00000001, \quad D2 = \$00000002, \quad D6 = \$00000001,$$

$$D7 = \$99000000, \quad A0 = \$00001000, \quad A6 = \$00001004.$$

1000	$33	$78
1002	$55	$44
1004	$AA	$FE
1006	$03	$00
1008	$00	$01
100A	$11	$00
100C	$FF	$FF

(a) MOVE.W #$F000,A0;   (b) MOVEQ #1,D0;   (c) ADD.W 8(A0),D0;

(d) MOVE.L D7,(A6)+;   (e) MOVE.L #0,D7;   (f) MOVE.W D6,$A(A0,D2.W).

9.6 Given the following state of an ARM RISC processor, what will be the effect of the following instructions? Each instruction is executed on the initial state.

$$R0 = 0, \quad R1 = 1, \quad R2 = 2, \quad R3 = 3, \quad R5 = 5.$$

(a) ADDS R1,R2,R3;

(b) ADD R2,R3,R1,LSL#1;

(c) SUB R0,R0,#1 (subtract operation);

(d) EORS R5,R0,R1 (bit-by-bit exclusive-OR operation);

(e) ADD R1,R1,R1,LSL#1;

(f) STR R0,[R2,#2].

9.7 What effect would the following operations have on the flags of the condition registers of the MC6809 and Z80 processors? (**Note**: This question requires the use of datasheets for both processors.)

(a) $10110110_2 + 10111111_2$;   (b) $28_{10} + 30_{10}$;

(c) $-10_{10} + -20_{10}$;   (d) logical shift left 20_{10}, carry = 0;

(e) rotate right 32_{10}, carry = 0;   (f) logical shift right 255_{10}, carry = 0.

9.8 If an MC68000 system has been designed such that one half of memory is devoted to user programs and the other half to supervisor programs, produce a circuit that will prevent a user program accessing supervisor data. Assume that A_{23} is used to distinguish between the two blocks of memory and that when $A_{23} = 1$ the supervisor RAM is being accessesed. (**Hint**: Table 9.2 will be useful.)

335

9.9 Describe how the MC68000 vector table can be used.

9.10 What circumstances would warrant the use of a single-chip microprocessor?

9.11 Most microprocessors have instructions that allow the programmer to call subroutines: for example, JSR (jump to subroutine) on the MC68000. Therefore explain why the MC68000 and ARM processors should require the TRAP and SWI instructions, repectively?

9.12 Explain how the MC68000 maintains compatibility with the older MC6800 peripheral devices.

9.13 A microprocessor architecture is described as being *orthogonal* if the instruction set is consistent in the way data types and addressing modes can be used. A truly orthogonal architecture would have a regular instruction set with very little duplication in the operations performed by each instruction. Using the processors presented in this chapter as examples, compare and contrast their instruction sets with regard to orthogonality.

Appendices

Appendix 1: NeMiSyS Software

There are two NeMiSyS programs supplied with this book; an assembler and a simulator. The assembler generates object files that can be loaded by the simulator for execution. Input to the assembler is in the form of a NeMiSyS assembly-language source file, which is a text file produced by any editor or wordprocessor with a non-document (ASCII file) mode. The NeMiSyS assembly language is described in Chapter 5, and the assembler and simulator can be used for all the assembly examples and exercises in this chapter. The simulator embodies the NeMiSyS processor described in Chapter 4. It has memory-mapped I/O for the keyboard, screen and a simple counter, plus a small amount of RAM. The simulator has a monitor that loads and executes the object files produced by the assembler. It has facilities for single-stepping through programs an instruction at a time and for displaying and modifying registers and memory. A micro-instruction step feature is provided so that the operation of the NeMiSyS processor can be investigated at the micro-code level. This uses the same register-transfer language notation as Chapter 4. The monitor uses a full-screen display with a simple menu for selecting its various options. Other areas on the monitor screen include register display, with temporary registers in micro-step mode, memory display and NeMiSyS screen output. There is an option to run the simulator in full-output-screen mode, which provides a 20-line display for memory-mapped output.

The NeMiSyS software is supplied on an IBM PC 360-Kbyte floppy disk. It should execute on any IBM PC microcomputer or true compatible using PC-DOS or MS-DOS, version 2.1 or later, with a mono, CGA, EGA, VGA or Hercules adaptor. However, the software may not work with some older PC-compatible microcomputers with a CGA adaptor.

The disk contains at least the following files.

ASSEMBLE.EXE	The NeMiSyS assembler.
NEMISYS.EXE	The NeMiSyS simulator.
NEMISYS.MIC	Microcode for the NeMiSyS simulator.
GUIDE.TXT	A short user guide to the software.

Before anything else is done, put a write-protect tab on the software disk. Then make a copy of it using the COPY or DISKCOPY command.

To run the assembler, prepare a source file using an editor or a suitable word processor. Place the software disk in the default drive and then run the assembler by using the following command.

ASSEMBLE [+S] [+L] [+W] [b:]filename[.filetype]

filename[.filetype] is the name of the source file to be assembled. If no file type is given it, a default of ASM will be used. [b:] is the drive that holds the source file. This can be omitted if the source file is on the disk in the default drive. The object file produced by the assembler will have the same name as the source file, but will have a type of OBJ. It will be on the same disk as the source file.

The option +S can be used to suppress the output listing written on the screen. When this option is used, only error messages will be printed on the screen. The options +L and +W will direct a full copy of the output listing to a disk file. This will be have the same name as the source file, but will have a type of LST. +L will produce an 80-column listing, while +W selects a wide listing format.

Once an object file has been produced the simulator can be used to run it. The simulator is executed by entering the following command.

NEMISYS [+T] [+V] [b:]filename[.filetype]

filename[.filetype] is the name of the object file to be run. If filetype is not given, a default of OBJ will be used. The drive that holds the file is given by [b:]. This can be omitted if the file is on the disk in the default drive. Alternatively, the object file can be selected, after the simulator has started, by using the load and go options.

Specifying +T will produce a file called NEMTRACE.LOG containing a list of all the instructions executed while the simulator's macro-step option is being used. When the micro-step option is used, this file will also contain the micro-code for each instruction executed. Specifying +V will produce a log file, called NEMVDU.LOG, of all the output appearing in the VDU section of the display.

For more information on the assembler and simulator, print the GUIDE.TXT files on the software disk.

Appendix 2: Character sets

Since only 1's and 0's are available within the microprocessor, combinations of these bits must be used to represent the characters (letters, numerals, etc.) that are used in the environment outside the computer. If computers are to communicate easily with each other then it is necessary that different machines use the same codes to represent the same characters. The two most widely used standards for character representation are listed below.

ASCII (AMERICAN STANDARD CODE FOR INFORMATION INTERCHANGE)

This is the more commonly used of the two systems. It is basically a 7-bit code, but often an eighth bit is added, either as a parity bit or sometimes just as a leading zero. The hexadecimal codes given below assume a leading zero.

Table A2.1. The ASCII character set.

Character	Code Binary	Hex	Character	Code Binary	Hex
NULL	000 0000	00	@	100 0000	40
SOH	000 0001	01	A	100 0001	41
STX	000 0010	02	B	100 0010	42
ETX	000 0011	03	C	100 0011	43
EOT	000 0100	04	D	100 0100	44
ENQ	000 0101	05	E	100 0101	45
ACK	000 0110	06	F	100 0110	46
BEL	000 0111	07	G	100 0111	47
BS	000 1000	08	H	100 1000	48
HT	000 1001	09	I	100 1001	49
LF	000 1010	0A	J	100 1010	4A
VT	000 1011	0B	K	100 1011	4B
FF	000 1100	0C	L	100 1100	4C
CR	000 1101	0D	M	100 1101	4D
SO	000 1110	0E	N	100 1110	4E
SI	000 1111	0F	O	100 1111	4F
DLE	001 0000	10	P	101 0000	50
DC1	001 0001	11	Q	101 0001	51
DC2	001 0010	12	R	101 0010	52
DC3	001 0011	13	S	101 0011	53
DC4	001 0100	14	T	101 0100	54
NAK	001 0101	15	U	101 0101	55
SYN	001 0110	16	V	101 0110	56

Table A2.1. *Continued*

Character	Code		Character	Code	
	Binary	Hex		Binary	Hex
ETB	0010111	17	W	1010111	57
CAN	0011000	18	X	1011000	58
EM	0011001	19	Y	1011001	59
SUB	0011010	1A	Z	1011010	5A
ESC	0011011	1B	[	1011011	5B
FSI	0011100	1C	\	1011100	5C
GS	0011101	1D	]	1011101	5D
RS	0011110	1E	^	1011110	5E
US	0011111	1F	_	1011111	5F
SP	0100000	20	`	1100000	60
!	0100001	21	a	1100001	61
"	0100010	22	b	1100010	62
#	0100011	23	c	1100011	63
$	0100100	24	d	1100100	64
%	0100101	25	e	1100101	65
&	0100110	26	f	1100110	66
'	0100111	27	g	1100111	67
(	0101000	28	h	1101000	68
)	0101001	29	i	1101001	69
*	0101010	2A	j	1101010	6A
+	0101011	2B	k	1101011	6B
,	0101100	2C	l	1101100	6C
-	0101101	2D	m	1101101	6D
.	0101110	2E	n	1101110	6E
/	0101111	2F	o	1101111	6F
0	0110000	30	p	1110000	70
1	0110001	31	q	1110001	71
2	0110010	32	r	1110010	72
3	0110011	33	s	1110011	73
4	0110100	34	t	1110100	74
5	0110101	35	u	1110101	75
6	0110110	36	v	1110110	76
7	0110111	37	w	1110111	77
8	0111000	38	x	1111000	78
9	0111001	39	y	1111001	79
:	0111010	3A	z	1111010	7A
;	0111011	3B	{	1111011	7B
<	0111100	3C	l	1111100	7C
=	0111101	3D	}	1111101	7D
>	0111110	3E	~	1111110	7E
?	0111111	3F	DEL	1111111	7F

Notes

1 Codes for the upper- and lower-case letters are the same except for bit 5: this is set for lower case and reset for upper case.

2 The numerals 0 to 9 are represented by 30_{16} to 39_{16}, respectively. Therefore the lower four bits of the code represent the numeral directly.

3 The control characters are generated by resetting bit 6 of the alphabet character. For example, control-H would produce BS (backspace).

EBCDIC (EXTENDED BINARY-CODED DECIMAL INTERCHANGE CODE)

This 8-bit code is less widely applied than ASCII and is used mostly in large IBM machines.

Table A2.2. The EBCDIC character set.

Character	Code Binary	Hex	Character	Code Binary	Hex
NULL	0000 0000	00	:	0111 1010	7A
PF	0000 0100	04	#	0111 1011	7B
HT	0000 0101	05	a	0111 1100	7C
LC	0000 0110	06	'	0111 1101	7D
DEL	0000 0111	07	=	0111 1110	7E
SOH	0000 1001	09	"	0111 1111	7F
STX	0000 1010	0A	a	1000 0001	81
ETX	0000 1011	0B	b	1000 0010	82
ENQ	0000 1101	0D	c	1000 0011	83
ACK	0000 1110	0E	d	1000 0100	84
BEL	0000 1111	0F	e	1000 0101	85
TM	0001 0011	13	f	1000 0110	86
RES	0001 0100	14	g	1000 0111	87
NL	0001 0101	15	h	1000 1000	88
BS	0001 0110	16	i	1000 1001	89
IL	0001 0111	17	j	1001 0001	91
DLE	0001 1000	18	k	1001 0010	92
DC1	0001 1001	19	l	1001 0011	93
DC2	0001 1010	1A	m	1001 0100	94
DC3	0001 1011	1B	n	1001 0101	95
DC4	0001 1100	1C	o	1001 0110	96
NAK	0001 1101	1D	p	1001 0111	97
SYN	0001 1110	1E	q	1001 1000	98
ETB	0001 1111	1F	r	1001 1001	99
DS	0010 0000	20	s	1010 0010	A2
SST	0010 0001	21	t	1010 0011	A3
FDS	0010 0010	22	u	1010 0100	A4
BYP	0010 0100	24	v	1010 0101	A5
LF	0010 0101	25	w	1010 0110	A6
EOB	0010 0110	26	x	1010 0111	A7
PR	0010 0111	27	y	1010 1000	A8
SM	0010 1010	2A	z	1010 1001	A9
VT	0010 1011	2B	A	1100 0001	C1
FF	0010 1011	2C	B	1100 0010	C2
CR	0010 1101	2D	C	1100 0011	C3
SO	0010 1110	2E	D	1100 0100	C4
SI	0010 1111	2F	E	1100 0101	C5
PN	0011 0100	34	F	1100 0110	C6
RST	0011 0101	35	G	1100 0111	C7
UC	0011 0110	36	H	1100 1000	C8
EOT	0011 0111	37	I	1100 1001	C9
CAN	0011 1000	38	J	1101 0001	D1
EM	0011 1001	39	K	1101 0010	D2
SS	0011 1010	3A	L	1101 0011	D3
ESC	0011 1011	3B	M	1101 0100	D4
FSR	0011 1100	3C	N	1101 0101	D5
GSR	0011 1101	3D	O	1101 0110	D6
RSR	0011 1110	3E	P	1101 0111	D7

Table A2.2. *Continued*

Character	Code Binary	Hex	Character	Code Binary	Hex
USR	0011 1111	3F	Q	1101 1000	D8
SP	0100 0000	40	R	1101 1001	D9
£	0100 1010	4A	S	1110 0010	E2
.	0100 1011	4B	T	1110 0011	E3
<	0100 1100	4C	U	1110 0100	E4
	0100 1101	4D	V	1110 0101	E5
+	0100 1110	4E	W	1110 0110	E6
I	0100 1111	4F	X	1110 0111	E7
&	0101 0000	50	Y	1110 1000	E8
!	0101 1010	5A	Z	1110 1001	E9
$	0101 1011	5B	0	1111 0000	F0
*	0101 1100	5C	1	1111 0001	F1
)	0101 1101	5D	2	1111 0010	F2
;	0101 1110	5E	3	1111 0011	F3
~	0101 1111	5F	4	1111 0100	F4
_	0110 0000	60	5	1111 0101	F5
^	0110 1010	6A	6	1111 0110	F6
,	0110 1011	6B	7	1111 0111	F7
%	0110 1100	6C	8	1111 1000	F8
-	0110 1101	6D	9	1111 1001	F9
>	0110 1110	6E	ERROR	1111 1111	FF
?	0110 1111	6F			

The following abbreviations are used for both codes.

ACK	Acknowledge		IL	Idle
BEL	Audible bell		LC	Lower case
BS	Backspace		LF	Line feed
BYP	Bypass		NAK	Negative acknowledge
CAN	Cancel		NL	New line
CR	Carriage return		NULL	All zero
DC	Device control		PF	Punch off
DC1	Device control1		PN	Punch on
DC2	Device control2		PR	Prefix
DC3	Device control3		RES	Restore
DC4	Device control4		RS	Record separator
DEL	Deletion		RSR	Record separator
DLE	Data line escape		RST	Reader stop
DS	Digit select		SI	Shift in
EM	End of medium		SM	Set made
ENQ	Enquiry		SO	Shift out
EOB	End of block		SOH	Start of heading
EOT	End of transmission		SP	Space
ESC	Escape		SS	Start of sequence
ETB	End of transmission block		SST	Significance starter
ETX	End of text		STX	Start of text
FDS	Field separator		SUB	Substitute
FF	Form feed		SYN	Synchronous idle
FS	File separator		TM	Trade mark
FSR	Field separator		UC	Upper case
GS	Group separator		US	Unit separator
GSR	Group separator		USR	Unit separator
HT	Horizontal tab		VT	Vertical tab

343

Index

2:4:2:1 code, 31
2-out-of-5 code, 42
4004 processor, 292
4×4 multiplier, 80
8008 processor, 292
8048 processor, 330
8049 processor, 330
8050 processor, 330
8080 processor, 292

abacus, 2
absolute addressing mode, 98, 133
absorption law in Boolean algebra, 53
access, direct memory, 234
access time for RAM, 157
ACE, 9
acknowledge, 227
  interrupt, 226
actual parameter, 272
actuator, 246, 251
ADA language, 15
ADC, 260
  ramp-converter, 260
  successive-approximation, 262
adder
  full, 59
  half, 59
  parallel, 59, 75
addition in binary arithmetic, 24
address
  decoder, 219
  decoding, 178
  effective, 300
  logical, 186
  physical, 186
addressing mode
  absolute, 98, 133
  auto-decrement, 302
  auto-increment, 302
  block-transfer, 310
  direct, 301, 309
  extended, 134, 301
  immediate, 99, 117, 134, 301, 309, 320
  implicit-references, 102
  implied, 309, 320
  indexed, 101, 117, 134, 288, 301
  indirect, 301, 321
  inherent, 300
  long absolute, 321
  modified base-page, 310
  PC-relative, 101, 134, 143
  permitted, 136
  postincrement register-indirect, 100, 136
  predecrement register-indirect, 100, 136
  register, 303
    direct, 99, 135, 320
    indirect, 99, 135, 321
  relative, 302, 310, 322
  short absolute, 321
Algol language, 15
aliasing, 252
alphanumeric display, 239
Altair, 8800 11
analog
  multiplexing, 255
  sensor, 247
analog-to-digital
  conversion, 247, 260
analysis
  lexical, 268
  semantic, 268
  syntax, 268
Analytical Engine, 5
AND
  gate, 58
  in Boolean algebra, 51
Apple, 12
arithmetic and logic unit (ALU), 73, 103
arithmetic instruction, 96
ARM processor, 323
array, 274, 283
ASCII, 47, 276, 340
assembler, 14, 125
  directive, 129
assembly code, 286
assign, 277
assignment instruction, 96
associative laws in Boolean algebra, 53
asynchronous
  counter, 68, 69, 71
  serial transmission, 229
auto-decrement addressing mode, 302
auto-increment addressing mode, 302
axioms and laws in Boolean algebra, 52

bank switching, 187
base pointer, 288
BASIC language, 12, 15, 267
BCD, 45, 63, 64
  to binary conversion, 29
  to XS3 code conversion, 64, 65
bi-directional
  databus, 73
  port, 224
biassed exponent, 38
binary
  arithmetic
    addition in, 24

344

division in, 26
  multiplication in, 25
  subtraction in, 25
counter, 68
number system, 21
point, 24, 36, 37
binary-coded decimal (BCD) number system, 29
begin, 268
Bisync, 231
bit, 21
  most significant, 23
bit-mapped display, 243
block-transfer addressing mode, 310
boolean, 274
Boolean algebra, 50, 53
  absorption law in, 53
  AND in, 51
  associative laws in, 53
  axioms and laws in, 52
  complement in, 51
  De Morgan's laws in, 53
  distributive laws in, 53
  exclusive OR in, 52
  inclusive OR in, 51
  NOT in, 51
  operators in, 51
  OR in, 51
bootstrapping, 171
branch instruction, 118
break, 285
buffer
  I/O, 226
  register, 104
  tri-state, 223
bus, 108
  instrumentation, 227

C language, 15, 268, 281
canonical form, 79, 80
carrier-sense, multiple-access, 246
case, 279
cathode-ray tube, 238
CD-ROM, 212
chain codes, 45
char, 283
char, 274
character
  codes, 47
  sets, 334
character-mapped display, 243
CISC processor, 293
clean zero, 39, 40
clear input for flip-flop, 68
clear-to-send, 230
clock, 228
  real-time, 233
CMOS
  NAND gate, 89
  NOR gate, 90
code
  assembly, 286
  chain, 45
  character, 47
  converter, 62, 64
  cyclic, 45, 46

distance, 43
Gray, 46, 250
Hamming, 43, 169
in-line, 287
position-independent, 302
pseudo, 267
redundancy, 42
reflected, 46
special, 42
2:4:2:1, 31
2-out-of-5, 42
unit-distance, 46
unweighted, 20, 31, 42, 45
weighted, 20, 31, 33, 45
collision detection, 246
Colossus, 7
combinational
  circuits, 65
  design, 58
comment, 270, 281
  lines, 131
communication, serial, 228
comparator, 260
compiler, 15, 266
  cross, 267
  native, 267
complement in Boolean algebra, 51
complementary metal-oxide semiconductor
    logic, 89
condition flag, 106
constant, 274, 282
contact bounce, 234
context, 227
control
  instruction, 97
  register, 225
  unit, 108
    hard-wired, 109
    micro-coded, 110, 117
conversion instruction, 97
counter
  interface, 232
  asynchronous, 68, 69, 71
  binary, 68
  ripple-through, 68
  synchronous, 71, 72
  three-stage, 68
  XS3, 72, 73
CRC, 232
cross compiler, 267
CRT interface, 242
CTS, 230
current loop, 254
current-mode logic (CML), 84
cycle stealing, 195
cyclic codes, 45, 46
cyclic-redundancy check, 232

DAC, 256
daisy-chaining, 226
data communication, packet, 245
data-access width, 127
data-link escape, 231
DC, 129
De Morgan's laws, 58, 61
  in Boolean algebra, 53

DEC, 10
decimal number system, 20
decimal–binary conversion, 23
decoder, 62, 63, 64
  address, 219
define, 282
define-constant directive, 129
define-storage directive, 130
demultiplexor (DMUX), 77
destination operand, 117
device, hard-copy, 243
Difference Engine, 4
differential signalling, 254
digital-to-analog converter, 256
digitizer, 237
direct addressing mode, 301, 309
direct memory access, 194, 234
directive
  assembler, 129
  define-constant, 129
  define-storage, 130
  END, 131
  equate, 129
  origin, 130
dirty zero, 39, 40
dispose, 277, 287
disk, 276
disk-drive interface, 244
displacement sensor, 248
display
  device, 237
  alphanumeric, 239
  bit-mapped, 243
  character-mapped, 243
  liquid-crystal, 238
  multiplexed, 239
  seven-segment, 239
distributive laws in Boolean algebra, 53
D-type flip-flop, 70
division in binary arithmetic, 26
"don't care" state, 62, 63, 65
DLE, 231
DMA, 234
DRAM controller, 170
DS, 130
dynamic
  RAM, 161
  storage, 287
  variable, 287

EBCDIC, 47, 342
ECL gate, 86
EDSAC, 9
EDVAC, 8
effective address, 300
else, 285
else, 279
END, 131
  directive, 131
end, 268
emitter-coupled logic, 84
encoder
  position, 250
  priority, 226
encoding
  horizontal, 111

vertical, 111
ENIAC, 8
enumerated type, 274
enumeration, 283
EQU, 129
equate directive, 129
error
  checking, 232
  detection and correction, 41
Ethernet, 246
ETX, 231
even parity, 43, 44
Ex-OR gate, 58
excess-three (XS3) number system, 30
exclusive OR
  in Boolean algebra, 52
  gate, 43, 45, 58
exponent, 37, 38, 39
  biassed, 38
extended addressing mode, 301
external adjacencies on Karnaugh map, 57

field, 275, 283
FIFO, 227
FILE, 283
file, 276
filter, 253
fixed-point notation, 36, 37
flag, 232
flash converter, 262
flip-flop, 65
  clear input for, 68
  D-type, 70
  J-K, 71
  preset input for, 68
  set-reset (SR), 65
  SRT, 67
  triggered, 67
float, 283
floating-point
  addition and subtraction, 39
  format, 41
  multiplication and division, 41
  notation, 36
  word structure, 38
flow
  control, 230
  sensor, 248
formal parameter, 272
fopen, 284
for, 286
for, 280
format, instruction, 94
framing
  code, 231, 245
  error, 230
FORTRAN language, 14, 267
fread, 284
full adder, 59
function, 272, 281
fwrite, 284

gate, 45
  symbology, 57
  AND, 58
  CMOS NAND, 89

NOR, 90
ECL, 86
exclusive-OR, 43, 58
NAND, 58
negative-input AND, 58
    OR, 58
nMOS NAND, 89
    NOR, 88
NOR, 58
NOT, 58
open-collector, 84
OR, 58
TTL NAND, 83
general-purpose register, 107, 126
global, 273, 282
    variable, 288
goto, 285
goto, 279
Gray code, 46, 250
grounding, 254
groupings in Karnaugh map, 57

half adder, 59
Hall effect, 237, 249
Hamming
    code, 43, 169
    distance, 43, 45
handshaking, 227, 245
hard-copy device, 243
hard-wired control unit, 109
Havard Mark I, 7
head parking, 203
heap storage, 287
hexadecimal number system, 28
hexadecimal–binary conversion, 28
high-level language, 15, 266
horizontal encoding, 111

IACK, 121
IBM, 6
identifier, 268
if, 285
if, 279
immediate addressing mode, 99, 117, 134,
    301, 309, 320
implicit-references addressing mode, 102
implied addressing mode, 309, 320
include, 283
inclusive OR in Boolean algebra, 51
indexed addressing, 288
    mode, 101, 117, 134, 301
indirect addressing mode, 301, 321
inherent addressing mode, 300
in-line code, 287
input, 276, 283
input/output, *see* I/O
instruction
    arithmetic, 96
    assignment, 96
    branch, 118
    control, 97
    conversion, 97
    cycle, 116
    format, 94
    jump, 118
    logic, 96

one-operand, 95
    register, 104
    two-operand, 95
    zero-operand, 95
instrumentation bus, 227
integrated circuit, 9
int, 281, 283
integer, 274
integrated injection logic, 86
interface, 218
    counter, 232
    CRT, 242
    disk-drive, 244
    operating-system, 269
    timer, 232
interference, 230
interpreter, 15, 266
interrupt, 120, 147, 150, 221, 300, 307, 316,
    326, 332
    acknowledge, 226
    mask, 228
    maskable, 121, 150, 153
    multiple, 226
    non-maskable, 121, 150
    vector, 121, 226
interrupt-service routine, 121, 122, 150,
    226
I/O (input/output)
    buffer, 226
    device, parallel, 225
    isolated, 220
    mapped, 220
    memory-mapped, 146, 219
    operations, 146
    port-mapped, 220
    register, multiple, 225
IREQ, 121
isolated I/O, 220
isolation, 249

J K flip-flop, 71
joystick, 237
jump
    instruction, 118
    to subroutine, 119

Karnaugh map, 54, 56, 60, 62, 63, 65, 72
    external adjacencies on, 57
    groupings in, 57
    multi-variable, 55
keyboard, 148, 234, 276
    multiplexed, 235
keyword, 268

label, 131, 279, 285
language
    ADA, 15
    Algol, 15
    BASIC, 12, 15, 267
    C, 15, 268, 281
    FORTRAN, 14, 267
    high-level, 15, 266
    Pascal, 15, 267, 270
    program-design, 137
large-scale integration (LSI), 73
laser printer, 244

INDEX

latency time, 201
LCD, 238
least significant bit, 36
LED, 238
level sensor, 248
lexical analysis, 268
library, 283
light pen, 237
light-emitting diode (LED), 238
line
  break, 230
  printer, 243
linear feedback shift register (LFSR), 45
linkage, object-code, 269
liquid-crystal display (LCD), 238
list, 277
local, 274, 282
locality of reference, 192
logarithms, 3
logic, 86
  (CML), current-mode, 84
  complementary metal-oxide semiconduc-
    tor, 89
  emitter-coupled, 84
  instruction, 96
  n-type metal-oxide semiconductor, 88
  symbols, 58
  transistor–transistor, 83
  tri-state, 108
logical address, 186
long absolute addressing mode, 321
loop, 279, 285

machine code, 14
macro-instruction, 110
magnetic, 157
main, 281
mainframe, 10
Manchester-encoded signal, 231
mantissa, 37, 38, 39, 40
mapped I/O, 220
mapper, 117
mapping table, 189
Mark-8, 11
mark/space ratio, 258
mask, interrupt, 228
maskable interrupt, 121, 150, 153
mask-programmed ROM, 173
MC6809 processor, 295
MC68000 processor, 289, 311
medium-scale integration (MSI), 73
memory
  address register, 104, 117
  addressing, 160
  buffer register, 105
  map, 180, 219
  random-access, 156
  read-only, 156, 170
memory-mapped I/O, 146, 219
micro-code, 110
micro-coded control unit, 110, 117
micro-instruction, 110
minicomputer, 10
mnemonics, 14
modem, 228
modified base-page addressing mode, 310

most significant bit, 23
motor, 251
  stepping, 252
mouse, 237
multi-variable Karnaugh map, 55
multiple
  I/O register, 225
  interrupt, 226
multiplexed
  display, 239
  keyboard, 235
multiplexing, analog, 255
multiplexor (MUX), 77
multiplication in binary arithmetic, 25

NAND, 58
  gate, 58
  implementation, 60
native compiler, 267
negative numbers
  signed binary, 32
  two's complement, 33
negative-input
  AND gate, 58
  OR gate, 58
NeMiSyS, 94, 102
  instruction statement, 131
network, 246
new, 277, 287
nil, 278
nMOS, 88
  NAND gate, 89
  NOR gate, 88
noise, 230, 253
non-maskable interrupt, 121, 150
NOR, 58
  gate, 58
  implementation, 60
normalization, 39
  post, 41
NOT
  gate, 58
  in Boolean algebra, 51
number system
  binary, 21
  binary-coded decimal (BCD), 29
  decimal, 20
  excess-three (XS3), 30
  hexadecimal, 28
  octal, 27
n-type metal-oxide semiconductor logic, 88
Nyquist limit, 253

object code, 14, 125
  linkage, 269
octal number system, 27
octal–binary conversion, 27
odd parity, 43, 60
origin directive, 130
one-operand instruction, 95
one's complement notation, 34
open-collector gate, 84
operand
  destination, 117
  source, 117
operating-system interface, 269

348

operational amplifier, 255
operators in Boolean algebra, 51
optimization, 267
OR
  gate, 58
  in Boolean algebra, 51
ORG, 130
otherwise, 279
out-of-range condition, 36
output, 276, 283
output-hold time for RAM, 165
overflow, 35, 36, 40
overrun error, 230

packet data communication, 245
parallel
  adder, 59, 75
  I/O device, 225
  interface, 227
parallel-to-serial translation, 70
parameter, 287
  actual, 272
  formal, 272
  passing, 144
    by reference, 144, 145
    by value, 144
parity, 230
  checking, 42
  even, 43, 44
  generator, 60
  odd, 43, 60
partial decoding, 180
Pascal language, 15, 267, 270
PC-relative addressing mode, 101, 134, 143
peek, 287
perfect induction, 53
peripheral, 218
permitted address mode, 136
phase control, 258
physical address, 186
pointer, 277, 282, 284
  base, 288
  stack, 288
pointing device, 237
poke, 287
pop, 139, 140
port, bi-directional, 224
port-mapped I/O, 220
position
  encoder, 250
  sensor, 248
    ultrasonic, 248
position-independent code, 302
post normalization, 41
postincrement register-indirect addressing
    mode, 100, 136
predecrement register-indirect addressing
    mode, 100, 136
pre-processor, 268, 282
preset input for flip-flop, 68
pressure sensor, 248
printer
  laser, 244
  line, 243
priority encoder, 226
procedure, 272

processor
  4004, 292
  8008, 292
  8048, 330
  8049, 330
  8050, 330
  8080, 292
  ARM, 323
  CISC, 293
  MC6809, 295
  MC68000, 311
  RISC, 293, 323
  Transputer, 294
  VL86C010, 323
  Z80, 295
program, 119
  counter, 104, 117, 120, 126
program, 270
program-counter register, 104
program-design language, 137
program-status register, 97, 106, 126
programmable logic array (PLA), 81
programmer's model, 126, 298
proximity sensor, 249
pseudo code, 267
pull-up resistor, 223
pulse overcrowding, 197
push, 139, 140

radix, 21, 22, 23, 27
RAM
  dynamic, 161
  output hold time for, 165
  random access for, 157
  read-access time for, 165
  refresh of, 167
  static, 161
ramp-converter ADC, 260
random access
  for RAM, 157
  memory, 156
re-entrantcy, 289
read, 276
read cycle, 106
read-access time for RAM, 165
read-only memory (ROM), 79, 156, 170
real, 274
real-time clock, 233
record, 275
recursion, 273
reed switch, 236
reflected codes, 46
refresh of RAM, 167
register, 104, 144, 219, 225
  addressing mode, 303
  buffer, 104
  control, 225
  general-purpose, 107, 126
  instruction, 104
  memory address, 104, 117
  memory buffer, 105
  program-counter, 104
  program-status, 97, 106, 126
  stack-pointer, 107
  status, 225
  temporary, 108, 117

INDEX

register-direct addressing mode, 99, 135, 320
register-indirect addressing mode, 99, 135, 321
relative addressing mode, 302, 310, 322
relay, 252
  relay, solid-state, 252
relocatability, 101
repeat, 280, 286
request, 227
request-to-send, 230
resistor, pull-up, 223
return from subroutine, 119
ripple-through counter, 68
RISC processor, 1, 293, 323
rollover, 236
ROM, mask-programmed, 173
RS232, 230
RTS, 230

sampling, 252
Schottky TTL, 85
scope rules, 273, 282
screen, 148
SDLC, 232
seek time, 201
semantic analysis, 268
semiconductor switch, 251
sensor, 246
  analog, 247
  displacement, 248
  flow, 248
  level, 248
  position, 248
  pressure, 248
  proximity, 249
  speed, 250
  temperature, 247
  ultrasonic position, 248
sequence number, 245
sequential circuit, 65
serial
  communication, 228
  transmission, asynchronous, 229
    synchronous, 231
serial-to-parallel translation, 70
service routine, interrupt, 226
servomechanism, 251
set, 274
set-reset (SR) flip-flop, 65
seven-segment display, 239
shift register, 45, 69, 70
  linear feedback (LFSR), 45
short absolute addressing mode, 321
sign, 32
  bit, 38
signed binary negative numbers, 32
sizeof, 284
small-scale integration (SSI), 73
soft error, 169
solenoid, 252
solid-state relay, 252
source
  code, 125
  file, 14
  operand, 117

special codes, 42
speed sensor, 250
SRT flip-flop, 67
stack, 102, 107, 120, 139, 145, 287
  frame, 288
  pointer, 119, 126, 288
stack-pointer register, 107
start bit, 229
static
  RAM, 161
  storage, 287
status register, 225
stepping motor, 252
stepwise refinement, 137
stop bit, 229
storage
  dynamic, 287
  heap, 287
  static, 287
strain gauge, 247
structure, 283
subrange, 274
subroutine, 97, 119, 141, 144
subtraction in binary arithmetic, 25
successive-approximation ADC, 262
supermini, 10
switch debouncing, 67
SYN, 231
synchronous
  counter, 71, 72
    formal design of, 72
  serial transmission, 231
switch, 285
syntax analysis, 268

Tabulating Machine, 5
tape, 276
temperature sensor, 247
temporary register, 108, 117
terminal, 276
three-stage counter, 68
then, 279
thyristor, 258
time-domain multiplexing (TDM), 77
timer interface, 232
toggle circuit, 70
token, 246
totem-pole output, 83, 84
tracking converter, 261
transformer, 249
transient protection, 255
transistor, 9
transistor–transistor logic, 83
Transputer processor, 294
tree, 277
tri-state
  buffer, 223
  drive, 73, 75
  logic, 108
  output, 84
triac, 258
triggered flip-flop, 67
truth, 79, 80
  table, 51, 53, 60
TTL NAND gate, 83
two-operand instruction, 95

two's complement
   arithmetic, 34
   negative numbers, 33
type, 274
   enumerated, 274
typedef, 283

ultrasonic position sensor, 248
union, 283
unit-distance codes, 46
universal logic module (ULM), 78
unweighted code, 20, 31, 42, 45
user interface, 234

var, 277
variable
   dynamic, 287
   global, 288
vector, interrupt, 226
vertical encoding, 111
virtual-memory management, 190
VL86C010 processor, 323

voltage-to-frequency converter, 262

weighted code, 20, 31, 33, 45
while, 286
while, 280
with, 275
wordlength, 28
WORM, 212
write, 276
write cycle, 106

XOFF, 230
XON, 230
XS3, 45, 64
   arithmetic, 31
   counter, 72, 73

Z80 processor, 295
zero
   clean, 39, 40
   dirty, 39, 40
zero-operand instruction, 95